Insurance under the ICE Contract

SLC

Books are to be returned on or before
the last date below.

17 AUG 1992

-3 FEB 2005

LIBREX—

Insurance under the ICE Contract

F.N. Eaglestone and C. Smyth

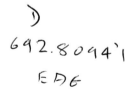

George Godwin
London and New York

George Godwin
an imprint of:
Longman Group Limited
Longman House, Burnt Mill, Harlow
Essex CM20 2JE, England
Associated companies throughout the world

Published in the United States of America
by Longman Inc., New York

First published 1985

British Library Cataloguing in Publication Data

Eaglestone, F.N.
    Insurance under the I.C.E. contract.
    1. Institution of Civil Engineers. Conditions
    of contract and forms of tender, agreement and
    bond for use in connection with works of
    civil engineering construction. 5th edition
    2. Civil engineering – Contracts and
    specifications – Great Britain
    I. Title   II. Smyth, C.
    624   TA181.G7
    ISBN 0-7114-5794-8

Printed in Great Britain at The Pitman Press, Bath

Contents

Foreword
Preface

Foreword

By Charles C. Timms
BSc, CEng, FICE, FIStructE, MIMechE, FIPHE, MConsE, FCIArb.

In common with many sections of business and commerce the construction industry suffers from the ills of watertight compartments. It is well known that
Engineers, Architects, Quantity Surveyors, Employers, Contractors and Academics
have suspicion of each other due mainly to lack of knowledge of the others' training
and experience. However, all the above categories are united in their ignorance and
fear of matters of insurance. They will all say that they do not understand insurance
and will all promptly deflect the whole matter to 'square leg'.

Events in recent years have, however, made it imperative that insurance in all its
forms be more widely studied and understood. A major contribution to this understanding has been made by Frank Eaglestone and Colin Smyth in this book which
is relative to insurance under the Institution of Civil Engineers' forms of contract.
All Civil Engineers in the construction side of the industry will soon realise after
reading just a few pages of this important and excellently written document that
their knowledge of insurance is very limited.

Even if the reader is familiar with the normal insurance clauses in civil engineering contracts he will realise that the whole subject is very much more complicated
than he would like to believe. A close study shows that there are risks which are
uninsurable but that also there are risks which are not generally recognised as such.

It is the hope of everyone who pays an insurance premium that the whole
operation will remain an academic one. However, every day the hand of fate takes
control and the other side of the coin requires attention. The insured is required to
make claim on the Insurance Company, and this aspect of the subject is covered in
great detail and with excellent guidance in Chapters 19 to 22 inclusive.

This publication is a most valuable addition to all the information already
available concerning contract matters. Civil Engineers and others will neglect this
book at their peril.

Charles C. Timms
Chairman of the Law and Contract Procedure
Examination Committee of the Institution of
Civil Engineers

Preface

Some years ago reviewers of my book *The RIBA Contract and the Insurance Market* (now out of print) suggested that a companion volume on the Institution of Civil Engineers' Conditions of Contract, would be welcome, and although this book was started a few years ago it has taken that time to produce it.

My co-author Colin Smyth has written the chapters on claims which we hope claims officials will appreciate. Claims men, including loss adjusters, complain that there is a dearth of written information dealing with their side (a very important side) of the insurance industry. This aspect will assist the book's claim to be included in the recommended reading lists for the relevant subjects of the Chartered Insurance Institute's and the Chartered Institute of Loss Adjusters' examinations.

The ICE Conditions were first issued in December 1945, having been agreed by the Institution of Civil Engineers and the Federation of Civil Engineering Contractors. The second, third and fourth editions (approved also by the Association of Consulting Engineers) were issued in January 1950, March 1951 and January 1955 respectively. The current edition, the fifth edition, was approved by the same three bodies in 1973. It was revised in 1979.

It is nearly twenty years since the Banwell Committee stated that a common form of contract for all construction work, covering England, Scotland and Wales, was both desirable and practicable. In fact, there has not been any move towards a common form of contract for building and civil engineering and there is strong support for retaining the ICE Conditions. Some arguments for and against were given in *Action on the Banwell Report* published by Her Majesty's Stationery Office in 1967. Therefore, it seems that there is little likelihood of a combined standard form of contract for both construction industries.

The major aim of this book is to supply the insurance official, engineer, employer, contractor and, dare we say it, the lawyer with an analysis of the clauses which directly and indirectly concern the insurance industry. Furthermore, it is intended to give details of those insurance policies and bonds which are involved in providing the protection required by the clauses just mentioned. At the ends of the chapters concerning the main insurance policies, the subcontract for use with the ICE Conditions and the FIDIC Contract we have endeavoured to indicate where the basic insurance policy cover, required by the contracts concerned, does not protect the contractor (subcontractor where involved) and the employer, or even falls short of the insurance requirements of those contracts, each of these parties being considered separately. In this connection the insurance cover considered is the normal cover provided in each case by the basic policy and not the special cover an insurance broker may be able to obtain for a particular client or for one particular contract. Moreover, this book does not attempt to explain any particular wordings which brokers specialising in the construction industry may persuade an insurer to give for their construction clients.

While in the main this book concerns insurance practice in the United Kingdom and deals with English law, the chapter on the International Civil Engineering Contract (FIDIC) compares that contract with the ICE Conditions, illustrates cases of forces of nature and, as mentioned earlier, indicates the risks remaining uninsured.

As 'insurance and bond in detail' (there are two chapters on sureties and performance bonds) are subjects in the ICE syllabus for the Civil Engineering Law and Contract Procedure examination it is hoped that engineering students as well as the qualified engineer will find the book useful in understanding these subjects which to some engineers, architects and quantity surveyors are rather a mystery.

We have tried to avoid too much repetition by means of cross references, but as a book of this nature will be used for reference purposes a certain amount of repetition is necessary and therefore acceptable.

We wish to express our grateful thanks to the Allstate Insurance Company Ltd for allowing us to use their contractors' combined policy and performance bond documents for illustration purposes, and to John Swarbrick, Humphrey Lloyd, Ken Cannar, John Gatward, and Roger Lewis who read some or all of the chapters and made many useful suggestions. John Swarbrick and Humphrey Lloyd did the bulk of this reading work and it was greatly appreciated as we were able to benefit from the insurance and the legal view.

Finally we acknowledge the hard work put in by Barbara Clegg and Margaret Smyth in deciphering and typing the authors' longhand.

F.N.E. and C.S.
September 1984

Chapter 1

Introduction and general points concerning contract conditions

Construction contracts frequently widen the contractor's liability at law in respect of damage to property (including the works) and for personal injuries arising from the carrying out of the works.

It is necessary to appreciate the extent of his liability in tort and his statutory liability in order to assess the extent to which the contract terms add to this liability.

The meaning of legal liabilities

In the United Kingdom there is only a right to compensation (called damages) for injury, death or damage to property in certain defined areas of liability. They are basically:

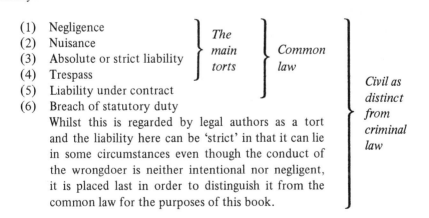

(1) Negligence
(2) Nuisance
(3) Absolute or strict liability
(4) Trespass
(5) Liability under contract
(6) Breach of statutory duty
Whilst this is regarded by legal authors as a tort and the liability here can be 'strict' in that it can lie in some circumstances even though the conduct of the wrongdoer is neither intentional nor negligent, it is placed last in order to distinguish it from the common law for the purposes of this book.

The main torts — *Common law* — *Civil as distinct from criminal law*

The distinction between crimes and civil wrongs does not depend on what is done but on the legal consequences that follow. If what are called criminal proceedings follow, the wrongful act is a crime. If civil proceedings follow, it is a civil wrong. If the act is capable of being followed by both types of proceedings, it is both a crime and a civil wrong. In civil proceedings the terms are different from those used in criminal proceedings. In the former a plaintiff sues a defendant, and successful proceedings result in a judgment for the plaintiff.

The phrase 'common law' can be used in various senses but always to point a contrast. Originally it meant the law that was not local law but was common to the whole of the country. While this may still be its meaning in certain contexts, it is not the usual meaning and in this book it means the law that is not the result of legislation, i.e. it is the law created by the custom of the people and the decisions of the judges. The latter is sometimes called case law. However, there are case law decisions on breaches of statutory duty which have been defined as exclusive of the common law, from which it will be appreciated that it is impossible to explain the law without defining as one proceeds. Even lawyers have difficulty in defining tort which is one type of civil wrong. From the above list it will be realised that a tort is a civil wrong independent of contract, i.e. it gives rise to an action for damages irrespective of any agreement giving a right to take the action concerned. However, the terms of a contract between two parties may override a tort action otherwise available between them. The meaning of 'tort' will become clear when each of the main torts are considered in turn.

It is necessary to record that the Pearson Committee, whose report was published in 1978, recommended a limited form of 'no fault' liability for injuries caused by motor vehicles. This is subject to statutory intervention which will take some time even if the recommendation is accepted, and in general the law concerning torts is retained. Probably the first statutory intervention introduced will be towards the strict liability of producers of products.

There are defences to the following torts of which the main ones are contributory negligence and consent, but the same defences are not always applicable to each tort. See chapter 19 for other defences and more detail.

(1) *Negligence*

In *Blyth* v *Birmingham Waterworks Co (1856)* negligence was defined as the 'omission to do something which a reasonable man, guided by those considerations which ordinarily regulate the conduct of human affairs, would do, or doing something which a prudent and reasonable man would not do'.

The following conditions must apply for the maintenance of an action for negligence:

    (a)   The defendant must owe a duty of care to the plaintiff.
    (b)   There must be a breach of that duty.
    (c)   The plaintiff must sustain injury or damage as a result.

Only by knowing the law can one be certain when a duty of care is owed. In driving a car the driver owes a duty of care to all those persons and their property with which the car might collide. On the other hand, the pedestrian strolling through the local park owes no legal duty to rescue the child drowning in the pond. In fact, legal liabilities are not necessarily moral liabilities.

Where the duty of care is owed, the standard is a reasonable one. What is reasonable is for the court to decide in each case, but it should be appreciated that this is an objective standard, i.e. how a hypothetical reasonable man would behave in a given set of circumstances, not what is reasonable for that particular defendant to do or not do in a particular case. Whilst the personal peculiarities of the individual whose conduct is in question are eliminated, the personality of the judge in applying the objective standard cannot be eliminated. It must be appreciated that in applying the objective standard the courts take into account the fact that the amount of care varies with the circumstances. Thus the contractor occupying property owes the same standard of care to both adults and children but he must exercise more care if the premises are used by children rather than solely by adults, e.g. the extension of the wing of a school where the main building is still in use by the pupils.

Another illustration is given by the case of *Haley* v *London Electricity Board (1964)* where it was decided that in view of the proportion of blind persons in the population (1 in 500), contractors who open a hole in a city highway must take reasonable care to prevent a blind person, using reasonable care for his own safety, from falling down the hole. While this liability is not a strict one, i.e. irrespective of negligence (see later), it obviously imposes a greater amount of care upon contractors and also local authorities.

In proceedings for negligence, as in most actions in respect of civil wrongs, the burden or onus of proof is borne by the plaintiff on the facts pleaded by him and not admitted by the defendant. In a civil case the **standard** of proof is proof on a preponderance of probability or on balance of probabilities, whereas in criminal law proof beyond reasonable doubt is required of the prosecution.

In certain cases in negligence where the maxim *res ipsa loquitur* (the thing speaks for itself) applies on proof of the occurrence of the incident, the onus is transferred to the defendant to disprove negligence. However, a word of warning is necessary as there are three requirements before this maxim applies:

(a)   The thing causing the accident must be under the defendant's exclusive control.
(b)   The accident is such as in the ordinary course of things does not happen if those who have the management use proper care.
(c)   There must be no explanation of the accident. This only means that if proof of the relevant facts is available, the question ceases to be whether they speak for themselves. The only question is whether, on the established facts, negligence is to be inferred or not.

If a contractor or his employees for whom he is responsible dropped a piece of scaffolding from a building in course of erection, onto a passer-by and injured him, this would be a matter of *res ipsa loquitur* as all the above requirements apply and it would create hardship on the injured plaintiff to establish negligence. In *Walsh* v *Holst (1958)* the defendants were demolishing a building adjacent to the highway. The plaintiff was injured by falling masonry, and sued in negligence. It was held that *res ipsa loquitur* applied as there was no evidence of how the masonry came to fall. On the facts the defendants escaped liability as they were able to show that all proper and reasonable precautions had been taken. However, it is not sufficient for the defendant to show that the accident could have happened without negligence on his part.

The leading case which deals with the duty of care is *Donoghue* v *Stevenson (1932)* where Lord Atkin said:

> You must take reasonable care to avoid acts or omissions which you can reasonably foresee would be likely to injure your neighbour.

Neighbours he described as:

> persons who are so closely and directly affected by my act that I ought reasonably to have them in contemplation as being so affected when I am directing my mind to acts or omissions which are called in question.

Cases extending this rule in the construction industry include:

(i) *Dutton* v *Bognor Regis UDC (1972)*, which was approved *(obiter)* by the House of Lords in *Anns* v *Merton London Borough Council (1977)*, held that an owner-builder was liable for negligence in constructing a house and this liability extended to a subsequent buyer who merely suffered depreciation in value of his house resulting from defects caused by the builder. Thus such cases are no longer an exception to the rule.

(ii) *Batty* v *Metropolitan Property Realisations Ltd (1978)*. A builder (non-owner) and a development company owning the land built houses on a steep slope. As a result of a severe landslide the court found the house concerned was doomed as within ten years it would slip downhill and there was imminent danger to the health and safety of the occupants. The builder and developer owed a duty of care to the purchaser from the developer and both were in breach of that duty.

(iii) *Junior Books Ltd* v *Veitchi Co Ltd (1982)*. This case extended the duty of care much further. The defendant negligently laid a new floor in the plaintiff's factory. They were subcontractors **and had no contractual relationship with the plaintiffs**. The House of Lords held that the cost of replacing the defective flooring and loss of profits in the resultant disruption was the responsibility of the subcontractors because of their negligence. This case decided that if a plaintiff is in sufficient close proximity to the defendant he may recover not only the cost of replacing work done for him which is defective but also foreseeable economic loss resulting from such work even though the defective work does not threaten danger to either persons or property.

(2) *Nuisance*

This tort was defined many years ago as a wrong done to a man by unlawfully disturbing him in the enjoyment of his property (a private nuisance) or, in some cases, in the exercise of a common right (a public nuisance).

Public nuisance A public nuisance is a crime and does not fall within the law of torts, unless it causes an individual special damage above that caused to the rest of the community.

Consequently, the builder who deposits a pile of sand on the highway and fails to light it at night would be guilty of a public nuisance because of the general inconvenience to the public in walking round the obstruction. This is a public nuisance and it is for the police to take action. However, if a motor cyclist collides with the unlit obstruction and suffers injury, he will have a right of action in tort against the builder for the public nuisance which caused him a specific injury over and above that caused to the rest of the section of the public concerned. Public nuisance includes not only obstruction of the highway but also the creation of dangers upon or near the highway and the matter of liability for the disrepair of premises abutting the highway.

With building works there is some tolerance with such work carried out adjacent to the highway. Hoardings, for example, are erected outside premises where building work is being carried out. These encroach upon the pavement, but they are for the public's protection and must be accepted. Incidentally, such erections can have statutory authority, e.g. under section 147 of the Highways Act 1959, the local authority controls this requirement. However, it may be a nuisance to obstruct the highway by placing builders' waste on it unless the required authority under section 146 of the 1959 Act has been obtained. In such cases sub-section 146(3) requires proper fencing and, by night, proper lighting. Furthermore, any negligence in fencing or erecting hoardings creates a nuisance in spite of the statutory authority.

The creation from building sites of dust and smoke which cross the highway and obscure the visibility of motorists causing an accident is another example of a public nuisance allowing those who sustain injury or damage to claim.

Private nuisance   Private nuisances were defined in *Cunard* v *Antifyre (1932)* as an 'interference for a substantial length of time by owners or occupiers of property with the use or enjoyment of neighbouring property'. They are of two kinds:

    (a)   Wrongful disturbance of rights attaching to land. The main rights, interference with which may be actionable, are rights to light and air, rights to support of land and buildings thereon, and rights in respect of water and rivers.

    (b)   The act of wrongfully causing or allowing the escape of noxious things, such as smoke, smells, noise, gas, vibration, damp and tree roots, into another person's property so as to interfere with his health, comfort, convenience, or enjoyment of, or cause damage to his property.

It should be noted that:

(i)   The element common to both forms of nuisance (public and private) is unlawful interference.

(ii)   It is said that a claim in private nuisance cannot be founded solely upon personal injury, at least no English case has been found on this point. Private nuisance is limited to actions based on injury caused in respect of the enjoyment of land and if personal injury is suffered without affecting the occupier's enjoyment of his land, the action will probably lie in negligence or trespass. Nevertheless, in principle, there seems to be no reason why a person who suffers an invasion of some proprietary or other interest in land should not be able to claim for a resulting bodily injury.

(iii)   Liability in nuisance may apply even though reasonable care has been taken, e.g. to avoid the emission of dust in building premises, which distinguishes this tort from negligence.

(iv)   It is not sufficient in nuisance for the plaintiff to show there is merely a prospective damage. In order to succeed in his action for nuisance there must be actual damage sustained by the plaintiff. This indicates one distinction between nuisance and trespass which is actionable *per se*, that is, taken alone. (See Trespass below).

(v)   In certain cases an action may lie in both nuisance and negligence. In the case of *Gold* v *Patman and Fotheringham Ltd (1958)* Gold was the owner of property which was to be modified and Patman and Fotheringham undertook to do the work. Piling operations took place **without** negligence on the part of the contractor but with the result that adjoining third-party property suffered damage due to the weakening of support. The third-party property owner succeeded in his action for nuisance against Gold and, had the contractor been negligent, there would also have been a liability on the part of Gold who would be vicariously responsible for the contractor's negligence in not taking precautions. This is one of those circumstances where a principal is liable for a contractor's negligence.

(vi)   Where a nuisance results from a positive act of misfeasance, the creator of that nuisance is liable for it and its continuance, even where he has lost the ability to abate it. Thus, where a building was erected so that it obstructed ancient lights (the right of access of light to any building enjoyed for the full period of twenty years without interruption), the builder was held liable in nuisance even though he had assigned his lease and was powerless to discontinue the nuisance. (See *Roswell* v *Prior (1701)*.) However, in such a case even though the landlord is liable for the creation of the nuisance, his liability is concurrent with that of the tenant, whose occupancy makes him liable for the continuance of the nuisance.

(vii)   The requirement of 'interference for a substantial length of time' in the definition of private nuisance given earlier is attained if a state of affairs is existing whereby the use of adjoining premises is rendered hazardous by a threat of fore-seeable damage likely to result from that state of affairs. However, there must be resulting damage likely to satisfy the requirement that some damage be caused. In *Andreae* v *Selfridge & Co Ltd (1938)* where a hotel proprietor complained of loss of custom through building operations, the Court of Appeal reversed an award of damages amounting to the full extent of the loss of custom, holding that some of the interference was reasonable and yet might have resulted in loss of custom. Consequently the court calculated the proportion of the business loss equivalent to the excess of noise and dust which alone was actionable.

It is appropriate to point out that the public liability policy does not cover consequential loss unless it results from injury or damage to property which is covered by the policy. (See the operative clause of this policy in chapter 5.)

(3) *Absolute or strict liability*

The latter term is considered more correct as it allows defences whereas the former probably does not.

Strict liability describes certain liabilities imposed by law making the defendant liable even though he has exercised reasonable care and did not intend the injury or damage. The main example under this heading is the rule in *Rylands* v *Fletcher (1866)* that the person who, for his own purposes, brings on his lands and collects and keeps there anything likely to do mischief if it escapes, must keep it at his peril, and if he does not do so, is *prima facie* answerable for all the damage which is the natural consequence of its escape.

There are certain defences:

(a) The escape was due to the plaintiff's default.
(b) It was due to an act of God.
(c) The incident results from natural use of the land.
(d) It was done with the plaintiff's consent.
(e) Statutory authority, depending on the wording of the Act concerned.
(f) It was the act of a stranger. The word stranger here denotes that it cannot include a servant, or contractor for whose acts the defendant would be responsible. For example, there would be no liability under the rule for the act of a trespasser.

In *H. and N. Emanuel Ltd* v *Greater London Council and Another (1971)*, the council employed a firm of contractors to demolish some prefabricated buildings on land that it owned. The contractor disposed of some of the debris by burning, and sparks from one of the bonfires set fire to surrounding buildings. The council were found liable for the damage, notwithstanding the contractors had been expressly forbidden to burn rubbish on the site. It was held that an occupier was liable for the escape of fire caused by the negligence, not only of his servant but also of his independent contractor and anyone else who was on his land with his leave and licence. The only occasion when the occupier would not be liable for negligence was when the negligence was the negligence of a stranger, although for this purpose a 'stranger' would include a person on the land with the occupier's permission who, in lighting a fire or allowing it to escape, acted contrary to anything which the occupier could anticipate that he would do. The council were 'occupiers' of the premises because they had a sufficient degree of control over the activities of the persons thereon and the contractor's men were not 'strangers' because they were present with the leave and knowledge of the LCC and, although they were forbidden to burn rubbish, it was their regular practice to do so. The council could reasonably have anticipated that the men would light a fire and ought to have taken more effective steps to prevent them. Ryland's rule was also applied.

Winfield on Tort (eighth edition) says:

In nuisance, a non-occupier of land on which the injury originates may be liable, e.g. the builder of a wall on land occupied by X had been held liable for thereby obstructing access to Y's private market: *Thompson* v *Gibson (1841)* . . . Despite *dicta* in the Court of Appeal that there is no case in which the doctrine of *Rylands* v *Fletcher (1866)* has been applied to an owner not in occupation, *St Anne's Well Brewery Co* v *Roberts (1928)*, it is submitted that anyone who introduces the mischievous thing and has control or management of the land at the time of the escape although he is not the owner or occupier of the land but has merely a licence to use it, may be liable under the rule, *Rainham Chemical Works Ltd* v *Belvedere Fish Guano Co (1921)*.

Consequently, it seems that whether or not the contractor is the legal occupier of the site he can be liable under the *Rylands* v *Fletcher* rule.

Other examples show that strict liability entails liability without proof of negligence.

At one time under common law but now under the Animals Act 1971 there is strict liability, regardless of fault, for damage done by an animal belonging to a dangerous species and for damage done by any other animal which to the keeper's knowledge has a mischievous propensity.

Again, there is strict liability when one person is vicariously responsible for the acts or omissions of another. This occurs in the following three relationships (the first named being responsible for the acts of the second named in each case):

(a) That of master and servant, generally speaking, in all circumstances.
(b) That of principal and independent contractor in certain circumstances only.
(c) That of principal and agent in particular situations only.

(4) *Trespass*

This is an unlawful act committed with 'force and violence' on the person or property of another. The violence may be only implied — e.g. a peaceable but wrongful entry on to the plaintiff's land. The injury or damage must be **direct**, i.e. it must not be consequential. For example, if a builder throws a brick into the road and it injures a passer-by it is a direct injury, but if the brick is thrown into the road and later a passer-by falls over it and is injured it is an indirect injury.

For practical purposes there is no significant difference between unintentional trespass to the person and the tort of negligence causing injury. (See *Fowler* v *Lanning (1959)* and *Letang* v *Cooper (1965)*.) It should also be borne in mind that liability policies do not cover **intentional** injury or damage, although this would be a trespass. The builder who intentionally digs a trench across a field and damages crops is guilty of a trespass among other wrongs, but the public liability policy would not cover such damage which is inevitable. Similarly, although there may be a trespass by merely going upon land and treading on the grass, there is no policy liability until actual damage is done, subject to the other terms and conditions of the policy.

(5) *Liability under contract*

A builder often increases his legal liability by entering into a written contract. The Standard Form of Building Contract and the Institution of Civil Engineers'

Conditions are examples of contracts where the contractor agrees to indemnify the employer (the principal), who is the other party to the contract, for damage to the works or in respect of any claim made against the employer by a third party as a result of injury or damage arising out of the performance of the contract, subject to exceptions.

(6) *Breach of statutory duty*

Certain Acts of Parliament create statutory liabilities and the following statutes are those most likely to be involved in the construction industry.

(a)   The Fatal Accidents Acts 1959 and 1976 give a right of action to the dependants of those killed in accidents to the extent of the former's pecuniary loss.

(b)   The Law Reform (Miscellaneous Provisions) Act 1934 has the main effect of providing that, on the death of any person, all causes of action subsisting against or vested in him at the time of his death survive against or for the benefit of his estate, including his claim for loss of expectation of life.

(c)   The Factories Act 1961 (which repeals and consolidates the previous Factories Acts) and other enactments relating to the safety, health and welfare of employed persons such as the regulations passed under the Health and Safety at Work etc, Act 1974 are also very important.

   The principal statute at present is the 1961 Act, but its provisions are being replaced by regulations issued under the Health and Safety at Work etc Act 1974. The Act of 1974 established the Health and Safety Commission and has brought together within the Health and Safety Executive five inspectorates: factories; explosions; mines and quarries; nuclear installations; and alkali and clean air; plus the operation of the Employment Advisory Service.

   It imposes new comprehensive statutory obligations on employers and others, and presents new duties and rights to employees, see sections 2 to 9, but breach of these sections by themselves carries no liability to a civil action, only to criminal prosecution.

   Ultimately, the 1974 Act will supersede all existing safety legislation and replace it gradually with regulations and codes of practice which will, according to the wording, give a civil right of action.

   The following regulations were published by virtue of the powers conferred on the appropriate Minister of the Government Department concerned under the current Factories Act then in force:

   The Construction (General Provisions) Regulations 1961
   The Construction (Lifting Operations) Regulations 1961
   The Construction (Working Places) Regulations 1966
   The Construction (Health and Welfare) Regulations 1966

Building and engineering operations are governed by these regulations. Employers are responsible under the civil as well as the criminal law to their own employees for breaches of these regulations, and contractors and users of plant or equipment also owe certain duties under them. Employees and self-employed persons are also required to comply with the regulations.

However, the Factories Act itself is not applicable to new building work but it will apply to the occupier of a 'factory' within which building work is being done. If the occupier is the employer (principal) and the building is a 'factory', the employer will be liable for injuries due to a breach of this Act. Furthermore, this liability is not limited to injuries caused to the principal's own employees. In *Whitby* v *Burt Boulton and Hayward Ltd (1947)* occupiers of a factory employed builders to execute repairs to the factory and were held liable to the builder's labourer for failing to provide a safe means of access under section 26(1) of the 1937 Act (now section 29 of the 1961 Act).

According to *Canadian Pacific Steamships Ltd* v *Bryers (1958)*, regulations issued under the Act apply to any person employed in the factory whether the direct servant of the occupier or a servant of an independent contractor, so long as he is employed on work in the factory. Also, under certain sections of the Act, the occupier owes the same duty to an independent contractor as he would to such a contractor's employees. For example, in *Lavender* v *Diamints (1949)*, it was held that the duty under section 29 (safe means of access and safe place of employment) includes a duty to an independent contractor. A window cleaner who provided his own tackle fell through an asbestos roof while making his way towards the window, and was awarded damages against the occupier for the injuries sustained.

(d)   The Occupiers' Liability Act 1957 reduces the categories of persons entering premises to visitors and trespassers. The common duty of care is owed to all visitors which is a duty to take such care as in all the circumstances of the case is reasonable to see that the visitor will be reasonably safe in using the premises for the purposes for which he is invited or permitted by the occupier to be there. In contracts for new building work the site and works are occupied by the contractor until they are handed over to the employer and thus the contractor is the occupier within the meaning of the Act. However, in *Wheat* v *Lacon (1966)* the manager of a public house and the brewery company who owned the premises were held to be occupiers and so they each owed the common duty of care to their visitor. In similar circumstances both employers and contractors can be occupiers of premises under the Occupiers' Liability Act.

The duty owed by an occupier to a trespasser is that the former is expected to act with ordinary humanity. In *Pannett* v *McGuinness & Co Ltd (1972)* the Court of Appeal held that an occupier/contractor was not absolved by previous warnings to a child trespassing on a demolition site which was left unprotected at a time when the site was at its most dangerous and alluring.

(e)   Building Regulations. The current regulations which govern building works throughout England and Wales except the inner London Boroughs, which have their own regulations, are made under the Public Health Act 1961. They provide *inter alia* minimum standards of construction and design; make provision for the deposit of plans; and among other things require adequate foundations.

Civil liability can occur under these regulations on the part of contractors:

    (i)   contractually by express provisions under the contract – clause 6.1.1 in the JCT form;

    (ii)   by breach of statutory duty by failure to comply with the regulations. Lord Wilberforce in *Anns* v *London Borough of Merton (1978)* said:

> Since it is the duty of the builder, owner or not, to comply with the by-laws, I would be of the opinion that an action could be brought against him in effect for breach of statutory duty by any person for whose benefit or protection the by-law was made.

(f)   The Defective Premises Act 1972. The Law Commission's report on the Civil Liability of Vendors and Lessors for Defective Premises (Law Comm No. 40) summarises their four recommendations for amendment of the law. The first of these deals with liability for defects of quality in newly built dwellings. The remainder deal with liability for dangerous defects in premises of any description sold or let. They are as follows:

(1) Those who build, or undertake work in connection with the provision of new dwellinghouses (whether by new construction, enlargement or conversion) should be under a statutory duty owed to any person who acquires a proprietary interest in the property, as well as to the person (if any) for whom they have contracted to provide the dwelling, to see that the work which they take on is done in a workmanlike or professional manner (as the case may be), with proper materials and so that the dwelling will be fit for habitation. The same duty should be owed by those who in the course of trade or business or under statutory powers arrange for such work to be undertaken.

Section 1 of the 1972 Act achieves the object recommended, but it only operates if section 2 does not apply, i.e. if an approved scheme is in operation in relation to the dwelling then no action can be brought in respect of the duty in section 1 of the Act. The National House-Building Council (NHBC) operates a scheme that has been approved by the Secretary of State in accordance with the Act. The House Purchaser's Agreement HB5 is the basis of the NHBC scheme and is a contract between the builder and purchaser of the dwelling and it contains a warranty on the part of the builder to build in an efficient and workmanlike manner and of proper materials and so as to be fit for habitation.

(2) The vendor's and lessor's immunity from liability for the consequences of their own negligent acts should be abolished; that is to say, *caveat emptor* should no longer provide a defence to a claim against a vendor or lessor which is founded upon his negligence.

Section 3 of this Act clearly attains this objective (as far as negligence in building or other work is concerned which is carried out on the land before the sale or letting) and sub-section 3(2) prevents the section operating retrospectively on contracts which have previously been negotiated.

(3) A person who sells or lets premises should be under the general duty of care in respect of defects which may result in injury to persons or damage to property and which are actually known to him at the date of the sale or letting. Section 3

also achieves this recommendation, subject to the statement in brackets mentioned in the previous paragraph.

(4) A landlord who is under a repairing obligation or has a right to do repairs to premises let should be under the general duty of care in relation to the risk of injury or damage arising from a failure to carry out that obligation or exercise that right with proper diligence; but a landlord should not, by reason of a right to enter and repair on the tenant's default, be liable to the tenant for injury suffered by the tenant's default. Section 4 achieves the desired effect here.

(g)   The Limitation Act 1980. This Act is operative from 1 May 1981 and consolidates the Limitation Acts from 1939 to 1980. There is no substantial change in the law. Thus claims in simple contract must generally be brought within six years from the breach of contract and in a contract under seal within twelve years. A claim in respect of personal injuries for breach of duty must normally be brought within three years from when the injury occurred. In all other tort actions the period is six years from the date the action accrues. To avoid the injustice which might otherwise result when a cause of action for personal injuries becomes statute-barred before the plaintiff knew of it, the limitation period ends three years after the date of the plaintiff's knowledge of the cause of action if that date is after three years from the accrual of the cause of action. The Act provides a complicated definition of 'knowledge' which involves interpretation by case law. The Act also gives the court an overriding discretion to allow the plaintiff to continue with his action which is out of time. See further details on page 250.

As regards causes of action other than those for personal injuries it is important to establish when the cause of action arises. In *Pirelli General Cable Works* v *Oscar Faber & Partners Ltd (1982)* the House of Lords held that the cause of action accrued when the damage came into existence. In this case the plaintiffs engaged the defendants to advise them concerning, *inter alia*, the building of a chimney. The defendants were responsible for the faulty design and building of the chimney in June and July 1969. Because of unsuitable materials damage appeared in the shape of cracks near the top of the chimney not later than April 1970. The plaintiffs discovered the damage in November 1977. A writ was issued in October 1978 and it was established that the plaintiffs could not have discovered the trouble before October 1972, so it would have been impossible for them to have issued their writ within the conventional six years limitation period from when the damage occurred. The House of Lords held that the cause of action was time-barred because it accrued in 1970 **when the cracks came into existence**; also, where successive owners of property are concerned if time runs against one it must also run against all his successors in title. It remains to be seen whether similar statutory authority will be produced to control the balance of justice between plaintiffs and defendants in this area as applies to personal injury cases.

Insurance law

The insurance contract has a set of legal rules in addition to and, sometimes, in place of those normally applying to the law of contract. Good faith must be observed in all legal contracts, that is to say, the parties must not act fraudulently. However, this does not mean that they are bound to reveal all they know or ought to know, about the transaction. The rule *caveat emptor*, let the buyer beware, applies to these contracts. If a person buys goods and finds subsequently that they have serious defects, he has little effective right of action at common law (as distinct from statute law) against the seller, unless the latter made representations or gave a warranty as to the non-existence of those defects. Goods can be inspected, and an intending buyer is able to look for himself, or employ a specialist to carry out an inspection on his behalf.

Contracts of insurance are different because one party to the contract alone (the proposer) knows, or ought to know, all about the risk proposed for insurance, and the other party (the insurer) has to rely mainly upon the information given by the proposer in his assessment of that risk. For this reason contracts of insurance are contracts *uberrimae fidei*, of the utmost good faith.

For example, the duty of utmost good faith is not discharged by the insured by replying truthfully to all the questions on the proposal form. If there are any other material facts which might affect the underwriters' mind in considering the risk they must be disclosed. This means any material facts not covered by questions on the proposal. See *Hair* v *Prudential Assurance Co Ltd (1982)*.

The following list, which is not exhaustive, should assist in the completion of the proposal form:

(a) Answer all questions fully and completely honestly. Although this may seem elementary it is often not complied with. Even if questions are not applicable this should be stated explaining why the question does not apply unless it is obvious.

(b) Disclose all matters which a prudent insurer would wish to consider in deciding whether to offer cover. If in doubt disclose the matter.

(c) Obtain a specimen of the policy required. Check that the exclusions and conditions are understood and are acceptable.

(d) Any special explanations required concerning the cover should be obtained in writing and the correspondence retained; similarly in the case of policy extensions.

(e) Sums insured and limits of indemnity must be adequate.

(f) Keep a copy of the completed proposal.

Throughout the policy period:

(a) Notify any change in risk which is basic to the cover provided, e.g. an extension of work which is undoubtedly more hazardous than indicated on the proposal.

(b) Notify claims immediately. If in doubt notify the circumstances concerning a possible claim.

(c) In the case of liability insurance claims make no admission, offer, promise or payment to the third party and pass relevant correspondence to the insurer immediately.

If renewal is intended: check that the proposal answers are still correct and notify changes or requirements which come to light.

The second principle of insurance is that the insured must have an insurable interest in the subject-matter of the insurance whether it is property, life or limb, or potential liability devolving upon the insured. That is to say, he must bear some relationship, recognised by law, to the subject-matter whereby he benefits by the safety of the property, life, limb or freedom from liability and is prejudiced by any loss, damage, injury or creation of liability.

Thirdly, all the policies mentioned in this book are policies of indemnity. This means that the object of these insurance contracts is to place the insured as nearly as possible in the same financial position after a loss as that occupied **immediately before** the happening of the insured event. It would be against public policy to allow the insured to make a profit out of the happening.

Fourthly, there is the principle of subrogation. This is the right which the insurer has of standing in the place of the insured and availing himself of all the rights and remedies of the insured, whether already enforced or not, but only up to the amount of the insurer's payment to the insured. This right of subrogation is exercisable at common law after the insurer has paid the claim. However, a condition of the policy may entitle the insurer to exercise the right before the payment is made.

Finally, the principle of contribution, like subrogation, is a corollary of indemnity. It only applies between insurers. Contribution is the right of an insurer who has paid under a policy to call upon other insurers equally or otherwise liable for the same loss to contribute to the payment. Before the principle can be applied, the insurances called into contribution must cover (a) the same interest, (b) the same subject-matter, and (c) the same peril.

The statements made above have recently been verified in the Scottish appeal case of *Steelclad Ltd* v *Iron Trades Mutual Insurance Co Ltd (1984)*. However, the interesting part of that case was the court's view of the words in the contribution condition of the policy of the defenders which refers to loss or damage insured by any other policy *'effected by the insured . . . on his behalf'* when the defenders would not be liable except in respect of any excess beyond the amount payable under such other policy. The pursuers were the insureds under the Iron Trades policy and were subcontractors on a project where the employer having the work done had arranged a project policy covering the employer, all contractors and all subcontractors and the contribution clause in that policy was worded exactly the same as the Iron Trades policy except for the phrase mentioned above in quotes. The court considered the project policy was not a policy effected by the insured *on his behalf*. The phrase was ambiguous and was therefore construed *contra proferentem* against the insurer who had refused to contribute. The project policy insurers had already paid almost a half of the claim and while the court did not have to go any further having come to a decision on this wording they did in effect say that even without this phrase or if the project policy had been effected on the pursuer's behalf they *would still not allow the two policy conditions to cancel the cover given* (by neglecting in each case the condition which was worded exactly the same way in the other policy) with the result that if the loss is covered elsewhere it is covered nowhere. The correct decision, and surely the intention, is that each policy should contribute subject always to a rateable proportion condition. Basically an insured must get an indemnity whether he insures once or twice the same subject-matter against the same peril, subject always to the other terms and conditions of the policy or policies. It is rather strange that the matter should ever have been in doubt.

If a policy contains an excess wording in the contribution condition stating that it does not operate until the second policy (which would otherwise be involved in contribution) is exhausted and the second policy does not contain such a clause, then the principle of contribution will *not* operate between the two policies. See the New Zealand Court of Appeal decision in *State Fire Insurance General Manager* v *Liverpool & London & Globe Insurance Co Ltd (1952)* NZLR 5. In the Steelclad case both policies contained such an excess wording.

The insurance contract

An insurance contract is an agreement whereby one party, the insurer, in return for a consideration, the premium, undertakes to pay to the other party, the insured, a sum of money or its equivalent upon the happening of a specified event which is against the insured's financial interest.

The contract is normally contained in a policy. This document is evidence of the contract which has usually come into existence before the policy is issued.

Policies vary with different classes of insurance and those dealt with in this book can be divided into the following sections:

(1) Recital clause, which refers to a schedule for details (see below). This clause also refers to the proposal form and its declaration as being the basis of the contract. However one qualifies 'the event' it must be uncertain.

(2) Operative clause, which describes the cover which is the subject-matter of the insurance.

(3) Exceptions which help to describe the cover by stating what is excluded.

(4) Signature clause on behalf of the insurers.

(5) Schedule, which contains the names of the parties, the address of the insured, the period of insurance, possibly the business of the insured, the sums insured or limit(s) of indemnity (in liability insurance), and the premium payable.

(6) Conditions, which limit the insured's legal rights, stipulate the various things the insured may or may not do and sometimes express or amend the common law or indicate an agreed state of affairs. It is usual to classify conditions (whether expressed or implied) as:

(a) Conditions precedent to the policy, e.g. all material facts must be disclosed during negotiations preceding the insurance contract.

(b) Conditions subsequent of the policy, e.g. notice by the insured of a change of risk during the period of the policy.

(c) Conditions precedent to liability, e.g. the notice of loss by the insured.

The following is the result of this classification:

(a)   Conditions precedent to the policy must be observed for the insurance to be valid from the beginning.

(b)   Conditions subsequent of the policy refer to matters arising after the contract has been completed and affect the validity of the policy from the date of breach of the condition.

(c)   Conditions precedent to liability only affect the claim which the breach of the conditions concerns, the policy remaining in force. See chapter 19 for further details.

The policy may be amended from time to time, with the agreement of the parties, by endorsements which are attached to the policy.

A condition by which the insured undertakes that some particular things will be done or not done or that a state of affairs exists or shall continue to exist is referred to as a warranty or a continuing warranty. Breach of such condition will probably allow the insurer to repudiate liability as a warranty must be complied with strictly and literally.

Occasionally a so-called condition in the policy has been held to be a mere stipulation. Thus it has been stated that the condition in the employers' and public liability policies requiring the insured to keep a proper wages book in order to return his annual wage roll on which the premium is based is a mere stipulation, a breach of which results in the insurer possibly obtaining damages as compensation but not allowing him to repudiate policy liability.

There is a rule of evidence among many concerning the interpretation of the insurance policy which is worth mentioning as it occasionally arises. It is called the *contra proferentem* rule and states that in the event of ambiguity a document will be construed strictly against the party who has drawn it up. Thus the interpretation less favourable to the insurer will be taken.

From what has been said under the last two headings concerning non-disclosure and breach of warranty, it will be appreciated that insurance law is tilted in favour of the insurer. However, the Law Commission's Working Paper No. 73 entitled 'Insurance Law, Non-disclosure and Breach of Warranty' recommends reforms to improve the insured's position.

The ICE Conditions

The liability and insurance provisions of the ICE Conditions, including responsibility for the works, concern mainly clauses 20 to 25 inclusive. However, clause 10 concerning sureties, sub-clause 54(3) relating to the vesting of materials and goods not on site, and clauses 58 and 59 on subcontracting are also involved.

These clauses are dealt with by quoting the wording and commenting on the features which concern the insurance industry. At suitable stages the appropriate policies or bonds will be considered.

Unless otherwise stated, only the fifth edition of the ICE Conditions, the third edition of the FIDIC Conditions and the ICE Form of Subcontract (as amended 30 March 1973) will be commented upon.

The word 'employer' is used in two senses in this book. Thus, when meaning the master in the master and servant relationship, e.g. the employers' liability policy, and when meaning the principal (the party to the contract who is having the work done), e.g. the employer in the ICE Conditions. When there is any likelihood of confusion the words 'master' or 'principal' will be used in parentheses after the word 'employer'.

There are five points concerning contract conditions generally. The first relates to the pattern of the main clauses under discussion. The second concerns the failure to print all the common policy exceptions in the contract conditions. The third indicates the lack of precision in drafting. The fourth concerns the effect of the Unfair Contract Terms Act 1977, and the fifth the effect of a common law requirement.

The pattern of the clauses   The usual pattern of these clauses in contract conditions is for a liability or responsibility clause to be followed by an insurance clause relating to that liability or responsibility. 'Liability clause' means a clause imposing a legal liability to indemnify the employer and 'responsibility clause' means a clause imposing a responsibility for loss or damage to the subject-matter of the contract, such as the works in a building contract, or plant in a plant agreement. Thus in the ICE Conditions clause 20 deals with the contractor's responsibility for any loss or damage to the works, and consequential reinstatement, from any cause other than those specified as excepted risks. Clause 21 is the corresponding insurance clause which requires the contractor to insure the works in the joint names of himself and the employer against all loss or damage from whatever cause for which he is responsible under the terms of the contract (and for this responsibility it is necessary to refer back to clause 20, bearing in mind the excepted risks).

Clause 22, which follows, is a liability clause for injury to persons or damage to property arising out of the construction or maintenance of the works. There are certain limitations on the contractor's liability to indemnify the employer. Then there is the complementary insurance clause 23 requiring the contractor to insure against the liability imposed by clause 22 and the policy is required to indemnify the employer in circumstances where the contractor is entitled to an indemnity. In the fourth edition sub-clause 24(1) referred to the liability of the contractor to indemnify the employer against the latter's liability for injuries to workmen in the employment of the contractor or subcontractor except when they are due to the employer's negligence, and sub-clause 24(2) required the contractor to insure against his liabilities under 24(1). In the fifth edition sub-clause 24(2) has been deleted because insurance under this heading is now compulsory but only for claims by employees of the insured contractor, as employees of subcontractors are third parties to the contractor and are not his employees. However, these clauses are still a good example of the usual pattern.

Lack of recognition of all common policy exceptions in contract conditions   The second general point concerns the failure to print all the common insurance policy exclusions in the contract insurance clauses, or to recognise the existence of policy conditions which can affect cover.

It seems that where one of the parties to a contract is able to dictate terms, inevitably the contract favours that party. However, where a standard contract is evolved in industry the wording is usually agreed between bodies representative of both parties to the agreement, and a contract fairer to both sides emerges — with one important exception. When the insurance clauses are imposed and accepted both parties to the contract seem to assume that insurance policies cover all risks and are without exclusions. This is very strange when one considers that these contracts are compiled by businessmen or their lawyers and even the man in the street is aware of the allegation made against insurers (however unfairly) that they use small print to set out their policy exceptions.

It is true that recently at least one or two comparatively successful attempts have been made after consultation with the British Insurance Association to match the contract insurance clauses to the usual type of cover granted by insurers, namely sub-clause 21(2)(1) and clause 21(3) of the JCT Contract and to some extent the fifth edition of the ICE Conditions. But as yet it is still common practice in most standard contracts to impose and accept a liability to insure which is impossible to fulfil completely, because the cover is not given by insurers without exceptions.

The British Insurance Association panel was consulted by those concerned with the preparation of the fifth edition of the ICE Conditions and while there has been an improvement in the wordings of the clauses mentioned, as will be seen in the following chapters, there is still room for improvement on this subject of printing insurance policy exclusions in the contract clauses concerning insurance. For example, nuclear risks are not mentioned as an exception in clauses 22 and 23 concerning liability (although found in clauses 20 and 21 concerning responsibility for the works) but this exception will always be found in the public liability policy which provides the cover required by these clauses; yet strictly speaking the contract conditions (in clauses 22 and 23) do not allow such an exception. Incidentally, clause 65 of the ICE Conditions protects the contractor in respect of damage or injury directly or indirectly due to war and kindred risks, which is an improvement on clauses 32 and 33 of the JCT Contract which only refer to war damage **to the works** and do not relate to kindred risks. The public liability policy excludes war and kindred risks.

Lack of precision in drafting It is surprising after so many years of standard forms of contract that words having a legal meaning are used in conjunction with general words without a legal meaning, thus leaving the reader in doubt as to whether the general words add anything or are superfluous. In the former case the drafter should be more specific by using additional words with a legal meaning to indicate his intention and in the latter event he should leave out the general words which are superfluous. See for example paragraphs (2) and (6) below.

Another fault is to use different phrases apparently to mean the same thing. See paragraph (1) below.

In the ICE Contract the following improvements in clauses 20 to 24 are suggested.

(1)   Omit the word 'injury' from clause 20(2) and use the same phrase in clause

20(2) as is used in the insurance clause 21. In this way both clauses will use exactly the same wording. This suggestion is on the assumption that nothing is intended to be added by the word 'injury' in clause 20(2). It cannot refer to bodily injury as clause 20 is the responsibility for the works clause.

(2)   Assuming in clauses 22(1) and (2) the phrase 'act or neglect' is only intended to mean common law negligence (according to the cases of *Hosking* v *De Havilland Limited (1949)* and *Murfin* v *United Steel Companies Limited (1957)* the phrase act or neglect' used in the indemnity clause 14(a) of the 1939 RIBA Form of Contract, which is similarly worded to clause 20(1) of the JCT 1980 edition, is limited to common law negligence and is not applicable to breach of statutory duty), then omit the word 'act' and leave the word 'neglect' or 'negligence' as the operative word. On the other hand, if the intention is to include the other torts, then they should be named, e.g. nuisance, trespass and, if desired, breach of statutory duty.

In *Strathclyde Regional Council* v *James Waddell and Son Ltd (1982)* a local authority contracted with the defenders for the execution of civil engineering works connected with the construction of a public sewer. The works were carried out under powers given by the Sewerage (Scotland) Act 1968, which Act also provided for the local authority making compensation to anyone suffering loss, injury or damage by reason of an exercise of those powers. The local authority met a claim for compensation by a tenant farmer who was alleged to have suffered loss as a result of certain failures by the contractors to carry out temporary fencing and water supply works. The contractors refused to repay the local authority on the grounds that: (a) the liability of the local authority under the 1968 Act was limited to circumstances where the exercise of the powers had not involved any negligence; and (b) anyway the indemnity contained in the ICE Conditions of Contract (fourth edition) under which the contractors had worked did not extend beyond damage to property arising out of an act or neglect by the contractors. It was held that (i) the claims under the 1968 Act were not confined to losses caused without negligence, and (ii) the indemnity clause covered a claim for compensation by a person affected by the works. It should be explained that in the fourth edition of the ICE Conditions there is no exception to the contractor's liability to indemnify the employer for breach of statutory duty by the employer (as in the fifth edition). However, it would seem that the court did **not** consider that the words in the fourth edition 'any act or neglect done or committed during the currency of the contract by the employer', for which the contractor did not have to indemnify the employer, included breach of statutory duty by the employer. This in effect is a similar decision to Hosking's case on the same phrase in the JCT Form mentioned earlier.

In sub-clauses 22(1)(a), 22(1)(b)(v) (where breach of statutory duty is included) and in clause 22(2) preferably the phrases used should be identical. In any event the word 'act' should be omitted.

(3)   It should be made clear that sub-clause 22(1)(b)(v) is overridden by sub-clause 22(1)(a) so far as it relates to contributory negligence or any contributory tort by the employer and those for whom he is responsible. Otherwise there is incon-

sistency between proviso 1(a) (reduction of contractor's indemnity where the employer contributes to the injury or damage) and proviso 1(b)(v) (no contractor's indemnity where the employer is only a contributor). However, it is arguable that the intention is to apply liability proportionately, as when the position is reversed (the employer indemnifying the contractor) in clause 22(2), proviso 1(b)(v) applies liability proportionately.

(4)   In clause 22 add another exclusion sub-clause namely 22(1)(b)(vi) excluding liability for nuclear risks which will cater for the exclusion in the public liability insurance policy. This assumes that war risks are considered to be adequately catered for in clause 65.

(5)   As specific provisions usually prevail over general ones at law clause 22 should exclude claims for accident or injury to employees of the contractor or his sub-contractors as these are dealt with by clause 24.

(6)   Unless the drafters have some other legal wrong(s) in mind in clause 24 the phrase 'act or default' should be replaced by the same phrase to be used in clause 22. If they have, then those legal wrongs should be stated.

The effect of the Unfair Contract Terms Act 1977 Negotiated forms of standard contract do not usually contain a term by which liability is avoided in the event of negligence by either party to the contract. A few do, and this Act, by section 2(1), invalidates such a term where it relates to negligence resulting in death or personal injury. If loss or damage results the term is subject to a test of reasonableness and the Act would abrogate the unreasonable term, by section 2(2). Furthermore, section 3 requires a test of reasonableness in deciding whether a term avoiding liability for breach of contract, in written standard terms of business, is valid.

The peculiarity of clause 20 of the ICE Conditions is that the excepted risks (the risks which affect the works where the employer accepts the expense of repair) do not include the employer's negligence. The effect was illustrated in *A.E. Farr Ltd v The Admiralty (1953)* which involved the General Conditions of Government Contracts for Building and Civil Engineering Works known as Form CCC/Works/1, Edition 5, which had a similar clause to clause 20 of the ICE Conditions. The plaintiffs were building a jetty for the Admiralty when, due to negligent navigation, an Admiralty vessel collided with it causing damage. The phrase 'any cause whatso-ever' (common to both the ICE and the CCC/Works/1 contracts) was given a wide meaning and no exception existed concerning the employer's negligence, although the current GC/Works/1 edition has such an exception.

Consequently the decision went against the contractors and they had to pay for the repair of the jetty and this is the present position under the ICE Conditions. However, see *Smith v South Wales Switchgear (1978)* mentioned under the next heading. The fact that the employer is indemnified anyway because of joint names policy cover required by clause 21 of these Conditions makes the point under discussion rather academic. See the remarks at the end of this chapter.

The question is whether clause 20 of the ICE Conditions, in so far as it makes the contractor responsible for loss or damage to the works caused by negligence on the part of the employer, satisfies the requirement of reasonableness. Is there a breach of section 2 of the 1977 Act? Possibly not, as the intention of the parties may well prevail in deciding reasonableness (for guidelines, see later). The representative bodies of both parties to the contract have had the opportunity to amend the contract and insert an exception in case of the employer's negligence (as the Government have done in their GC/Works/1 contract) because there have been two editions since Farr's case was heard in 1953 but no such alteration has been made.

The reference in section 3 of the 1977 Act to a contracting party who 'deals . . . on the other's written standard terms of business' is vague as to whether it extends, for example, to any employer under the ICE Conditions, who is not a member of one of the bodies sponsoring those conditions. Presumably section 3 does include the client of an architect or engineer where the contract incorporates the JCT or ICE Conditions.

Fortunately the ICE Conditions in clause 22 follow the normal pattern of making each party to the contract responsible for his own negligence. Thus no major alteration by the 1977 Act is anticipated. However, those employers who might consider altering the wording of the Conditions should bear in mind the terms of this Act which is effective from 1 February 1978.

There are two other aspects to consider regarding this Act and the first concerns the fact that, apart from the case where one of the parties to the contract is a consumer, the Act does not appear to apply to indemnity clauses. The argument in support of this statement is that section 4 of the Act states that a person **dealing as consumer** cannot by reference to any contract term be made to indemnify another for his negligence or breach of contract except in so far as the contract term satisfies the requirement of reasonableness. Incidentally a party 'deals as a consumer' in relation to another party if:

(a) he neither makes the contract in the course of a business nor holds himself out as doing so; and

(b) the other party does make the contract in the course of a business; and

(c) in the case of a contract governed by the law of sale of goods or hire-purchase, or by section 7 (which refers to contracts involving work and the supply of materials, e.g. the ICE Conditions) the goods passing under or in pursuance of the contract are of a type ordinarily supplied for private use or consumption.

Thus in the great majority of cases where the ICE Conditions apply neither party to the contract is dealing as a consumer because (a) above does **not** apply, i.e. usually as well as the contractor making the contract in the course of a business the employer does so as well. The conclusion must be that unless section 4 is superfluous, which must not be presumed, indemnity clauses in contracts between non-consumers are not affected by the 1977 Act. Therefore clause 22 which is an indemnity clause in the large majority of cases will not be caught by the Act, even if altered, because the employer usually makes the contract in the course of a business, and the contractor always does so.

The final aspect concerning this Act and the ICE Conditions is that the test of reasonableness is stated by Schedule 2 of the Act to depend upon certain 'guidelines'. Now while these guidelines apply particularly to certain sub-sections they operate in respect of the whole Act whenever the reasonableness test is applicable. The first guideline and the one used most often in practice to prove the term concerned is reasonable reads 'the strength of the bargaining positions of the parties relative to each other . . .' which implies that business companies and firms (certainly those of similar size) are capable of looking after themselves or can obtain legal advice to protect themselves and if they do not they must take the consequences, i.e. the term even though *prima facie* harsh will be considered reasonable. It is significant that little or no case law has been produced concerning this aspect since the Act came into force on 1 February 1978, which lends force to the argument that when companies and firms contract it is not easy to show that a term is unreasonable*.

The other guideline to reasonableness which applies particularly to standard forms reads 'whether the customer knew or ought reasonably to have known of the existence and the extent of the term (having regard, among other things, to any custom of the trade and any previous course of dealing between the parties)'.

It is difficult for a member of the representative bodies named on the front of the ICE Conditions to argue that he did not know of the terms of the standard form approved by his own representative organisation particularly if he is using the form frequently. See also the quotation from *Photo Production Ltd* v *Securicor (1980)* at the end of the next section.

The common law position The position outside the Unfair Contract Terms Act 1977 is governed by the House of Lords decision in *Smith and Others* v *South Wales Switchgear Ltd (1978)*.

In Smith's case the standard contract terms applicable in an overhaul contract contained an indemnity clause as follows:

> In the event of . . . the carrying out of work by the Supplier (South Wales Switchgear) . . . on land and/or premises of the Purchaser (Chrysler), the Supplier will keep the Purchaser indemnified against . . . (b) Any liability, loss claim or proceedings whatsoever under Statute or Common Law (i) in respect of personal injury to, or death of, any person whomsoever . . . arising out of or in the course of or caused by the execution of this order (the contract works).

The injured man was an electrical fitter employed by South Wales Switchgear at Chrysler's factory. The trial judge held that the accident had been caused by negligence and breach of statutory duty by Chrysler and this decision was not appealed. The judge also decided that Chrysler were entitled to be indemnified by South Wales Switchgear and this was appealed in Scotland and affirmed. On appeal to the House of Lords it was decided on a proper and strict construction of the indemnity clause wording that it was not wide enough to cover negligence on the part of Chrysler.

---

* However, see *Phillips Products* v *Hampstead Plant Hire (1983)* where in the circumstances clause 8 of the CPA Conditions was considered unreasonable.

The House of Lords disagreed that the phrase 'all claims or demands whatso-ever' (a similar phrase to that quoted above) was wide enough to include negligence as stated in *Gillespie Brothers Ltd* v *Roy Bowles Ltd (1973)*. They said that an express agreement to accept liability for negligence required the word 'negligence' or a synonym for it. It was held in Smith's case that the clause only applied to liabilities incurred by the factory owners for the acts or omissions of other parties in connection with the contract work. Moreover, the clause could also be read as covering the far from fanciful liability of the factory owners at common law for the acts or omissions of the other parties, as occupiers of the factory or as employers.

The reason the court referred to the application of the clause in Smith's case to liabilities other than Chrysler's own direct negligence is that they adopted the test put forward by Lord Greene M.R. in *Alderslade* v *Hendon Laundry Ltd (1945)* and quoted by Lord Merton in *Canada Steamship Ltd* v *The King (1952)*. He indicated three principles, viz:

(1)   If the clause contains language expressly exempting a party from the consequences of his negligence effect must be given to that provision.

(2)   If there is no such express reference, the court must consider whether given their natural meaning the words are wide enough to cover negli-gence. If there is a doubt the clause is construed against the person in whose favour it is made.

(3)   If the answer to (2) is 'yes' then the court must consider whether the head of damage may be based on some ground other than that of negligence. Such other ground must not be fanciful or remote, but, subject to such qualification, the existence of a possible head of damage other than negligence is fatal to any attempt to apply the clause to negligence.

Consequently, although the decision in Smith's case is sometimes said to be that to contract out of negligence requires the use of the word 'negligence' or a synonym for that word, which is what the House said about the first principle mentioned above, the other principles must not be forgotten. For example, in the *'Raphael'* (1982) there was no mention of negligence or a synonym but the Court of Appeal decided that the words 'any act or omission' were wide enough to comprehend negligence, and the main point was that the third general proposition above applies, i.e. the words could only apply to negligence.

The ICE Conditions, as already mentioned, do not give an indemnity to the employer for his own negligence in clause 22 but if the contract was amended to this effect it would possibly fall foul of Smith's decision. See Lord Keith's remarks below. As regards *Farr* v *The Admiralty* mentioned earlier, Smith's case now appears to alter the decision in Farr's case. Clause 20 uses the words 'any cause whatsoever' in making the contractor responsible for the works and Farr's case decided that such a phrase was wide enough to include negligence. While the decision in Smith's case disagrees with Farr's decision on this point and being a House of Lords decision would override Farr's case, this assumption has yet to be tested in the courts and it must be borne in mind that if Smith's decision is followed the result would be contrary to the intention of the representative bodies of the parties to the contract, who have been content not to alter the effect of Farr's decision by amending the wording. In this connection it should be noted that in Smith's case Lord Keith said:

The matter is essentially one of ascertaining the intention of the contracting parties from the language they have used, considered in the light of surrounding circumstances which must be taken to have been within their knowledge.

While Smith's case was heard before the Unfair Contract Terms Act 1977 came into force, it was said by Lord Wilberforce in the House of Lords in *Photo Production Ltd v Securicor (1980)* (which also concerned a contract entered into before the 1977 Act was passed):

After this Act [the 1977 Act], in commercial matters generally, when the parties are not of unequal bargaining power, and when risks are normally borne by insurance, not only is the case for judicial intervention undemonstrated, but there is everything to be said, and this seems to have been Parliament's intention, for leaving the parties free to apportion the risks as they think fit and for respecting their decisions.

Therefore it is possible that the 1977 Act will have no effect on standard forms agreed by the representative bodies of both parties which makes the parties of equal bargaining power and where the risks are normally borne by insurance. However, this still leaves open the question of the effect of Smith's case on Farr's decision.

Perhaps it is timely to mention that the above statutory and common law points are to some extent academic as far as the ICE Conditions are concerned, not only because clause 22 does not give the employer an indemnity in respect of his negligence, but also because it is the intention that the insurance required by clause 23 should cover the employer's negligence to the extent that the claim is one 'in respect of which the contractor would be entitled to receive indemnity under the policy'. Furthermore, by clause 21 the insurance of the works would indemnify the employer even in respect of his negligence as the policy is required to be in the joint names of the contractor and the employer, and the employer's negligence is neither an excepted risk under the contract nor an exception under the policy, which in this case is the contractors' all risks policy.

Chapter 2

Responsibility for the works: clause 20

Care of the Works

20. (1) The Contractor shall take full responsibility for the care of the Works from the date of the commencement thereof until 14 days after the Engineer shall have issued a Certificate of Completion for the whole of the Works pursuant to Clause 48. Provided that if the Engineer shall issue a Certificate of Completion in respect of any Section or part of the Permanent Works before he shall issue a Certificate of Completion in respect of the whole of the Works the Contractor shall cease to be responsible for the care of that Section or part of the Permanent Works 14 days after the Engineer shall have issued the Certificate of Completion in respect of that Section or part and the responsibility for the care thereof shall thereupon pass to the Employer. Provided further than the Contractor shall take full responsibility for the care of any outstanding work which he shall have undertaken to finish during the Period of Maintenance until such outstanding work is complete.

Responsibility for Reinstatement

(2) In case any damage loss or injury from any cause whatsoever (save and except the Excepted Risks as defined in sub-clause (3) of this Clause) shall happen to the Works or any part thereof while the Contractor shall be responsible for the care thereof the Contractor shall at his own cost repair and make good the same so that at completion the Permanent Works shall be in good order and condition and in conformity in every respect with the requirements of the Contract and the Engineer's instructions. To the extent that any such damage loss or injury arises from any of the Excepted Risks the Contractor shall if required by the Engineer repair and make good the same as aforesaid at the expense of the Employer. The Contractor shall also be liable for any damage to the Works occasioned by him in the course of any operations carried out by him for the purpose of completing any outstanding work or of complying with his obligations under Clauses 49 and 50.

Excepted Risks

(3) The 'Excepted Risks' are riot war invasion act of foreign enemies hostilities (whether war be declared or not) civil war rebellion revolution insurrection or military or usurped power ionising radiations or contamination by radio-activity from any nuclear fuel or from any nuclear waste from the combustion of nuclear fuel radioactive toxic explosive or other hazardous properties of any explosive nuclear assembly or nuclear component thereof pressure waves caused by aircraft or other aerial devices travelling at sonic or supersonic speeds or a cause due to use or occupation by the Employer his agents servants or other contractors (not being employed by the Contractor) of any part of the Permanent Works or to fault defect error or omission in the design of the Works (other than a design provided by the Contractor pursuant to his obligations under the Contract).

Care of the works: sub-clause 20(1)

This clause was defined at the beginning of this book as a 'responsibility' clause indicating where responsibility for the subject-matter of the contract (the works) lies. 'Works' are defined in clause 1(1) as meaning 'the permanent works together with the temporary works'. The 'permanent works' are defined as 'the permanent works to be constructed completed and maintained in accordance with the contract', and similarly the 'temporary works' are stated to mean 'all temporary works of every kind required in or about the construction completion and maintenance of the works'. It is relevant to note that the form of tender* commences with a short description of works which reads:

All Permanent and Temporary Works in connection with [and then follows a space for completion as appropriate].

This form of tender is clearly an offer to construct and complete the whole of the said Works and maintain the Permanent Works in conformity with the said Drawings, Conditions of Contract, Specification and Bill of Quantities for such sum as may be ascertained in accordance with the said Conditions of Contract. Also clause 8(1) requires the contractor to construct complete and maintain the works. The additional category of 'temporary works designed by the engineer' is introduced by clause 8(2) and 'temporary works specified or designed by the engineer' in clause 26(2)(c). As a result of these two clauses the contractor is not responsible for the design or specification of the permanent works (except as may be expressly provided in the contract) or of any temporary works designed by the engineer nor for obtaining planning permission in respect of such permanent or temporary works.

Taking the wording of clause 20(1) in stages the clause opens with the statement that:

'The Contractor shall take full responsibility for the care of the Works'
Subject to what has been said already bearing in mind the contractor's duty to hand over completed works to the employer, this phrase makes the contractor liable to make good at his own expense (irrespective of insurance cover) any damage to the works before the completion date next specified.

'from the date of the commencement thereof until 14 days after the Engineer shall issue a Certificate of Completion for the whole of the Works pursuant to Clause 48.'
This statement makes it clear that the date of the issue of the certificate of completion is relevant and the employer now has fourteen days in which to arrange his insurance cover for the completed works. When issuing the certificate the engineer should at the same time warn the employer accordingly. The contractor's all risks policy must cover the fourteen days just mentioned. Clause 48 states that the whole of the works must be substantially completed and have passed any final test

* See the end of the ICE Conditions for the form of tender and appendix which forms part of the tender.

prescribed by the contract. The contractor's notice in writing to that effect to the engineer or his representative must be accompanied by a written undertaking to finish any outstanding work during the period of maintenance (stated in the appendix to the form of tender and running from the date of completion of the works or any section or part thereof certified by the engineer in accordance with clause 48 as the case may be, see clause 49(1)).

If in the opinion of the engineer the works are not substantially completed he shall give instructions in writing to the contractor specifying the work to be done before the issue of a certificate. If the works are substantially completed the engineer is required to issue a certificate of completion (with a copy to the employer) within twenty-one days of delivery of the contractor's notice or completion of the necessary work specified by the engineer. Clause 48 also makes provision for sectional completion and clause 20(1) makes provision for the contractor's responsibility for the care of the works to cease in accordance with that sectional completion fourteen days after the engineer shall have issued the certificate of completion in respect of the section concerned and responsibility for the care thereof shall thereupon pass to the employer. Consequently this sub-clause connects the cessation of responsibility for the care of the works with the certification of completion procedure under clause 48.

The contractor's obligation to complete the works is set out in clause 8 and this means that in the event of loss or damage before completion the expense of repair and replacement falls on the contractor. This situation was illustrated in *Charon (Finchley) Ltd* v *Singer Sewing Machine Co Ltd (1968)* where the defendants employed the plaintiffs to convert premises into a shop. When work on the shop was almost complete vandals caused damage. Repairing the damage involved doing again work already done and paid for. The plaintiffs sued for the cost of this extra work. The court held that Singer were entitled to have the work completed and the builders had to bear the cost. The court quoted *Hudson's Building and Engineering Contracts*, ninth edition, page 223:

Indeed, by virtue of the express undertaking to complete (and, in many contracts maintain for a fixed period after completion) the contractor would be liable to carry out work again free of charge in the event of some accidental damage occurring before completion even in the absence of any express provision for protection of the work.

Apparently this was not a JCT or ICE Contract but the decision indicates the legal position.

A possible exception is the law of frustration. Lord Radcliffe in *Davis Contractors Ltd* v *Fareham UDC (1956)* said:

Frustration occurs whenever the law recognises that without default of either party a contractual obligation has become incapable of being performed because the circumstances in which performance is called for would render it a thing radically different from that which was undertaken by the contract.

In such circumstances the rights of the parties are decided under the Law Reform (Frustrated Contracts) Act 1943. This Act applies under English law when a contract has become impossible of performance as described in the above quotation and the parties to the contract have, for that reason, been discharged from the further performance of the contract. The Act *inter alia* allows a party to make a claim for expenses incurred or for benefits conferred on the other party.

There are not many cases where it can properly be said that a construction contract may have been truly frustrated in law and there are few cases where such an allegation has succeeded. Therefore the opinion of the Judicial Committee of the Privy Council in restating and applying established principles to a modern contract is worth reading in *Wong Lai Ying and Others* v *Chinachem Investment Co Ltd (1979)*.

One effect of the responsibility of the contractor for the works from commencement until fourteen days after the certificate of completion of the whole works is that the responsibility to insure in conjunction with the provisions concerning certification of completion of sections and parts of the works (see clause 48) avoids any repetition of the case of *English Industrial Estates Corporation* v *George Wimpey (1973)* which concerned the Standard Form of Building Contract and not the ICE Conditions.

The responsibility to insure in the case of sectional completion was considered in this case, although this concerned a matter of alleged sectional completion. A factory owned by English Industrial Estates Corporation but leased to Reed Corrugated Cases Ltd was damaged by fire. In 1969 Reed wanted to extend the factory, while continuing to make corrugated cardboard, to install a large new machine and have storage space for hundreds of reels of paper. George Wimpey & Co were awarded the contract which was arranged on the standard JCT Form incorporating the bills of quantities. Wimpey had a blanket contractors' all risks policy which provided the cover required by sub-clause 20[A](1).

By January 1970, a great deal of the work had been carried out by Wimpey's and Reed had installed the machine and had stored 1500 reels in the new reel warehouse. On 18 January 1970 much of the new factory was gutted by fire, the damage being estimated at £250,000. At that time the contractors had not finished their work.

The Corporation argued that, as the works had not been completed, it was Wimpey's duty to insure and that the loss, therefore, should fall upon them or their insurers. Wimpey, however, argued that the Corporation, through Reed, had taken possession of several parts of the works and that, under clause 16, the risk had passed to the Corporation in respect of those parts.

Clause 16 provides that:

If at any time or times before practical completion of the works the employer, with the consent of the contractor shall take possession of any part or parts of the same, then, notwithstanding anything expressed or implied elsewhere in the contract such parts shall, as from the date on which the employer shall have taken possession thereof be at the sole risk of the employer as regards any of the contingencies referred to in Clause 20[A].

The matter was complicated by special provisions in the bills.

Lord Denning was prepared to consider clause 16 without placing any reliance on the provisions just mentioned. In his opinion the words 'taking possession' of a part of the works must be so interpreted as to give precision to the time of taking possession and in defining the part, because of the important consequences which followed on it.

To achieve this precision, the parties themselves had evolved suitable machinery to determine it by way of a definite handing over of the part by the contractors to the employers. The practice was for the contractors to tell the architect that a part was ready for hand-over. The architect would inspect it and, if satisfied, would accept it on behalf of the employers. He would give a certificate defining the part, its value and the date of taking possession. The hand-over was thus precise and definite. It was the accepted means of defining hand-over.

In Lord Denning's view, the contractors at the time of the fire had not handed over to the employers the responsibility for the board machine house, the reel warehouse or the two extensions. Although Reed were using those places, it was the contractor's responsibility to insure them until actual hand-over. The risk remained with them and they, or their insurers must bear the loss. While the other two judges in the Court of Appeal came to the same conclusion as Lord Denning they merely considered that some formality was required in interpreting clause 16.

However, the ICE Conditions make it clear that the issue of a certificate of completion is essential for the contractor's responsibility for the works to cease fourteen days after the issue of the certificate not the date of completion stated in the certificate. This means that the insurance may have to continue for as much as thirty-five days after actual completion, because the completion certificate may only be issued within twenty-one days of completion of the whole, section or part of the works under clause 48. See this clause for details.

The proviso to sub-clause 20(1)

There is an extension of clause 20(1) the proviso of which makes the employer liable for damage from any cause to any part of the permanent works certified complete by the engineer, with clause 20(3) making the employer responsible for 'excepted risks', because one of these risks reads 'a cause due to use or occupation by the employer his agents servants or other contractors (not being employed by the contractor) of any part of the permanent works' and this applies whether that part is certified complete or not.

The question of occupation and responsibility must be matched by insurance taken out by the person responsible for the works at any one time. Unless the position is cleared with the insurers they could rightly adopt the attitude that partly occupied buildings or other works are not buildings or other works in course of construction (apart from the operation of a policy exclusion) but a far more hazardous contract. Thus not only would the insurers regard themselves as not on cover for the part occupied (which applies in any event) but also possibly not on cover for the remainder of the works which are unoccupied because of the hazardous effect the occupied part has on the unoccupied part. Admittedly, normally the part occupied would have to be put to a more hazardous use than construction work, nevertheless all other factors have to be considered. The point being made is that the policy exception only excludes loss or damage to the part taken into use or occupation by the principal (employer), therefore the contractor should clear the position with his insurers to avoid the insurers using their common law rights as far as the unoccupied part is concerned.

If the policy follows this clause, as it should, in wording the cover to continue for fourteen days after the issue of a certified completion date of the whole or part of the works then the employer, as already stated, has a built-in protection for that period. However, unless the same thing is endorsed in the policy for part 'use or occupation' without certification of completion (an excepted risk) it is essential for the employer to be told to insure **before** 'use or occupation' takes place. Otherwise several days can pass before the employer can arrange his cover, during which time a fire, for example, would be a catastrophe against which there would be no protection.

The further proviso to sub-clause 20(1)

The further proviso to sub-clause 20(1) is presumed to mean that the contractor is liable for each item of outstanding work included in an undertaking under clause 48 (concerning completion) until that item in the undertaking is complete. This involves the engineer in explaining in detail each item of outstanding work.

Responsibility for reinstatement: sub-clause 20(2)

Clause 20(2) commences as follows:

> 'In case any damage loss or injury from any cause whatsoever (save and except the Excepted Risks defined in sub-clause (3) of this Clause) shall happen to the Works or any part thereof while the Contractor shall be responsible for the care thereof the Contractor shall at his own cost repair and make good the same so that at completion the Permanent Works shall be in good order and condition and in conformity in every respect with the requirements of the Contract and the Engineer's instructions.'

In the phrase 'damage loss or injury' the word 'injury' seems to be not only superfluous but possibly confusing. Perhaps it is because the insurance world associates the word 'injury' with personal or bodily injury and not with damage to property, whereas the contract wording in the construction industry seems to use this word as meaning damage to property. Therefore in the phrase concerned the word 'injury' is at least superfluous as this sub-clause only applies to property, i.e. the works.

Whether the phrase 'any cause whatsoever' includes the negligence of the employer or his agents has been considered in chapter 1 under the headings 'The effect of the Unfair Contract Terms Act 1977' and 'The common law position'. While the decisions in Farr's and Smith's cases seem to clash, which leaves the latter (as a House of Lords case) overriding the former, one cannot help feeling that the apparent intention of the parties to accept Farr's decision (by virtue of the passage of time and publications of new editions of the ICE Conditions since that case without amendment on this point) may prevail. In any event the insurance policy will indemnify the employer as well as the contractor (see clause 21 regarding joint names cover). However, use or occupation by the employer his agents, servants or other contractors is an 'excepted risk' and a commentary on this exception appears under the next heading. Furthermore, the risk of faulty design is similarly excepted and may be due to the negligence of the employer's agent. See further comment under the next heading. The general meaning of the phrase 'any cause whatsoever' resulted in the birth of the contractors' all risks policy as will be seen in the next chapter when discussing clause 21.

The statement '(save and except the excepted risks . . .)' indicates the express qualifications upon the contractor's obligation to complete and the second sentence of sub-clause 20(2) makes it clear that if loss or damage occurs due to any of the 'excepted risks' the employer meets the expense of repair or replacement.

The second sentence reads:

> 'To the extent that any such damage loss or injury arises from any of the Excepted Risks the Contractor shall if required by the Engineer repair and make good the same as aforesaid at the expense of the Employer.'

By the way clause 20 is worded, with sub-clauses 20(1) and 20(2) putting responsibility for the care of the works on the contractor and sub-clauses 20(2) and 20(3) indicating the exceptions, the onus of proof of the damage arising from an excepted risk is on the contractor. The problem is to decide responsibility when both an excepted risk operates and a risk for which the contractor is responsible. The insurance cases illustrate the operations of the rules concerned in deciding the proximate cause of the incident, sometimes called the 'dominant or effective' cause, as this is what is considered in deciding whether the insurer is liable, i.e. whether it is an insured peril or an excepted peril under the policy or under the contract conditions whether the employer or contractor pays for the loss or damage.

In *Pawsey* v *Scottish Union and National (1908)* a clear definition was given as follows, although the practical application of the doctrine is sometimes difficult:

Proximate cause means the active, efficient cause that sets in motion a train of events which brings about a result, without the intervention of any force started and working actively from a new and independent source.

It is necessary to consider whether:

(i) there is a single cause or an unbroken chain of events; or
(ii) there is a broken chain of events; and, if so, whether a new and independent cause exists;
(iii) the predominant peril in the case of concurrent causes (i.e. the proximate cause), is insured or uninsured or in the case of the contract conditions one for which the employer or contractor is responsible.

Bearing in mind that excepted risks under the contract are usually excepted perils under the policy – under (i), policy liability exists where:

(a) the single cause, or
(b) one of a natural chain of events (but not necessarily the last of them because a cause may be proximate in efficiency but not necessarily in time),

is an insured peril.

Under (ii), there can be various series of events. Subject to case law, loss or damage by an insured peril is covered to the extent proved to have been caused by that peril, even if subsequently a new and independent cause arose from an excepted peril. Similarly, if an excepted peril is followed by the occurrence of an insured peril, as a new and independent cause there is a valid claim caused by the happening of an insured peril. Thus, if in the course of removing goods from a house set on fire by rioters, a fireman accidentally carries things which happen to be burning into a neighbouring building and sets it on fire, the fire caused thereby is not a fire occasioned by riot.

Under (iii), the insurers are liable if the predominant event is caused by an insured peril. If there is an excepted peril concerned and its effects can be separated from the results of the operation of the insured peril, there is liability for the latter but not for the former, but, if not, there is no policy liability at all.

Wayne Tank & Pump Co Ltd v *The Employers' Liability Assurance Corporation Ltd (1973)* concerned a public liability policy issued to contractors (not a material damage policy — for the difference, see chapter 4), but it illustrates the doctrine of proximate cause. Contractors supplied and installed electrically operated equipment for a factory owner. The equipment was unsuitable for its purpose and, as it turned out, a fire hazard. Harbutts' Plasticine factory was burnt to the ground when a pipeline installed by the plaintiffs disintegrated after an employee of the contractor switched on the equipment, although it had not been tested, and it was left unattended throughout the night. It was held that there were two causes of the loss, namely the dangerous nature of the equipment and the negligence of the employee. However, the former cause was held to be the dominant one by the Court of Appeal, therefore the insurers were not liable under their policy which excluded cover for any liability consequent on 'damage caused by the nature or condition of any goods . . . supplied by' the plaintiff contractors.

The statement 'the contractor shall if required by the engineer repair and make good the same . . . at the expense of the employer' raises the question of consequential loss suffered by the contractor. *Prima facie* the contractor is entitled only to payment for repairing or replacing loss or damage caused by the excepted risks to the works and not for consequential losses due to delay to the remainder of the works while the damage is made good. However, at common law the employer would be liable for such consequential loss if he were negligent or in breach of his contract. It will be recollected that there is a doubt as to whether the employer's negligence, not being an excepted risk, is still the contractor's responsibility (see the last two headings of chapter 1). Furthermore the phrase 'at the expense of the employer' is deemed to include normal cost of repair and replacement to the contractor plus a reasonable percentage for profit.

The real problem regarding 'consequential loss' is that the legal interpretation of what is consequential loss in the contract wording may be different from the exclusion of consequential loss in the contractors' all risks policy, depending on how the insurers word their exclusion and/or the operative clause of their policy. In *Croudace Construction Ltd* v *Cawoods Concrete Products Ltd (1978)* the Court of Appeal considered the extent of a contractual stipulation which excludes 'consequential loss or damage' upon which there is a lack of modern authority. The court followed the decision in the case of *Millar's Machinery Co Ltd* v *David Way*

& *Son (1934)* in deciding that 'consequential loss or damage' meant that loss or damage which did **not** result directly and naturally from the breach of contract concerned. What is meant by 'direct loss or damage'? Presumably it is the loss or damage which flows naturally from the breach without an intervening cause and independently of special circumstances. In Millar's Machinery case Maugham L.J. said: 'on the question of damages the word "consequential" has come to mean "not direct".'

Unfortunately in the insurance industry there is a tendency to regard consequential loss or damage as all loss or damage which is not physical or tangible loss or damage to property. Furthermore, economic loss is regarded by insurers as a consequential loss (at least by liability insurers) and the business interruption policy is probably wrongly referred to as a consequential loss policy. Yet in the majority of cases all these losses are direct losses in that they flow naturally from the breaches concerned, therefore legally they are not consequential losses. The court said in Croudace's case that: 'the word "consequential" provides a fertile ground for philosophical debate as do other words involving the concept of causation'. In this case there was late delivery of and defects in masonry blocks. Part of the main contractor's claim included the costs of loss of production and additional costs of delay in executing the main contract works plus an indemnity against a claim by a subcontractor for delay in that subcontractor's work caused by the absence of the defendants' materials. Because this was a preliminary hearing there was no finding that any losses of which the plaintiffs complained were or were not recoverable. However, there is little point in having such a preliminary issue tried on the meaning of the exclusion 'consequential loss or damage' if, on the facts, there is a possibility that it would be held that the losses did not directly and naturally arise from the breach of contract. Probably the defendants accepted that the plaintiffs' losses would be recoverable unless the word 'consequential' included that which was the direct and natural result of the alleged breaches. Also in Millar's case a machine supplied was not in accordance with the contract and an exclusion of consequential damages was held not to exclude the buyer's expenses incurred in obtaining other machinery to replace the defective machine.

Furthermore, in *Saint Line Ltd* v *Richardson, Westgarth & Co Ltd (1940)* a breach of contract concerning the supply of engines for a ship resulted in a claim for loss of profit during the time the buyers were deprived of the use of the ship, for the expenses of wages, stores etc., and for fees paid to experts for superintendence. It was held that all these items of damage were recoverable, in spite of an exclusion of liability for any indirect or consequential damages whatsoever, as they were considered direct and a natural result of the breaches and were not indirect or consequential.

As the ICE Conditions in clause 20 do not exclude consequential loss specifically it could be argued that 'loss or damage' includes consequential loss however interpreted and thus avoid the above discussion. However, this is doubtful as elsewhere in the contract where consequential losses are included other phrases are used, for example in clause 22 where an indemnity to the employer is given, the phrase used is 'all losses and claims for injuries or damage to any person or property whatsoever' and later in the same clause the phrase used is 'against all claims demands proceedings damages costs charges and expenses whatsoever'. Thus in clause 20 it seems that these excepted risks concern repair or replacement of loss or damage to the works and not other costs charges and expenses, as already mentioned, except the contractor's profit. Also, under clause 12 the contractor could claim for unforeseen physical conditions or artificial obstacles causing consequential loss in the shape of delay and additional cost. Similarly, clause 13 makes provision for the cost of unforeseen delay and disruptions. This clause deals with the engineer's requirements and this could involve repair of damage or replacement of loss due to excepted risks.

In practice clause 12 may not be easy for the contractor to invoke, as clause 11, which deals with 'inspection of site', puts the responsibility on the contractor to satisfy himself about the site and its surroundings and the nature of the ground and subsoil (so far as is practicable and having taken into account any information in connection therewith which may have been provided by or on behalf of the employer) and in general to have obtained for himself all necessary information as to risks, contingencies and all other circumstances influencing or affecting his tender. However, if the information provided by the employer is misleading this may be one factor in assisting the contractor to establish a claim under clause 12.

By way of illustration, if the water authority employs a contractor to supply pre-stressed concrete tanks and provides the contractor with a detailed soil mechanics report, based on bore-holes, but nevertheless the hillside concerned (which according to the report was stable) collapsed onto the site, the contractor may well have a valid clause 12 claim.

The last sentence of sub-clause 20(2) reads:

> *'The Contractor shall also be liable for any damage to the Works occasioned by him in the course of any operations carried out by him for the purpose of completing any outstanding work or of complying with his obligations under Clauses 49 and 50.'*

Clause 49 deals with 'maintenance and defects' and clause 50 with the contractor's responsibility (if required by the engineer in writing) to carry out such searches and tests or trials as may be necessary to determine the cause of any defect imperfection or fault under the directions of the engineer.

It will be recollected that the further proviso to sub-clause 20(1) is taken to mean that the contractor is liable for each item of outstanding work included in an undertaking under clause 48 (concerning completion) until that item of work in the undertaking is complete. That is to say, when an item has been so completed that item is no longer outstanding and thus there is no question of the whole undertaking having to be completed before any item can be considered as dealt with. This involves the engineer in explaining in detail each item of outstanding work.

To summarise the position: the works are at the risk of the contractor until completion plus fourteen days, and after this, under clause 49 he will receive additional payment for work of repair during the maintenance period, which is not due to a breach of contract by him. However, under the last sentence of sub-clause 20(2) quoted above the contractor remains responsible for any damage caused by him during the maintenance period when completing outstanding works or maintenance obligations.

The normal contractors' all risks policy makes provision for covering outstanding work where such loss or damage arises out of the performance of the contract and occurs during the period of maintenance (see chapter 5).

Excepted risks: sub-clause 20(3)

Clause 20(3) reads:

> 'The "Excepted Risks" are riot war invasion act of foreign enemies hostilities (whether war be declared or not) civil war rebellion revolution insurrection or military or usurped power ionising radiations or contamination by radioactivity from any nuclear fuel or from any nuclear waste from the combustion of nuclear fuel radioactive toxic explosive or other hazardous properties of any explosive nuclear assembly or nuclear component thereof pressure waves caused by aircraft or other aerial devices travelling at sonic or supersonic speeds or a cause due to use or occupation by the Employer his agents servants or other contractors (not being employed by the Contractor) of any part of the Permanent Works or to fault defect error or omission in the design of the Works (other than a design provided by the Contractor pursuant to his obligations under the Contract).'

The expression 'any cause whatsoever' is qualified by the exclusion of 'excepted risks' from the contractor's responsibility and these 'excepted risks' listed in clause 20(3) can now be considered in detail. Incidentally, they form the basis of the exclusions in the contractors' all risks policy, except for riot in the United Kingdom.

(a)   *Riot* is defined in *Halsbury's Laws of England* as:

a tumultuous disturbance of the peace by three or more persons who assemble together, without lawful authority, with an intent mutually to assist one another, by force if necessary, against any who shall oppose them in the execution of a common purpose and who actually execute, or begin to execute, that purpose in a violent manner displayed not merely by demolishing property but in such a manner as to alarm at least one person of reasonable firmness and courage.

In *Field* v *Receiver of Metropolitan Police (1907)* it was laid down that all the following elements must be present to constitute a riot, viz.:

(a)   no less than three persons;
(b)   a common purpose;
(c)   execution or inception of the common purpose;
(d)   intent on the part of the persons assembled to help one another by force, if necessary, in face of opposition;
(e)   force or violence, not merely used in and about the common purpose, but displayed in such a manner as to alarm at least one person of reasonable firmness and courage.

In *J. W. Dwyer Ltd* v *Receiver of Metropolitan Police District (1967)* it was indicated that an essential element of a riot enabling a claim to be made under the Riot (Damages) Act 1886 was 'tumult'. A comment on the recovery aspect under this Act appears in the next paragraph.

While this definition indicated that a more organised clash with authority was required than that, for example, displayed by hooligans leaving a football ground after their club's defeat, on the other hand in *Munday* v *Metropolitan Police District Receiver (1949)* a similar situation did arise when private property was damaged after invasion by a football crowd endeavouring to see a match when the ground was full. It was clear from the uncontradicted evidence of the plaintiff's gardener, who had met with violence when he tried to recover his ladder from those who had climbed on to a garage, that there was a riot. This case is of interest to insurers who occasionally find it necessary to claim under the Riot (Damages) Act 1886 if property insured by them is damaged in a riot. Incidentally, those claiming have only fourteen days from the occurrence in which to record an application for recovery against the appropriate authority.

It is a short step from Munday's case to visualise a similar situation where a new uncompleted football stand is being built and unauthorised entry in defiance of authority could result in damage to the erection by rioters. Now while the ICE Conditions do not require the contractor under his all risks policy to insure against riot it is the practice of insurers not to exclude this risk in England, Scotland and Wales and as the policy provides cover in the joint names of the employer and contractor the employer's responsibility under the contract for riot may be covered by this policy. Some policies exclude risks which are the responsibility of the principal under conditions of contract. As already stated, all the other 'excepted risks' are excluded from the policy, and strictly speaking insurers should require notification of damage, forming a recovery claim from the police, within seven days in the case of riot in order to get their recovery claim made within fourteen days against that authority.

(b) *War* was defined in *Pesquerias y Secaderos de Bacalao de Espana SA* v *Beer (1949)* as 'the state or condition of governments contending by force'. In fact this and the following terms of 'invasion, act of foreign enemies hostilities (whether war be declared or not)' imply the existence of a state of war with a foreign power, and the pattern of international events during and since the last war is reflected in the qualification 'whether war be declared or not'. The exclusion has been worded in such a way that it applies not only to losses due to enemy acts, but also to those incurred through the steps taken to resist the enemy, as shown in the Falklands Islands crisis.

As one of the excepted risks is war it is necessary to refer to clause 65 on this subject. While the kindred risks of invasion, act of foreign enemies, and hostilities are listed in sub-clause 20(3) they are not included in clause 65, but a fair presumption would be to assume that clause 65 also applies to these kindred risks.

Clause 65 provides for work to continue for twenty-eight days if this is possible and if the work is then complete the contractor is to be paid without necessarily carrying out all maintenance work. If, after twenty-eight days, the work is not complete the employer has the option of determination at any time with the usual procedure for paying the contractor. For work following the outbreak of hostilities a special variation of price clause applies.

(c) *Civil war*   See under 'Civil commotion' later.

(d) *Rebellion*   In *Lindsay and Pirie* v *General Accident etc.*, *Corpn Ltd (1914)* rebellion was defined as the taking up of arms traitorously against the Crown whether by natural subjects or others when subdued. It can also mean disobedience to the process of the law by the courts. Revolution is very similar.

(e) *Insurrection* is a stage between civil commotion and rebellion and is defined in Jowitt's *Dictionary of English Law* as a rising of the people in open resistance against established authority with the object of supplanting it.

(f) *Military or usurped power*   The term 'military power' includes acts done by the Crown's military forces in opposition to subjects of the realm in open rebellion and organised as a military force. 'Usurped power' applies to an organised rebellion which is acting under some authority and has assumed the power of government by making laws and enforcing them.
    In *Curtis* v *Mathews (1919)* Banks, L.J. said:

Usurped power seems to me to mean something more than the action of an unorganised rabble. How much more I am not prepared to define. There must probably be action by some more or less organised body with more or less authoritative leaders.

It has been suggested that the last four words of the kindred risks to war, namely 'military or usurped power' come within the *ejusdem generis* (meaning 'of the same kind') rule. This legal rule of construction applies where there is a particular list of precise words sufficient to identify the intention, followed by some general words or 'omnibus' description. The latter are confined to the same class or kind as the former. The assumption is that the general words are only intended to preserve some accidental omission in the list of the kind mentioned and are not intended to extend to words of a completely different kind.
    An example will make the position clear. A ship was to be relieved from liability for not delivering cargo at a certain port or ports if it was in the opinion of the master unsafe to do so 'in consequence of war, disturbance or any other cause'. The question arose whether a port inaccessible through ice was within the exception. It was held not to be so. 'Any other cause' must be read as applying to causes *ejusdem generis* or similar to 'war or disturbance'. See *Tillmans and Co* v *SS Knutsford (1908)*. Possibly in the ICE Conditions the words military or usurped power are already sufficiently specific that it is not necessary to apply the rule. In any event the intention is clearly to include in the excepted risks both war and kindred risks.

All the above excepted risks follow exclusions required by the insurance market with the exception of riot, which, with civil commotion, is usually covered except in Northern Ireland, the Republic of Ireland and probably elsewhere abroad.

(g) *Civil commotion*   Why this has been omitted from the list of excepted risks is a matter for conjecture. Civil commotion is insurrection of the people for the purposes of general mischief and directed against law and authority but not amounting to a rebellion or usurped power. Civil commotion is probably a stage between a riot and civil war. Civil war implies something in the nature of acts of organised warfare.

(h)   The next item under the list of 'Excepted Risks' is:

> *'ionising radiations or contamination by radio-activity from any nuclear fuel or from any nuclear waste from the combustion of nuclear fuel radioactive toxic explosive or other hazardous properties of any explosive nuclear assembly or nuclear component thereof'*

The wording used here is the same as the exception applicable to all insurance policies covering property in the United Kingdom. By the provisions of the Nuclear Installations Acts 1965 and 1969 the holder of a nuclear site licence must by insurance or in some other way make provision against third party (including employees) injury or property damage claims to an aggregate amount of £5 million (increased by the Energy Act 1983 to £20 million with the Government taking liability for the next £190 million). Illustrations of the type of event considered are the accidental dropping of nuclear bombs in Spain in 1966 (unarmed in that case) and the possible use of nuclear exposives for civil engineering. It is understood that Parliament has provided for compensation for the damage caused by incidents due to materials in explosive devices of this kind. A better example is the escape of radioactive material from Windscale in 1957 when compensation had to be paid to farmers for the loss of the sale of milk.

The exception does not preclude radioisotopes or equipment such as X-ray machines as they are considered capable of being covered by the insurance market. These articles can be used to test welding and in this respect the construction industry may be involved.

(i)   The next excepted risk reads:

> *'pressure waves caused by aircraft or other aerial devices travelling at sonic or supersonic speeds'*

and is an exclusion in all material damage insurance policies but not in liability policies (see chapter 4). The United Kingdom Government has stated that it will pay compensation if such damage results from Concorde's test flights but as such test flights are now over there seems to be an assumption that by implication such a statement applies to its routine flights. In any event section 40 of the Civil Aviation Act 1949 provides that where damage is caused to persons or property by an aircraft or anything falling from it, damages shall be recoverable without proof of negligence. This is provided that the claimant is not responsible or partly responsible, and also that the operator has complied with the provisions of the Act, the most important of which are the Air Navigation Orders and Regulations.

(j) This exception, and the next one, are the most important in the list of excepted risks from a practical point of view.

> 'a cause due to use or occupation by the Employer his agents servants or other contractors (not being employed by the Contractor) of any part of the Permanent Works'

This exception has been considered, when discussing the employer's responsibility for the works after completion, as an additional liability of the employer for the works used or occupied by the employer etc., irrespective of completion. Furthermore, the exception quoted above does not exactly fit in with the equivalent part of the exclusion under the usual contractors' all risks policy which often merely excludes damage to parts taken into use or occupancy. Thus the policy exclusion is wider in that it is not concerned with cause but narrower in that it only excludes the parts taken into use. See further discussion in chapter 5.

A case concerning this exception (although involving the JCT Form) was *English Industrial Estates Corporation* v *George Wimpey & Co Ltd (1973)* and details will be found earlier in this chapter and chapter 20.

The words 'or other contractors (not being employed by the contractor)' mean that the risk attaches to the employer once subcontractors employed directly by the employer use or occupy the site and cause loss or damage arising from that use or occupancy. Thus the loss or damage is not limited to the part occupied; it can happen to any part of the works as long as the loss or damage was caused by the use or occupancy 'of any part of the permanent works' by the employer etc. However, it is more likely that subcontractors directly employed by the employer will cause such loss or damage than the employer himself or his servants or agents, particularly before substantial completion, for two reasons. In the first place there are usually more subcontractors than employers, servants or agents and secondly they are usually longer in occupation than the employer, his servant or agents, whose visits to the site tend to be transient, although this is not applicable if the latter are in permanent occupation.

The period of fourteen days after the issue of the completion certificate stipulated in sub-clause 20(1), which (as mentioned under that sub-clause) can mean thirty-five days after actual completion because of the additional period for the issue of the completion certificate in clause 48, does mean that for thirty-five days this excepted risk might be invoked. In this period it is more likely that the employer, his servants or agents might use or occupy the works, although specialist contractors might be employed to install special equipment.

(k) The final excepted risk reads:

> 'to fault defect error or omission in the design of the Works (other than a design provided by the Contractor pursuant to his obligations under the Contract).'

Presumably the introductory words to the previous risk, 'a cause due', apply to this risk as they are connected by the conjunction 'or'.

In the first place the limitation in brackets should be noted, the point being that if the contractor submits a design for part of the works or is requested to do so by the engineer and the contract does not allow for this the employer is still responsible for that design under this clause. The limitation mentioned in brackets is important as it saddles the contractor with responsibility for his own design work and this type of cover is normally only available in a limited form under the contractors' all risks policy, although a few insurers will not exclude the consequences of design but the actual defective design work itself is always excluded. It is normally a matter for the professional indemnity policy to cover but usually this policy is only issued to proposers who are qualified in their profession. The point to remember is that many contractors' all risks insurers exclude all claims arising from defective design. When it is essential for the contractor to obtain some cover for his liability for design work he should approach an insurance broker specialising in the construction industry, but the contractor should be prepared for the type of cover which leaves him responsible for a financial proportion of any claim either in the shape of an excess (deductible) or a direct percentage of the loss. Cover might be offered on an excess of loss basis, i.e. the insurer only covering claims which exceed a fairly substantial sum, e.g. £50,000.

The legal and insurance aspects are discussed in some detail in chapter 5 when considering the policy exclusion based on this contractual exception. The word 'design' in construction contracts will include any requirement which may regulate or control the suitability of the completed work for its intended purpose (not only structural designs, calculations and dimensions but also the choice of materials and control of work processes); see *Hudson on Building and Civil Engineering Contracts*, tenth edition, page 274. It appears that the excepted risk applies either to a design defect of the permanent works or of the temporary works.

Because the words 'solely due to the engineer's design of the works', included in the fourth edition of the ICE Contract, have been dropped from the fifth edition the design can now be wholly or partly the fault of any person (subject to the limitation in brackets at the end of clause 20(3)). There seems little doubt that this excepted risk has been widened by the removal of the words 'solely due to the engineer's design' and this allows the contractor to invoke the excepted risk in many cases of serious damage to the works. Consequently, contractors' insurers and those of the engineer, and possibly the insurers of the employer as well, have found themselves involved in lengthy disputes, if not litigation.

In the case of *Norman Hughes & Co Ltd* v *Ruthin Borough Council (1971)* counsel for the employer conceded that if the arbitrator concluded that settlement of a sewer was not due to bad workmanship on the part of the contractor, it followed by a simple process of elimination that the engineer's design was the sole cause of the settlement that had taken place, in that the ground had rendered that design unsuitable.

The relevance of this case is arguably that the removal of the words 'solely due to the engineer's design' from the fifth edition makes it easier for a court to come to the conclusion reached in Norman Hughes' case, decided before the fifth edition was published, because the excepted risk has been widened to include wholly or in part any person's design other than that of the contractor. Thus, if the contractor is eliminated as the designer and the fault is not due to his bad workmanship then the excepted risk must apply to any other faulty design.

In conclusion, as this book will probably be used for reference purposes, it is reasonable to indulge in a certain amount of repetition with the contents of chapter 5, by explaining that it is not absolutely clear whether the phrase 'fault defect error or omission in the design of the Works' is not limited to negligent design as the words 'faulty design' are not so limited in the exception in the contractors' all risks policy. See the Queensland Government case in chapter 5 where it was said that the design concerned was outside the policy cover notwithstanding that the design complied with the standards that would be expected of designing engineers according to engineering knowledge and practice at the time of their design. The mere fact that the design proved to be inadequate made it a 'faulty design'. Probably the tendency will be to interpret the contract in the same way as the policy because of the case law just quoted. See *Keating on Building Contracts*, fourth edition, pages 471–72. On the other hand Duncan Wallace in *The ICE Conditions of Contract, Fifth Edition* says:

The words 'fault defect error or omission in the design' may, incidentally, be somewhat narrower than the simpler 'due to the engineer's design' wording in the fourth edition, in those not uncommon cases where the design, while not negligent in the current state of knowledge, was the actual cause of damage.

In the fifth edition of the ICE Contract, as compared with the fourth edition, there is also in this excepted risk a limitation to 'fault defect error or omission' in design but whether this has the effect of confining the exception to negligent design will be discussed in the chapter on the 'Contractors' all risks policy' when considering the case of *Queensland Government Railways* v *Manufacturers' Mutual Insurance (1969)*. In the case of *C.J. Pearce & Co Ltd* v *Hereford Corporation (1968)* the word 'design' was said to mean:

that documents of the nature of the plans, and so forth, are handed to the contractor, showing the precise detail of the work the contractor is to carry out. In other words the word 'design' deals with how the work is to be carried out and not which work is to be carried out.

For further discussion of the design exclusion in the contractors' all risks policy see chapter 5 on that policy and the case of *Pentagon Construction (1969) Co Ltd* v *United States Fidelity and Guarantee Co (1978)* which is unsatisfactory on the meaning of 'design' because two judges disagreed and the third judge abstained.

Chapter 3

Insurance of the works: clause 21

Insurance of the Works, etc.

21. Without limiting his obligations and responsibilities under Clause 20 the Contractor shall insure in the joint names of the Employer and the Contractor:

    (a)   the Permanent Works and the Temporary Works (including for the purposes of this Clause any unfixed materials or other things delivered to the Site for incorporation therein) to their full value;

    (b)   the Constructional Plant to its full value;

against all loss or damage from whatever cause arising (other than the Excepted Risks) for which he is responsible under the terms of the Contract and in such manner that the Employer and Contractor are covered for the period stipulated in Clause 20(1) and are also covered for loss or damage arising during the Period of Maintenance from such cause occurring prior to the commencement of the Period of Maintenance and for any loss or damage occasioned by the Contractor in the course of any operation carried out by him for the purpose of complying with his obligations under Clauses 49 and 50.

Provided that without limiting his obligations and responsibilities as aforesaid nothing in this Clause contained shall render the Contractor liable to insure against the necessity for the repair or reconstruction of any work constructed with materials and workmanship not in accordance with the requirements of the Contract unless the Bill of Quantities shall provide a special item for this insurance.

Such insurances shall be effected with an insurer and in terms approved by the Employer (which approval shall not be unreasonably withheld) and the Contractor shall whenever required produce to the Employer the policy or policies of insurance and the receipts for payment of the current premiums.

This is the insurance clause which is complementary to clause 20.

Probably the most important sentence in clause 20, it will be recollected, is the first one in sub-clause 20(2) making the contractor responsible for loss or damage to the works until they are officially completed plus fourteen days. If there is any loss or damage to the works other than by the excepted risks the contractor must pay for the repair or replacement himself.

Obviously the employer wishes to be satisfied that the contactor has sufficient money to carry out those responsibilities under the clause expeditiously, otherwise any delay will result in loss to the employer. This is the reason for the insurance clause 21. It does not alter in any way the effect of clause 20 as it opens with the following statement: 'without limiting his . . . responsibilities under clause 20 . . .'. Consequently, if there is a failure of the insurance policy to operate in whole or in part (for example, the sum insured may not be sufficient to meet the loss — see later) the contractor's responsibility will continue to apply. However, apart from the failure of the insurance policy to provide the cover required, it will be remembered that in the previous chapter when discussing the phrase 'loss or damage' it was considered that this phrase was limited to the cost of repair or replacement of physical or tangible loss or damage and did not include any other costs, charges and expenses other than the profit of the contractor. In other words the expression 'loss or damage' is read as excluding all consequential loss.

While normally it is advisable to avoid the term consequential loss because of the difficulties of definition (see the discussion in the previous chapter) it can be taken that in the following discussion consequential loss means all loss which flows directly or indirectly from the physical or tangible loss or damage other than the profit of the contractor in repairing or replacing such physical loss. This interpretation leaves the party to the contract who suffers the consequential loss, subject to any tortious remedy, responsible for all such consequential losses because clause 20 and thus clause 21 do not cater for them.

While a number of examples of the contractor's consequential losses were mentioned in the previous chapter he is not the only one to suffer in this way. As such losses by both parties to the contract are considered from an insurance point of view in chapter 5 on the 'Contractors' all risks policy', only one consequential loss suffered by the employer will be mentioned here. In the construction of a building to house machinery with an anticipated start-up date, damage by fire to both the building and machinery will delay the completion date for several months. The employer will thus suffer a loss of production and loss of profit because of this delay. This could be a large sum.

Returning to the wording of clause 21, the main requirements are shown below. The contractor is required:

(1) to insure in the joint names of the employer and the contractor;   } The insured

(2) the permanent and temporary works (including unfixed materials and other things on site for incorporation therein) to their full value;

(3) the constructional plant to its full value;   } The subject-matter of the insurance and the sum insured

(4) against all loss or damage from whatever cause arising (other than the excepted risks) for which he is responsible under the terms of the contract;   } The perils insured against

(5) both during the construction period and the maintenance period arising from a cause occurring during the construction period or through damage caused by the contractor during operations carried out in accordance with the maintenance clause plus 14 days in each case.   } The period of insurance

The clause is considered in the following stages, which do not follow exactly the item numbers above.

(1) *Insurance in the joint names of the employer and contractor*

From the employer's point of view insurance in the joint names may seem necessary to prevent the contractor's insurers recovering from the employer after paying a claim in cases where the contract indicates that the employer is responsible (the excepted risks). See *MacGillivray's Insurance Law*, sixth edition, 1975 where he says in section 1916: 'The fact that the policy is in joint names will almost invariably mean that both parties are intended to benefit and that there is no scope for subrogation.'* However, when it is appreciated that the contractors' all risks policy excludes the excepted risks, other than riot in England and Wales (but see p. 188), this argument becomes irrelevant. Nevertheless, it does give the employer the right to claim directly against the contractor's insurers under the policy in respect of damage by any risk which is covered and there may be a good reason for taking this shorter route to recovery rather than claiming against the contractor under the contract. In the case of the contractor's insolvency the employer would probably have the right under the Third Party (Rights against Insurers) Act 1930 if the insurance was in the contractor's name only. Also in these cases of insolvency a performance bond in accordance with clause 10 would provide some protection for the employer. See chapter 16 concerning sureties. However, the bond protection is at best only 10% of the contract price in the ICE Conditions and the employer's interest in the contractor's insurance of the works will be greater than the 10% cover of any bond. Additionally, the joint names insurance will also avoid the possibility of the contractor claiming under the policy for work for which he has already been paid by the employer.

(2) *The subject-matter of the insurance*

'Permanent works' is defined in clause 1 as meaning 'the permanent works to be constructed, completed and maintained in accordance with the contract'.

'Temporary works' is defined in the same clause as meaning 'all temporary works of every kind required in or about the construction, completion and maintenance of the works'. On to these definitions is grafted 'any unfixed materials or other things delivered to the site for incorportion therein', which is more precise than the fourth edition of this contract which reads 'materials and other things brought on to the site'. It is now clear that the temporary works do not include temporary buildings on construction sites. See also the drafting discrepancy between clauses 20 and 21 mentioned under heading (5) below.

(3) *The construction plant*

This is defined in clause 1 as meaning:

all appliances or things of whatsoever nature required in or about the construction, completion and maintenance of the Works but does not include materials or other things intended to form or forming part of the Permanent Works.

This definition is wide enough to include the temporary buildings on construction sites. It is a fact that a high proportion of fires on construction sites start in the temporary buildings. This type of fire usually means that several structures are destroyed including offices containing important drawings and stores with key

---

* This is repeated in the seventh edition 1981, section 1213 and verified in *Petrofina (UK) Ltd and Others* v *Magnaload Ltd and Others (1983)*.

components. It only takes a fire in a hut a few minutes to spread to the roof of a new factory in course of erection. A typical example was the destruction by fire of a nearly completed brewery amounting to about £3.5 million. The fire started after the working day among the contractors' huts. Note also the drafting discrepancy between clauses 20 and 21 mentioned under heading (5) below.

(4) *The sum insured*

An important aspect is choosing an adequate sum insured, bearing in mind the words used in clause 21 are 'to their full value' or 'to its full value' as the case may be, which, ignoring such matters as debris removal costs and professional fees, must cover the reinstatement cost of the works, so that the insured has to make a calculation:

(a) of the increases in cost of materials, goods and labour charges during the period of the contract, i.e. price fluctuations, and

(b) following damage during the construction period, of the effect of inflation on those costs incurred:

    (i) in the reinstatement of the damaged works, and

    (ii) in completion of those parts of the works that have not been started at the time of damage and which are delayed as a consequence.

It is the usual practice for contractors' all risks policies to provide for an automatic increase in the indemnity limit (in the policy schedule) for contract works by an amount not exceeding 20% subject always to an adjustment of premium but this type of clause can only cater for (a) and (b)(i) above. However, see chapter 5 on the 'Contractors' all risks policy' for further details and an example of an extension cover to this policy to cover consequential loss of the nature of (a), (b)(i) and (ii), although cover for (b)(ii) is not called for by the contract wording.

(5) *The perils insured against*

The words 'all loss or damage from whatever cause other than the excepted risks for which he is responsible under the contract' are clear enough in themselves. However, in this clause the words apply to permanent and temporary works plus unfixed materials and constructional plant while in clause 20 reference is only made to the works, i.e. no reference is made to unfixed materials or constructional plant. Furthermore 'works' is defined in clause 1(1) as meaning 'the permanent works together with the temporary works', and neither of the definitions of permanent or temporary works includes unfixed materials etc. The ambiguity is whether the contractor is required:

(a) to insure constructional plant and unfixed materials even against the excepted risks because he is contractually responsible for this property in the absence of any qualification in clause 20, or

(b) not to insure such plant and materials against the excepted risks in accordance with clause 21.

The difficulty can also be expressed by comparing the words 'other than the excepted risks' with the words 'for which he is responsible'. If the former override the latter then no insurance against the excepted risks is required but if the former are unnecessary and the latter override them then (a) above applies. The tendency in practice when arranging the insurances is to assume that (b) above applies, but it is by no means certain.

A possible guide to the intention of the drafters (which supports the practice just mentioned) is given in clause 53(9) which states that the employer shall not be liable for the loss or injury to any of the plant goods or materials which have been deemed to become the property of the employer when on the site 'save as mentioned in clause 20'. This seems to assume incorrectly that clause 20 qualifies the liability of the contractor in some way but as just explained it does not as far as plant and materials are concerned. However, clause 53(9) may indicate that the intention of the parties to the contract is to include unfixed materials within the word 'works' in clause 20, although this is hardly acceptable in the case of constructional plant.

It has been suggested that the matter should be resolved by obtaining the engineer's instructions under clause 5. Clause 5 states that ambiguities and discrepancies shall be explained and adjusted by the engineer. What if the engineer instructs that the contractor is required to insure unfixed materials etc., and constructional plant even against the excepted risks? The short answer is that the standard contractors' all risks policy does not cover the excepted risks except 'riot' assuming the site is not in Northern Ireland. In fact it is doubtful whether an underwriter would be prepared to cover any of the excepted risks at least without considering each contract individually; even then some excepted risks are clearly uninsurable, e.g. war risks.

(6) *The period of insurance*

The cover is stated to be for the period stipulated in clause 20(1) which is the construction period starting with the commencement of the works until fourteen days after the engineer shall have issued a certificate of completion for the whole of the works pursuant to clause 48. Cover is also to be provided during the maintenance period for any delayed loss or damage arising from an insured cause or event occurring prior to the period of maintenance and loss or damage actually caused by the contractor while working on the site (that is, in carrying out his maintenance obligations under clauses 49 and 50). Incidentally, fourteen days after the issue of the certificate of completion or partial completion may in fact result in the cover continuing for thirty-five days after actual completion — see the commentary on clause 20(1) in the previous chapter.

(7) *The proviso (concerning no insurance for the repair or reconstruction of any work . . . not in accordance with the contract)*

The proviso is limited to the actual repair or reconstruction of the defective work itself. Defective work resulting in structural or other failure causing considerable destruction or damage to other parts of the works, not themselves defective, is not within the proviso. Such consequential loss must be insured and in fact the contractors' all risks policy only excludes the cost of repairing or replacing the defective work or materials. It has been suggested that the words 'workmanship not in accordance with the requirements of the contract' might include failure by the contractor to comply with any express provisions in the bills or specifications requiring the contractor to take certain measures to protect the works during construction such as shoring or pumping for drainage.

However, if the policy exclusion was worded as mentioned above and not in the specific wording of the contract it would not be possible for the insurers to use the wording of the policy exclusion to repudiate liability but it might be argued by insurers that a reasonable precautions condition had been violated. Such a condition is usually worded as follows:

The Insured shall take reasonable precautions to prevent loss . . . and to safeguard the property insured.

Presumably the purpose of this proviso in not asking for insurance is that it was thought by the drafters that difficulty might be experienced in obtaining such cover.

(8) *Approval of insurer*
The contractor is required to see that the insurance is effected with an insurer and in terms approved by the employer (which approval must not be unreasonably withheld). The only satisfactory way for the contractor to comply with this stipulation is to submit the policy of the insurer to the employer so that he may satisfy himself on the matter. This is suggested in spite of the fact that the final paragraph of clause 21 goes on to require the contractor to produce the policy or policies and receipts for the premiums **whenever required**. The stipulation regarding approval in clause 21 goes a little further than the requirement in clause 25, which will be dealt with later, as the latter only operates if the employer requests evidence that there is in force the insurance referred to in clauses 21 and 23. It has been suggested that the contract documents (the specifications or bills of quantities) should set out the exact terms of the policy (including any permitted exceptions). In fact clause 21 goes quite a long way towards achieving this ideal because of the excepted risks set out in that clause but underwriters will always add their own exceptions (see chapter 5).

Clause 21 makes no provision for the operation of an annual contractors' all risks policy (as distinct from a policy issued for the specific contract) but normally if such a policy provides the cover required by clause 21 it would be accepted.

The engineer will inform the employer of the certified completion of the works (and usually prior to the estimated date thereof) in order that the employer may make suitable arrangements for insurances in his own name from the date the contractor's policy ceases to provide the cover the employer requires.

Chapter 4

Classes of insurance policies which give the cover

While the previous chapter dealt with clause 21 of the ICE Conditions, which requires a contractors' all risks policy, before discussing that policy it is necessary to explain to those outside the insurance industry the classes of policies which construction contracts normally require.

The classes usually concerned are liability policies and material damage policies.

Liability policies

There are two:

(1)   employers' liability; and
(2)   public liability.

These policies cover the insured's legal liability to third parties, that is, liability to any person who is not a party to the insurance contract subject to exceptions.

The employers' liability policy covers the liability of an employer (the master) to his employees (those persons under a contract of service or apprenticeship with the employer) for bodily injury or disease arising out of and in the course of their employment.

The public liability policy provides an indemnity against personal injury claims by the public (other than employees), and property damage claims by any third party including employees. If the contractor has a design (or engineers or architects) department and enters into 'design and construct' contracts he will need a professional indemnity policy to cover the professional negligence risk. See more about this at the end of this chapter under 'Project insurance'.

Material damage policies

This type of policy covers loss or damage to property in which the insured has an insurable interest which may arise through ownership, possession or contract. A material damage policy only covers the property specified. The contractors' all risks policy is a material damage policy.

Composite and combined policies

These policies cover in one document risks which are frequently covered by separate policies.

In the case of a contractors' combined policy the risks covered are material damage and legal liability. A specimen policy will be found in appendix 1. It will be seen that section 1 covers employers' liability, section 2 public liability and section 3 loss or damage to the contract works, (thus section 3 gives the cover provided by the contractors' all risks (CAR for short) policy mentioned earlier). Similarly the CAR insurance as well as offering comprehensive protection against loss or damage to the contract works (as explained earlier), to the construction plant and equipment and/or construction machinery, can include third party claims in respect of bodily injury or property damage.

A very few insurers go even further in their CAR cover mentioned in the last paragraph, by including a consequential loss section and even fewer include a limited professional indemnity cover for architects, engineers and professional advisers restricted to 'site activities'. At this stage the insurers are getting nearer to providing project insurance assuming 'the insured' in the policy schedule includes the:

(a)   employer;
(b)   architects, engineers and professional advisers;
(c)   contractors (under contract to the employer);
(d)   subcontractors (under contract to the contractor).

However, more of this aspect will be mentioned in the next chapter.

These composite or combined policies have an advantage over separate policies because only one proposal form is required, one policy is issued, and, assuming an annual basis, one renewal notice and one renewal premium suffice. Another advantage is that there is one declaration of turnover for the purpose of adjusting the premium.

Miscellaneous policies

Some insurers have introduced a liability policy which goes a stage further than the composite policy described above since, although the cover is confined to the employers' and public liability risks and does not include loss or damage to the contract works, it does not separate the liability risks into sections with their own exclusions but contains a combined operative clause covering (a) injury to any person, and (b) damage to property.

In this type of policy it is necessary to indicate where an exception applies only to personal injury or only to damage to property, or does not apply to an employee of the person indemnified.

Project insurance

Particularly in the case of large civil engineering projects the owner who commissions the project often wastes money as a result of the system of making every participant arrange separate insurance policies for his part of the work, his own plant and equipment, and for the liabilities to workmen and other third parties which arise from performance of his part of the work.

Comprehensive project (sometimes referred to as wrap-up or omnibus) insurance, preferably arranged by the owner, has been frequently advocated. When this is done, however, it is usually confined to conventional risks only, i.e. loss or damage to the works and to other property at the site, and liability to third parties, including cross liabilities between contractors. For an explanation of cross liabilities see chapter 7.

Theoretically there is no reason why the consulting engineer's professional liability or that of any other professional man should not be included. Why is this not done as a general practice?

The normal method used is for all risks insurance of the works to be taken out by the employer or contractor. If the contractor takes it out the employer is usually a joint insured. The consulting engineer has his professional indemnity insurance. Each of the other parties takes out his own material damage and liability insurances except that occasionally the main contractor takes out insurance for his subcontractors.

The professional indemnity insurance market is limited, i.e. it is quite a small one. As the majority of insurers do not transact this class of business they would not be prepared to give professional indemnity cover to the professional men involved in a construction project.

Even in the case of insurance of the works and public liability insurance (which are not generally speaking profitable classes of business from the insurers' viewpoint), the large contractor will often maintain that the cover provided by his annual policies (especially where they are tailor-made) is wider than that which can be provided for him by the owner (employer). This is not always true, but if so it is an argument against project insurance, although the owner can insist upon the project insurance applying.

There should be no difficulty in omitting from the yearly returns to the contractor's annual insurers the information on which the premium is calculated if the contractor is to come within the owner's project insurance in certain single project cases, thus avoiding a duplication of cover and premium.

The above reasons against comprehensive project insurance do not avoid the criticism that the insurance policies do not meet the requirements of the construction industry in this respect.

At least one insurer now offers to go further to help overcome the problems arising and disputes which frequently occur between contractors and designers/consultants with regard to liability and responsibility for accidents arising from defects.

The contractors' all risks policy can be widened in the first place to include:

(1)   the owner (employer) and all the contractors he employs directly;
(2)   the main contractor;
(3)   all subcontractors;
(4)   consulting engineers, architects and quantity surveyors.

Secondly the policy provides design damage cover (as described above) so that the professional men in (4) above are covered not only for their site activities but also for their office work in connection with the project.

This cuts across the cover provided by these individuals' professional indemnity policies but only so far as such negligence:

(a)   causes damage to the works (not defects without such damage);
(b)   arises during the construction period (not the maintenance period).

This cover helps to overcome arguments as to who is to blame in tort or contract, and the resultant delays and extra costs which so easily occur, bearing in mind that the interests of all the parties mentioned in (1) to (4) above are jointly covered. Reference to the case of *Commonwealth Construction Co* and the remarks in chapter 20 are pertinent.

Finally, it is essential that the terms and conditions of such a comprehensive project insurance are disclosed to bidders in the tender documents, so that all bidders will be able to cost their tenders accordingly.

The advantages and disadvantages of project insurance

A discussion of the advantages and disadvantages of project insurance tends to become one of employer-arranged against main contractor-arranged insurances for the simple reason that the employer will include all contractors he directly employs as well as the main contractor, and all subcontractors in the policy he arranges for the project, while the main contractor will not include all these parties in his policy. However, before listing the points for and against project insurance it must be clear what cover is being discussed.

In project insurance the professional indemnity cover must be ignored for the reason mentioned earlier (limited cover) plus the fact that professional indemnity policies are usually arranged on a claims made basis; thus cover virtually ceases when the policy lapses, i.e. when the contract is completed. While the CAR policy with a public liability extension will also often cease on completion of the contract, plus a limited form of cover during the maintenance period, this policy is on a losses occurring basis consequently the cover is wider. Professional indemnity claims, particularly in the construction industry, have a long tail, i.e. defects can arise over many years after the contract is completed. See for example the cases following *Anns* v *The London Borough of Merton (1978)* and as a result the Government's suggestions in the draft Regulations for the Housing and Building Control Bill 1983 where fifteen years' professional indemnity cover is suggested for the approved inspectors who may take over from the local authority inspectors.

Normally in project insurance it is only the CAR and public liability risks that are included. This leaves as a residue the insurances of employers' liability and motor outside the project cover. As these two are legally compulsory in the United Kingdom it is probably better to leave such insurances for the participants in the project to make their own arrangements.

Contractors will give the following drawbacks in employer-arranged insurances:

(1)   The employer's insurers cannot know, at the time when the covers are being arranged, even the identity of the contractor who is going to carry out the work, and certainly nothing about the efficiency of the contractor's organisation and his insurance claims record. Therefore it is difficult to see upon what basis the risks insured can be rated. Consequently most underwriters would load premium rates and excesses on CAR insurance arranged by the employer, where the identity of the successful contractor is unknown.

(2)   If, because of the foregoing, the employer's insurance is not arranged until after the contract is let, then there are going to be delays in effecting cover which do not occur where the contractor is responsible for his own insurances because the contractor will almost invariably (except for overseas projects) have an annual policy on a post-declaration basis so that cover is always in force in respect of any contract undertaken.

(3)   Contractors carry out many contracts and therefore are in the best position to know the risks against which they should insure and the contractor's policy will be tailored to his needs and the activities which he undertakes. See for example the consequential loss extensions available to the CAR policy in chapter 5. Furthermore the contractor is constantly in the market and dependent on his claims experience thus he is in a better position to obtain competitive rates in respect of the risks which are insured. The employer, on the other hand, will frequently only be undertaking a once in a lifetime construction project and will lack the 'muscle' in the insurance market, quite apart from the lack of knowledge or where to go for the specialist and best advice. In any event it is doubtful whether the employer will have sufficient detailed knowledge of such factors as the temporary works as these will only become apparent when they have been designed by the contractor (e.g. cofferdams and falsework to bridges etc. are usually the contractor's responsibility) and according to the Bragg Report on Falsework*, temporary works failures have been the cause of large insurance claims.

(4)   Frequently there are residual risks which are stated not to be included in the covers effected by the employer and these have to be insured by the contractor, usually under his annual policy, and the resultant saving by the contractor between the full risks and the residual risks can be marginal. In fact it has been known for employer-arranged CAR cover purporting to be project insurance not to cover the

---

* Full title *Final Report of the Advisory Committee on Falsework*, published under the auspices of the Health and Safety Commission in 1976 and available from HMSO.

temporary works or the constructional plant, temporary buildings and other property of the contractor. While, on the face of it, this may not seem unreasonable as he owns most of these items, if the contractor has to insure this property separately the cost can be prohibitive as it is the hazardous part of the risk. Similarly some insurers are unable to add a public liability section to the CAR policy particularly on overseas projects as their reinsurance facilities do not allow this, and again the premium cost would be heavier.

Furthermore the overall premium (employer's and contractor's) is higher than if one policy handled all the risks.

(5)   There is an argument for saying that as the contractor is usually responsible for the works under the contract he should also be empowered to take steps to protect himself including effecting insurance, whereas when the employer seeks to arrange the contract insurances he still leaves responsibility for the works with the contractor. Similarly if the employer insures the project the contractor should be relieved of responsibility for any inadequacies in the insurance cover to the extent that it falls short of the contract requirements. A contractor usually has a long-standing relationship with his insurers and the loss adjusters his insurers use and claims are expeditiously and often generously settled, therefore a contractor does not take kindly to an employer-arranged package which falls short of those arranged by the contractor. Consequently it will often be necessary for the contractor to arrange a 'Difference in Conditions' policy in order to avoid exposure to any risk attaching to him and normally covered under his own insurances. For example, the JCT Form and the ICE Conditions require insurances for off-site goods certified and paid for but it is doubtful whether an employer-arranged CAR cover would always cater for these risks or for the transit cover of such goods, and as mentioned in (4) above, separately arranged off-site goods and transit cover can be expensive for the contractor.

In spite of what is said by contractors certain advantages do exist for the employer-arranged project insurance.

(1)   The very strong argument is that mentioned at the beginning of this consideration of project insurance, namely, it avoids the time and trouble spent by all parties, involved in a loss, damage or liability claim, blaming the other parties to the contract or subcontract.

(2)   In very large projects, particularly where civil or mechanical engineering as well as building work is involved with many subcontractors being used project insurance covering all interested parties avoids gaps in each of the individual parties' cover.

(3)   The employer, in one way or another, pays the premium, even when the contractor arranges insurance cover, therefore the employer has the right to protect himself by buying project insurance, but the contractor regards this as a theoretical argument as the employer is not often in a position to provide more adequate

insurance than the contractor or at least to match his cover. However, assuming the employer is properly advised it could apply particularly in the case of the inadequately insured subcontractor. In any event the cost of the insurance to the employer can only be defined in any detail when the main contractor is known. The cost can only be broadly established before the contractor is selected.

(4)   Subject to the employer obtaining the best advice:
    (a)   The use of excesses can be controlled.
    (b)   Loss or damage by design failures can be minimised as the employer would control the overall cover and the settlement of claims would be quicker and easier as it would be a matter of negotiation between the employer's brokers and the insurers or their loss adjusters, with the consequent improvement in overhead costs and cashflow on insurance claims.
    (c)   Use and occupancy problems can be catered for easily.

Example of project insurance

An illustration of what can be achieved by project insurance for the very large contract can be seen from one of the contracts covering the Thames Barrier. This involved a consortium of contractors based upon a construction period of four years in 1973 at a contract price rising from £33 to £38 million during the year between submission and acceptance of the tender. Incidentally the contract took 8½ years to construct and the final cost had then leapt to over £240 million.

The contract was based upon the Greater London Council's General Conditions of Contract, amended so as to be similar to the ICE Conditions of Contract current at the time. This resulted in the possible exposure of the consortium of contractors to third party liability from flooding caused by say temporary blockage of the river during construction. Therefore a limitation of liability was arranged at a figure of £10 per occurrence. The GLC agreed to indemnify the consortium for amounts in excess of that figure.

The CAR insurance was placed with a leading office and eight other UK insurers plus Lloyds. The policy covered the contract works, temporary works, materials and non-marine construction plant (marine hull insurance had to be effected elsewhere) tools and equipment, together with a public liability section.

The main points of this insurance were:

(1)   All subcontractors were covered as joint insureds.
(2)   Cover was for all operations anywhere in the UK including storage and transit risks whether by land or water.
(3)   Indemnity for loss or damage caused by defective design, materials or workmanship **without** exclusion of the defective part.
(4)   Indemnity for increased cost of working sustained in expediting the reinstatement of loss or damage to the insured property in order to avoid or reduce a delay in completing the contract, subject to a limit of £2 million.

(5)   Cover for the cost of professional fees, removal of debris, shoring up etc., without separate lower limits.

(6)   Cover for permanent buildings on the site used by the contractors in relation to the contract.

(7)   Deductibles (Excesses) of:
  £25,000 each claim for defective design, materials or workmanship.
  £10,000 each claim for property in the river or dock.
  £500 each claim for all others.

Bearing in mind that the insurance market when this contract was placed was less competitive that it is a decade later, it is felt that at the time of writing it would be easier to place the insurances with possibly lower rates and wider coverage. For example it is possible that a separate Marine hull insurance could be arranged in one omnibus policy with the CAR etc. cover.

It is interesting to note that in this contract there were problems with ground conditions but the costs of overcoming these were paid for under the equivalent clause to clause 12 of the ICE Conditions of Contract dealing with adverse physical conditions and artificial obstructions which could not reasonably have been foreseen by an experienced contractor. Thus no payment had to be made for this by the insurers.

Further details of this example can be found in an article in *Policy Holder Insurance News* dated 30 September 1983, page 19.

Chapter 5

The contractors' all risks policy

The birth of the contractors' all risks policy is said to be due to the ICE General Conditions of Contract. Probably the advent of the second edition of the ICE Conditions in January 1950 requiring the contractor to insure the works in the joint names of the employer and the contractor on an all risks basis really gave this policy a boost, although it had been produced even before the last war on rare occasions. A policy was issued in 1929 in connection with the construction of Lambeth Bridge.

Therefore, gradually since the war 'all risks' cover has become usual rather than the more limited fire and special perils cover, even where the contract only asks for insurance in this limited way, as in the JCT Form of Contract. This is because of the duty of the contractor to complete the contract which imposes a responsibility for all risks apart from the contractual insurance requirement.

Because of the connection between the contractors' all risks policy and the ICE Conditions it was almost inevitable that the excepted risks listed in clause 20(3) of those Conditions (and dealt with in considerable detail in chapter 2) should form the basis of the policy exceptions. However, most insurers will include other exceptions for reasons which will be discussed later.

The contractors' all risks policy basically indemnifies the insured for loss, damage or destruction of the contract works, which includes the materials and goods on site to become part of the works and often includes materials and goods certified for payment although not on site but elsewhere or in transit.

While the construction or contractors' all risks (CAR) department handles risks connected with the erection of structures and civil engineering works, the installation in these works of mechanical or electrical plant is normally excluded. The engineering department, on the other hand, deals with risks connected with the installation of such plant, and with the constructional plant used to install it.

Consequently the term 'contract works' or 'erection all risks' in its usual form denotes the installation of machinery compared with CAR which deals with the ordinary building and civil engineering contracts. It should be noted that the overseas market tends to use the term 'erection all risks policy', not 'contract works', to denote the installation of machinery. Nevertheless, there is no clear-cut line as quite often the contract works policy will cover both the building or civil engineering as well as the installation of machinery risk, and even the CAR policy can be extended to cover the installation risk.

The main point to note about the CAR policy is that it is an 'all risks' policy. Thus it covers loss, damage to or destruction of the works etc., by any cause whatsoever but subject to certain exceptions. In this book the terms 'contractors' all risks' and 'contract works' are synonymous and do not include the installation of machinery unless otherwise stated.

The works

It has already been mentioned in chapter 3 that 'works' are defined in clause 1(1) of the ICE Conditions as meaning 'the permanent works together with the temporary works'. 'Permanent works' means the permanent works to be constructed completed and maintained in accordance with the contract. 'Temporary works' means all temporary works of every kind required in or about the construction completion and maintenance of the works.

In *Rowlinson Construction Ltd* v *Insurance Company of North America (1980)* the meaning of the contract works under an annual contract works policy was considered, although it did not involve the ICE Conditions. The plaintiffs, who were the insured under the policy, entered into an agreement with the Manchester City Council to lease land bounded by the River Medlock where there was a retaining wall. Under clause 5 of the agreement the insured were to build six industrial units on the land. The 99-year lease was to take effect from the date when the last unit was completed or 25 March 1982. Clause 11 of the agreement provided that the tenant would at his expense maintain, repair and renew to the council's satisfaction the retaining wall and the bank of the River Medlock.

There was a draft lease attached to the agreement and by clause 2(iii) of this lease the insured covenanted to repair, maintain and cleanse and in all respects keep in good and tenantable repair and condition the said buildings . . . including the retaining wall adjoining the River Medlock.

The wall collapsed and water accumulated on the inside of the retaining wall due to heavy rainfall, and frost prevented this water escaping. The pressure of the water forced the wall to overturn on its foundations which resulted in two items of claim under the contract works policy.

Firstly, £42,054 was claimed for the repair of the part of the wall which collapsed, and secondly £58,481 was claimed for the cost of strengthening the remaining sections of the wall not yet affected. The latter claim was clearly not covered as it involved the prevention of anticipated damage which is not a function of the policy.

The policy covered loss or damage to the permanent and temporary works at any contract site in the name of the insured and their principals as required by the terms of any contract within the scope of the policy. There was also a clause regarding special contracts which stipulated that unless the contract was below £10,000, to obtain cover the insured was required to provide full details to the insurers of any contract involving:

(a)   bridges or other work in or over water or tidal or coastal areas;
(b)   any other bridges;
(c)   tunnels or dams or weirs.

The policy described the contracts covered as any contract carried out by the insured in the normal course of his business whether by way of construction, alteration, repair or maintenance.

The insurers raised three items in their defence of the claim:

(1)   The damage did not arise out of the performance of any contract within the policy wording.
(2)   The retaining wall did not form part of the permanent or temporary works.
(3)   If the agreement for the lease was a contract within the policy wording then it was a special category contract and not covered.

Item 3 was eliminated as no part of the wall was built in or over the river so the special category allegation failed. Item 1 also failed as the court held that although the lease was not an ordinary building contract, nevertheless it came within the spirit and the letter of the policy and the work was thus covered.

Item 2 was more difficult to resolve and in order to decide whether the retaining wall was included in the permanent works the court considered what the insured were actually going to do under the contract. The insured were not going to build the retaining wall, thus in this respect it was not part of the permanent works. At the time of its collapse the insured had no intention of doing anything to the wall other than improve its appearance. Consequently it was wrong to describe the wall as part of the works.

The insured maintained that the policy applied to contracts by way of alteration, repair or maintenance and by clause 11 of the agreement for the lease the insured was obliged to 'maintain, repair and renew to the satisfaction of the council the existing retaining wall'. The court held that what the insured might have to do was irrelevant. What was relevant was what they actually intended to do and this did not include the rebuilding of the wall. Furthermore, this clause 11 obligation applied not only during the period of construction but also under clause 2(iii) of the lease for the whole of the 99-year period of the lease. Therefore, *prima facie* the insurers would continue on risk in respect of the wall so long as they renewed the policy from year to year and the court did not think this was the intention of the parties.

The policy wording

A specimen wording of this policy can be seen in appendix 1, section 3 of the contractors' combined insurance policy.

The operative clause indicates that loss or damage to the following are covered from whatsoever cause occurring during the period of insurance:

(a) the contract works consisting of the permanent and/or temporary works to be erected in performance of the contract and all the materials (some wordings say – all other things) for use in connection therewith;

(b) tools, plant, equipment and temporary buildings the property of the insured or for which he is responsible;

(c) personal effects and tools of the insured's employees in so far as such items are not otherwise insured.

This particular policy not only covers the property just listed while on the contract site, but also in transit thereto or therefrom by road, rail or inland waterway, plus any liability which falls on the insured under clause 16(2) of the Standard Form of Building Contract or clause 54(3) of the standard ICE Conditions in respect of loss or damage to materials and goods for incorporation in the works while temporarily stored away from the contract site. (See appendix 1, section 3, paragraph 4.) This latter cover is important. Without it, the insured would have to arrange cover for materials off site which are to be certified for payment by the engineer or architect and for transit to the site. The contractor really does not want to be bothered with this arrangement in each case and therefore it is important that the contractors' all risks policy provides this additional cover. Not all insurers' policies do so. On the other hand, some insurers give a wider cover (no clauses are mentioned) and insure property for the contract 'anywhere in the United Kingdom'.

There are limits of indemnity specified in the policy schedule for each of the items mentioned in the operative clause, that is to say (a), (b) and (c) above with the exception that item (b) is divided into two parts, namely tools, plant equipment and temporary buildings belonging to the insured on the one hand and such items hired in on the other.

In accordance with appendix 1, section 3, paragraph 1, liability for loss and/or damage to the contract works can also happen within the period of maintenance and arise from a cause occurring prior to the commencement of the period of maintenance. Alternatively it can be caused by the insured during the period of maintenance in the course of any operations carried out for the purpose of rectifying any defects but excluding the actual defects. The period of maintenance is the period stated in the contract conditions but not exceeding twelve months.

Paragraph 2 of appendix 1, section 3, provides cover for architects', surveyors', legal, consulting engineers' and other fees necessarily incurred in the reinstatement following loss or damage to the property insured by any peril covered by the policy, at a percentage in accordance with the authorised scale of the appropriate body or institute concerned. Fees for preparing any claim as distinct from fees incurred in the reinstatement of the loss or damage are not covered.

Paragraph 3 provides cover for costs and expenses incurred by the insured with the consent of the company in: (i) removing debris; (ii) dismantling and/or demolishing; and (iii) shoring up or propping of the property insured under the policy following loss or damage by any peril it covers.

Proviso (i) of section 3 of the policy provides for an automatic increase in the indemnity limit (in the policy schedule) for contract works by an amount not exceeding 20% subject always to adjustment of the premium in accordance with condition 7 (see later). Some wordings do not tie this clause to the contract value. An automatic escalation clause of this nature is essential in these days of inflation, especially with large contracts which are going to take a number of years. After all, the original contract price can increase tremendously over two or three years and, if there is a large claim towards the end of the completion of the contract, the sum insured could be found to be inadequate. While the contract usually insists upon insurance for full value, full value can only mean reinstatement value, that is, rebuilding value.

An important aspect of any material damage policy is choosing an adequate sum insured which, ignoring such matters as debris removal costs and professional fees, must cover the reinstatement cost of the works, so that the insured has to make a calculation:

(a)   of the increases in cost of materials, goods and labour charges during the period of the contract, i.e. price fluctuations; and

(b)   following damage during the construction period, of the effect of inflation on those costs incurred:

(1)   in the reinstatement of the damaged works; and

(2)   in completion of those parts of the works that have not been started at the time of damage and which are delayed as a consequence.

The type of automatic escalation clause mentioned above in proviso (i) can cater for (a) and even (b)(1) above, but better protection against inflation can be obtained by an extension of the policy with a separate sum insured in respect of the inflation factor which is geared to a rate of interest suggested by the insured and agreed to by the insurers. This extension is not a type of consequential loss policy which could be added to those mentioned later under exception 8; whereas, in the case of (b)(2) above, it is such an addition. During the last decade the monthly rate of inflation in building costs has fluctuated between 1 and 2½% (see example on page 214).

It is appreciated that the last two paragraphs are to some extent a repetition of the remarks under the heading 'Sum insured' in chapter 3 but as this book is intended as a work of reference a degree of repetition is desirable.

Proviso (ii) caters for the reinstatement of the sum insured after a loss. As the provision of the reinstatement of the sum insured is not always an automatic matter (in some policies it is), it is essential for the insured or their agents that this request is made to the insurer after a loss. This reinstatement of the sum insured is subject to the payment of an additional premium.

Proviso (iii) merely ensures that the structure which is the subject matter of the contract is to be built with standard construction materials, and thus controls premium in the case of annual policies.

Proviso (iv) imposes an average clause on the sum insured in respect of tools plant equipment and temporary buildings. That is to say, if the value of these items at the time of any loss or damage insured by the policy is of greater value than the sum insured, then the insured is considered as being his own insurer for the difference and has to bear a rateable share of the loss accordingly. The effect of this clause is to alter the contract, for instead of being a contract to pay up to a specified amount in the case of loss, it becomes a contract to pay not the amount of the loss, but such proportion of that loss as the sum insured bears to the value at the time of the loss of the property insured, not exceeding in all 100%, i.e. amount payable under the policy equals:

$$\frac{\text{Sum insured}}{\text{Full value}} \times \text{Loss}$$

Example

	(£)
Sum insured	1,500
Full value	2,000
Loss	500

$$\text{Amount payable } \frac{£1,500}{£2,000} \times £500 = £375$$

If the property is insured for more than its full value, the average clause is inoperative, as the insured cannot recover more than the loss he has actually sustained. Some policies may not have a separate sum insured for plant etc., or an average clause which is not easy to operate.

Policy exceptions

The position under clause 20(3) of the ICE Conditions is that the employer is responsible for the excepted risks and these form the basis of the policy exceptions. Most of them are uninsurable and are now considered in some detail.

(1)   Riot is uninsurable only in certain areas, e.g. Northern Ireland, for the obvious reason that the risk is excessively hazardous in that area.

(2)   War (including kindred risks) and nuclear risks are by past practice the responsibility of the Government and by statute the responsibility of the licensee of the nuclear site respectively. See appendix 1, general exception 3 of the contractors' combined insurance policy. See more detail of nuclear risks on page 38.

(3)   Sonic waves can cause considerable damage and the United Kingdom Government announced that it will pay compensation if damage results from Concorde's test flights and by implication all flights. Insurers never intended to cover loss, destruction or damage directly occasioned by pressure waves under their material damage policies and have made the position clear by an exclusion which would apply to the contractors' all risks policy. See more detail on page 67, exception 9.

(4)   Fault, defect, error or omission in the design of the works (other than a design provided by the contractor pursuant to his obligations under the contract). Loss arising from defective design is not usually covered as it is a professional indemnity risk — thus a more specific insurance is available. According to an Australian case cited below, it is wrong to confine faulty design to professional negligence. In the past, to meet competition, a few insurers have included cover against the consequence of defective design in approved cases. Damage to the item of the works proved to be defective was not insured. While in 1968 a number of insurers agreed that all consequences of design defects should be excluded, the following wording, giving a wider cover, is used by some insurers:

The Policy excludes claims for the cost of replacement or rectification of that part of the permanent works rendered necessary by its own defective design or specification but this insurance shall cover the cost of making good any damage to other parts of the Property Insured hereunder resulting from such defective design plan or specification.

A case on the meaning of 'faulty design' as used in the exclusion to this policy is *Queensland Government Railways* v *Manufacturers' Mutual Insurance, Ltd (1969)*.

A railway bridge in Australia was being constructed by EPT for QGR (railway authority) to replace the bridge built in 1897 which had been swept away by flood waters. Prismatic piers (similar to the original piers, but strengthened) were being erected when they were overturned by flood waters after exceptionally heavy rains. EPT and QGR claimed to be indemnified by the insurers under a contractors' all risks policy, which provided, *inter alia*:

. . . this insurance shall not apply to or include:
   (vii)   cost of making good faulty workmanship or construction . . .
   (xi)   loss or damage arising from faulty design and liabilities resulting therefrom.

The insurers denied liability, contending that the loss was due to faulty design of the piers. The arbitrator found that, in the state of engineering knowledge at that time, the design of the new piers was satisfactory, but investigations into the cause of failure of the piers showed that during floods the piers were subjected to greater transverse forces than had been realised and that the loss was not due to faulty design, in that 'faulty design' meant that 'in the designing of the piers there was some element of personal failure or non-compliance with the standards which would be expected of designing engineers', and that, therefore, the insurers were liable. The insurers applied to have an award set aside or remitted on the ground that the arbitrator misconstrued the term.

It was held by the Supreme Court of Queensland that, in the context, 'faulty design' implied some element of blameworthiness or negligence, which had been negatived by the arbitrator's findings; that subsequently acquired knowledge revealing that the piers were not strong enough, could not convert the design, which would at the time have been accepted by responsible and competent engineers, into a 'faulty design' and that, therefore, the insurers were liable.

On appeal the decision of the Queensland Supreme Court was reversed. It was held that 'faulty design' did not imply an element of blameworthiness or negligence; the loss of piers through the inadequacy of their design to withstand an unprecedented flood was outside the policy notwithstanding that the design complied with the standards that would be expected of designing engineers according to engineering knowledge and practice at the time of their design.

The exclusion concerning defective workmanship and material is limited to the cost of replacement of the defective part or parts and does not extend to exclude liability for damage to other parts of the insured property caused by the defect. A common wording is: 'Excluding . . . the cost of repairing replacing or rectifying property which is defective in material or workmanship.'

In the Canadian case of *Pentagon Construction (1969) Co Ltd* v *United States Fidelity and Guarantee Co (1978)*, the interpretation of the following wording was considered:

This insurance does not cover: (a) Loss or damage caused by: (i) faulty or improper material or (ii) faulty or improper workmanship or (iii) faulty or improper design.

In this case Pentagon were building contractors engaged to construct a sewage treatment plant which included a concrete tank. The plans and specifications required a number of steel struts to be laid across the top of the tank with each end welded to a plate let into the concrete wall beneath it. The purposes of the struts were to strengthen the tank by holding the sides together and to hang equipment from them. The contract required Pentagon to test the tank. Pentagon insured the work under a contractors' all risks policy.

After the concrete work of the tank was completed, and the struts laid across the tank but before the ends of the struts had been welded, the tank was tested by filling with water. The tank bulged and a claim was lodged under the policy and repudiated by the insurers who relied on the above exclusion.

At the court of first instance it was held that the design of the tank was not faulty or improper and there was no faulty or improper workmanship. The insurers appealed, and argued that the word 'design' included the plans and specifications and that they were faulty in that they omitted to state that the tank should not be tested until after the struts had been welded.

All three judges in the appeal court decided that as the evidence clearly established that the wall of the tank failed because of the failure to weld the steel struts to the top of each side of the wall before testing, this amounted to improper workmanship or to put it another way, testing before welding was improper workmanship, which led one judge to decide it was unnecessary for him to consider the question of faulty or improper design. The other two judges reached different opinions on the meaning of 'design'.

The conclusions to be drawn from this case are that:

(a)   Workmanship is not limited to the work or result produced by a worker. It includes the combination or conglomeration of all the skills necessary to complete the contract including, in this case, the particular sequence necessary to achieve the performance of the contract. Failure to follow that sequence could constitute faulty or improper workmanship and in this case did so.

(b)   We do not know whether:

(i)   Detailed instructions in the plans and specifications on how the work of construction is to be carried out are not part of the design, which was one judge's view (he added that if he were wrong he did not think it was necessary for the plans and specifications to warn that the tank should not be filled with water before the struts were welded); or

(ii) Design includes the drawings and specifications, which was the other judge's view.

Thus on the meaning of 'design' with the third judge abstaining, the case is unsatisfactory.

As was explained at the end of chapter 2, while there is a slight doubt as to whether the contract phrase 'fault defect error or omission in the design of the works' includes non-negligent design, there is no doubt that a similar phrase (if used), namely 'faulty design', in an exclusion of the contractors' all risks policy does include non-negligent design. See the Australian case mentioned above. Some insurers repeat the actual wording used in the ICE Contract, as illustrated in the policy given in appendix 1, and in the case of policy wording any ambiguity must be construed in the insured's favour so it is just possible that the longer phrase may carry sufficient doubt to call the *contra proferentem* rule into play. In this event the more detailed phrase used in a policy exclusion would not include non-negligent design which would be covered. Furthermore, the contractor is still left unprotected by insurance (even if only responsible for negligent design unless professional indemnity cover is arranged), in respect of a design he provides pursuant to his obligations under the contract, for which he is responsible. See the 'Summary of risks remaining uninsured' at the end of this chapter.

Defective workmanship is a contract hazard normally accepted by contractors. It is important to note that while the contractor is expressly liable to repair damage to the works due to defective materials or workmanship 'during the progress of the works' under sub-clause 39(1)(c) and during the maintenance period under sub-clause 49(3), these liabilities are excepted from the obligation to insure by the proviso to clause 21 unless the bill of quantities provides otherwise. The reason may be that there is some difficulty in obtaining such insurance cover, at least so far as the cost of replacement of the defective part or parts is concerned. Thus the policy exclusion concerning defective workmanship and materials, unlike the usual design exclusion, **does not extend to exclude liability for damage to other parts of the insured property caused by the defect**. Defective workmanship is a contract hazard normally accepted by contractors, as already mentioned. The cost of doing such work twice is not properly a matter for insurers and if the use of defective materials is not due to the negligence of the contractor he will probably have a remedy against the suppliers. See appendix 1, section 3, exception 1(a) and (b).

The following are other wordings for defective design materials and workmanship exclusion:

(a) *Outright defect exclusion*

This Policy excludes all loss of or damage to the Property Insured due to defective design, plan, specification, materials or workmanship.

This wording even excludes liability for damage to other parts of the insured property caused by defective materials and workmanship.

(b) *Limited defect exclusion*

This Policy excludes the costs necessary to replace, repair or rectify any of the Property Insured which is in a defective condition due to a defect in design, plan, specification, materials or workmanship, but this Exclusion shall not apply to the remainder of the Property Insured which is free of such defective condition but is damaged as a consequence of such defect.

This wording includes the wider design cover mentioned earlier (protecting other parts of the insured property caused by defective design).

(5)   Loss or damage due to use or occupation by the employer his servants agents or other contractors (not employed by the contractor) of any part of the permanent work. The contract does not require such cover as it is an excepted risk. In any event once the situation in this exclusion arises the contractor probably no longer has an insurable interest in that part of the works. See appendix 1, section 3, exception 5.

The above risks appear as exclusions under a contractors' all risks policy plus the following which is not an exhaustive list:

(a)   Wear, tear and gradual deterioration. Probably this cover would not be expected by the contractor as material damage policies are not maintenance contracts. See appendix 1, section 3, exception 2.

(b)   Loss of or damage to plant due to its own explosion, mechanical breakdown, or derangement. This risk can be covered by an engineering policy, but the exception does not exclude damage to other property caused by the explosion, mechanical breakdown or derangement. See appendix 1, section 3, exception 3.

(c)   Loss of or damage to mechanically propelled vehicles (other than certain equipment) locomotives, aircraft or watercraft exceeding twenty feet in length or hovercraft. These risks are the subject of more specific insurances such as motor, engineering, aviation and marine, although special arrangements can be made where individual types of such plant are required to be covered by this policy. See appendix 1, section 3, exception 6 which does not include mechanically propelled vehicles.

(d)   Loss or damage due to confiscation, nationalisation, requisition or order of the government or public or local authority. This is not a risk for insurers to cover, and in practice mainly concerns overseas contracts.

(e)   In respect of each occurrence giving rise to loss or damage, it is the practice to make the insured responsible for an excess which applies to all perils. Insurers are tending to increase excesses and £100 is often considered a minimum figure. The trend is for higher excesses especially in civil engineering projects.

(f)   Consequential or economic loss whether caused directly or indirectly. The term 'consequential loss' can have more than one meaning and this is best illustrated by the story of the man who was asked 'How is your wife?', and he answered 'Compared with what?'. Thus 'consequential loss', like the term 'common law', is always used to make a contrast.

Most fire insurance officials use the term to denote what is now more usually called 'business interruption insurance' as distinct from material damage insurance, e.g. the standard fire policy. Most material damage policies do not cover consequential loss.

The liability underwriter on the other hand, uses the term 'consequential loss' to denote economic or financial loss which flows from an event or liability. Most public liability policies cover consequential loss which flows from loss or damage to third party property or bodily injury to third parties.

Finally, a lawyer follows the interpretation by the court of 'consequential loss' as meaning loss or damage which does not result directly and naturally from the breach alleged. See *Croudace Construction Ltd* v *Cawoods Concrete Products Ltd (1978)*, details of which are given in chapters 2 and 20. Clearly the courts take a more limited view of the meaning of 'consequential loss'. Thus, provided the policy indicates exactly what is covered without using the term 'consequential loss' (or, if used, defining it), there should be no difficulty. However, if a contractors' all risks policy used the term as an exception intending to exclude loss of profit for example, the insurer could find himself giving this cover on the court's interpretation as it flows directly from the loss or damage. See *Saint Line Ltd* v *Richardson Westgarth and Co Ltd (1940)* and *Wraight Ltd* v *P.H. and T. (Holdings) Ltd (1968)*.

Some consequential losses may be covered under special policies; sometimes the employer (principal) is the insured. These policies normally give cover against the same perils as the material damage policy required for the protection of the works and are subject to such a policy operating. Consequential loss policies can be obtained covering:

(i)   *Advance profits*   The protection provided by this policy is in respect of financial loss through delay by damage to the works or to important plant or equipment at the supplier's premises or during transit. Payment under this policy does not start until the date the business would have commenced but for the damage, and is in respect of the anticipated income, i.e. the gross profit of a manufacturer or the rent of a property developer which is not earned at the estimated date.

(ii)   *Additional cost of working*   This is an expense incurred by the contractor and can be an extension of the contractors' all risks policy as it follows a claim under that policy and involves costs beyond those incurred in making good that damage. For example, the basis of claims settlement in the case of payment to workmen by reason of guaranteed time or such agreements is by calculating the difference between the amount paid to the workmen and the amount which would have been payable had no such agreement been in force. In the case of plant

standing idle, the calculation for hired plant is the amount payable for
the affected period and for the contractor's own plant is based on an
allowance in respect of loss of working time, say 66²/3% of the rates for
such plant in either of the publications applicable (*Definition of Prime
Cost of Daywork carried out under a Building Contract*, published
jointly by the RICS and NFBTE, or the *Schedules of Dayworks carried
out incidental to Contract Work*, issued by the Federation of Civil
Engineering Contractors). Sometimes the expense is difficult to identify
especially when time has also been lost on the contract before the
damage.

(iii)   *Fines and damages*   This is also a possible extension of the contractors'
all risks policy covering fines and damages payable by the contractor
under the construction contract, following loss or damage due to some
or all of the perils covered by this all risks policy. This cover is not
easily obtainable and the rate of premium is high.

(iv)   *Additional or extended interest charges*   Where the subject matter of
the contract is to be sold on completion, in the case of damage there
could be delay in receipt of the money from the sale. The cover can
give the agreed amount of interest on the net amount of the sale to the
extent that it is delayed subject to a limitation of the indemnity period.
On the other hand, the actual interest on a loan could be insured so far
as it is extended due to the damage.

(g)   This exception refers to the damage to tyres by application of brakes or by
punctures cuts or bursts. Possibly the excess which always operates in this kind of
policy would ensure that this type of claim is excluded, but this is not necessarily
so as tyres on large types of plant can be very expensive, hence the exception. See
appendix 1, section 3, exception 4, but some insurers do not apply this exception
especially if there is a large excess on their basic policy.

(h)   Loss of or damage to tools, plant and equipment belonging to subcontractors
or their employees are excluded because the insured contractor has no knowledge
of the plant brought on to the site by the subcontractors and thus the sum insured
is unknown and therefore it is undesirable for insurance. See appendix 1, section 3,
exception 7. This exception can be amended if subcontractors' tools etc have to be
covered.

(i)   Material damage and consequential loss policies were never intended to cover
loss, destruction or damage directly occasioned by sonic waves and insurers took
the opportunity to make the point clear by applying the wording of exception 9 of
section 3, appendix 1. The UK Government at the time of Concorde's test flights
announced that it would pay compensation if damage resulted from such flights
and by implication the same would apply to normal schedule flights. While the
clause is included in the own damage cover given by the motor policy, it is not
included in the other sections and is not included in the employers' and public
liability policies. It was understood that if Concorde's tests showed that there

was a real risk of claims arising under certain liability policies there could be a reassessment of the position. However, this has not become necessary.

(j)   This exception excludes cash, banknotes, cheques, stamps or security for money, and is the subject-matter of a money policy. It may be thought that the wording of the operative clause would be limiting enough to exclude cash, etc., as it is confined to contract works and the like. However, there are likely to be large sums on the site at the time of payment of wages apart from the normal float for day-to-day expenses and consequently some insurers consider it is as well to make it quite clear that these items are not covered by the policy. See appendix 1, section 3, exception 10.

(k)   The exception excluding property more specifically insured mentioned in appendix 1, section 3, exception 11, is not always included as it merely emphasises condition 9 of the specimen policy in the appendix. In any event the exclusion of property more specifically insured is ineffective if the other insurance also excludes liability in the event of a more specific insurance. Each insurer has to share the risk: see *Gale* v *Motor Union* and *Loyst* v *General Accident (1928)*. A few insurers avoid sharing such losses by stating that their policy only operates after the other insurance has paid to the full extent of its liability: see *State Fire Insurance General Manager* v *Liverpool and London and Globe Insurance Co Ltd (1952)* before the New Zealand Court of Appeal. Note the situation in *Steelclad* v *Iron Trades (1984)* mentioned in chapter 1 when discussing contribution.

(l)   There is usually an exception the object of which is to exclude shortages not arising out of a specific incident but due to regular pilfering over a period of time. Sometimes this exclusion uses the wording 'disappearance or shortage which is only revealed when a routine inventory is made' or as in appendix 1, section 3, exception 12, which relates the occurrence to one notified to the insurer under the terms of condition 1 of appendix 1. Condition 1 requires notice as soon as possible after the occurrence of any accident. Obviously the policy is not intended to cover shortages discovered at the time of stocktaking, i.e. unexplained shortages.

(m)   This exception concerns loss or damage which is the responsibility of the principal (employer) under the conditions of contract, which in the case of the ICE Conditions are those 'excepted risks' listed in sub-clause 20(3). It might be wondered why, if the exceptions take care of such excepted risks individually, it is necessary to have an overall exclusion applying to such risks. In the first place 'riot', as already mentioned, is not the subject of a specific exception in the policy although it is an excepted risk in the contract. However, in England, Scotland and Wales no doubt insurers would be prepared to cover this risk if requested to do so. Secondly, when insurers word their policies they have to have in mind contracts other than the ICE Conditions and it is possible that when an employer is made responsible the intention is that there should be no cover for the employer under the contractor's policy even though the perils concerned are otherwise covered by the contractor's policy.

Thus in the JCT Standard Form of Building Contract clauses 22B and 22C leave only the employer responsible for the listed perils. The intention is for the employer to arrange his own policy for these risks. There is no request for joint named cover in the JCT Form under clauses 22B and 22C. On the other hand, in the ICE Conditions *prima facie* the request for insurance cover in the joint names of the employer and the contractor in clause 21 can cause confusion in the light of the exclusion under discussion. However, it will be seen when examining the 'indemnity to principals' extension of the contractors' all risks policy that it is subject otherwise to the terms and conditions of the policy in so far as they can apply. Thus the contractors' all risks policy would cover the employer when the ICE Conditions apply but only for the risks which are **not** 'excepted risks', as only these excepted risks are the responsibility of the employer. As to when the employer would need such cover, see the commentary under the specific heading 'Indemnity to principals' mentioned later.

Clauses applicable to all sections of the contractors' combined policy

(1) *Excess clause*
This has already been mentioned when discussing the exceptions to section 3, 'Loss of or Damage to Contract Works'. Note that the expression 'claim' means a claim or series of claims arising out of one event.

(2) *Indemnity to personal representatives*
This is the usual extension to cover the insured's personal representatives after his death provided that they observe the terms exceptions and conditions of the policy as far as they can apply.

(3) *Indemnity to principals and joint names cover*
An example of this clause is given in appendix 1. Many contracts including the ICE Conditions (in clause 21) stipulate that the employer shall be a joint insured with the contractor in respect of the required insurance of the works.

Whether a clause worded as that in appendix 1 is used or the employer (principal) is named as an insured in the policy schedule makes no difference (assuming the interpretation given below concerning proviso (c) is correct), as the clause in appendix 1 also requires the principal to be named in the policy schedule, but not as an insured because the clause has this effect. However, it is understood that the insurers concerned intended the phrase '(named in the schedule hereto)' to read '(described in the schedule hereto)', as in that schedule under the word 'principal' the words 'any person, firm, ministry or authority for whom the insured is undertaking work' appear; therefore the intention and practice is to give cover to any principal without naming the principal in the schedule. At least this is the position in the case of annual policies. For individual contract policies the principal is named in the policy. However, (subject to the interpretation of proviso (c) being correct) there is no difference between this clause as practised and naming a principal in the schedule as a joint insured without such a clause. In the former case the clause is stated in an overall heading to apply to all sections of the policy, therefore it must apply to section 3 (loss of or damage to contract works). Also the clause states that once a contract condition so requires the benefits of the policy are stated to apply jointly to the insured and the principal who will be indemnified within the policy terms for any claim resulting from loss or damage as defined in the policy. Provisos (a) and (b) merely emphasise that the principal is subject to the limitations terms and conditions of the policy so far as they can apply. This is automatic when the principal is named as an insured in the policy schedule. It is suggested that proviso (c) only applies to sections 1 and 2 of the policy which concern the employers' and public liability covers where the intention of proviso (c) is to have the effect of a cross liabilities clause allowing the principal to claim against the contractor, for example, for loss or damage to property belonging to the employer (which is not the subject-matter of the section 3 insurance, i.e. the works etc.), as such a claim is otherwise excluded from the public liability policy. A similar example can be given under the employers' liability cover, and this aspect is discussed further in the appropriate chapters concerning the liability policies. There does not seem to be any point in including the subject-matter of the section 3 cover in proviso (c) as the effect would be to cancel the cover intended to be provided by this clause, the intention being to allow the principal to claim under the policy as a joint insured. Whether the insured is shown as a joint insured in the policy schedule or an 'indemnity to principal' clause as explained is used, it is clear that any loss or damage due to the use or occupation of the works is excluded by a term of the cover. See appendix 1, section 3, exception 5.

It is not always appreciated by insurers why the employer should require to be a joint insured. Although the works usually remain the sole responsibility of the contractor until final completion (in the ICE Conditions it is until fourteen days after the engineer has issued a certificate of completion for the whole of the works pursuant to clause 48; see clause 20(1)), nevertheless it is a common provision in contracts that the title in all goods delivered to the site for incorporation in the works is transferred to the employer. In the ICE Conditions clause 53(2) makes all plant, goods and materials owned by the contractor when on the site the property of the employer. Plant for the purposes of this clause means constructional plant, temporary works and materials for temporary works but excludes any vehicles engaged in transporting any labour, plant or materials to or from the site. The point is that by these contractual terms the employer has an insurable interest in all the property which becomes the subject of the contractors' all risks policy.

Petrofina (UK) Ltd v *Magnaload Ltd (1983)* was the first English case to decide that insurers cannot sue a joint insured in the name of the other joint insured in order to recover their outlay under a material damage policy, although there have been Canadian and American cases to this effect.

General exceptions

In appendix 1, general exception 1, the intention is not to cover injury, illness, loss or damage caused elsewhere than in Great Britain, the Isle of Man or the Channel

Islands.

While the ICE Conditions are largely limited in their use to these areas, the International Contract (FIDIC) based on the ICE Conditions and considered in a later chapter, is by its very nature used elsewhere than in the areas mentioned.

The number of insurers prepared to provide insurance cover for civil engineering projects is limited and even more so when the site is situated abroad. This aspect will be considered further in the chapter dealing with the FIDIC Contract.

General exceptions 2 and 3 in the same appendix exclude nuclear risks and war and kindred risks.

War risks for property on land are such that no insurer could undertake them. They are a matter for the Government, as happened in the First and Second World Wars.

Liability in respect of radioactive contamination – this is necessary because legislation directs liability to the nuclear reactor owner and the Government. Similar responsibility for loss or liability arising from explosive nuclear assemblies would probably fall upon the UK government department concerned or the Atomic Energy Authority.

This exception does not include radio-isotopes or equipment such as X-ray machines as they are considered capable of being covered by the insurance market. These articles can be used to test welding and in this respect the construction industry may be involved. The proposal form usually includes an appropriate question.

It is clear that insurance policies were never intended to cover the effects of nuclear explosives as being excessively hazardous and insurers have decided to make this clear by the clause or a similar one which appears as mentioned earlier. The same exception includes the radioactive contamination wording.

Finally, exception 4 excludes any sums payable under the contract as fines or penalties for delay or non-completion of the works. This is not normally a risk insurers are prepared to accept as it involves a guarantee of completion of the works, and may be considered by some as against public policy.

Policy conditions

The conditions will be taken in the order set down in the policy (see appendix 1).

Condition 1 requires the insured to give notice of any occurrence which may give rise to a claim. This wording, to some extent, leaves the decision to the insured as to whether a claim may arise, but the intention of the insurer is that if there is any doubt at all the claim should be reported to the insurer.

The remainder of the condition leaves no decision to the insured. He must send to the insurer immediately every letter, claim, writ, etc. and immediately notify the insurer of any impending prosecution, inquest or fatal inquiry.

Condition 2 incorporates the usual wording by which insurers can take over the conduct and control of the claim, and prosecute or defend it in the name of the insured. No admission, offer, promise or payment may be made without the consent of the insurer in writing.

Condition 3 is the reasonable precautions condition and is self-explanatory.

Condition 4 merely restates the common law concerning non-disclosure or mis-statement of material facts and/or alteration or material increase of the risk.

Condition 5 increases slightly the common law rights (an insurer has to stand in the shoes of the insured and claim in his name after indemnifying the insured) by allowing such rights before indemnifying the insured.

Condition 6 does not apply to the contract works risk.

Condition 7 sets out the requirements concerning returns a company requires in order to estimate its premiums, and the right to examine the insured's records concerning turnover or wages on which the premium is based.

Condition 8 sets out the period of notice of cancellation by the insurer and varies from insurer to insurer from seven to thirty days according to the practice of the insurer concerned. There is provision for the adjustment of premium in the event of cancellation.

Condition 9 is the usual contribution clause (see chapter 1). If a company says it is not to be called upon to contribute under this insurance if there is another insurance in operation that is covering the same subject matter against the same risk in respect of the same insured, the effect is the same as if it has said it would only pay its rateable proportion assuming both insurers say they are not going to contribute and both their policies are in operation. Note the Steelclad case in chapter 1.

Condition 10 contains facilities for arbitration in the event of there being a difference as to the amount to be paid under the policy.

Condition 11 makes the other conditions precedent to liability as far as they can be so. Condition 7, for example, could not be so regarded because a condition precedent to liability is one which must be observed in respect of any particular loss in order that such loss may become payable by the insurer. The requirement concerning the calculation of the premium is nothing to do with any particular loss.

Types of policy cover (period of insurance)

While the usual practice in major civil engineering works is to issue a policy for each contract and rated for that contract, an annual policy is available where the work normally undertaken is of a similar nature. However, with civil engineering work unless the contractor is involved only in specialist work uniformity is unlikely to exist and as the usual annual policy would exclude hazardous operations, e.g. tunnelling, bridges, and work in water which the contractor may well undertake, a policy issued contract by contract is more likely to be the normal procedure. If it is possible to issue an annual policy the initial premium is often rated on 50% of the estimated value of the annual turnover, which is subject to adjustment at the end of the insurance year according to the actual value of the completed work. From the information supplied to the insurer concerning the type, amount, period of construction and maintenance and completion date of contracts undertaken a contract limit is usually imposed, and often a cancellation clause is included in the policy.

The annual, blanket (or floater) policy has the following advantages:

(a) As the cost is known it makes tendering simpler.
(b) There is a minimum number of policies and amount of administrative work.
(c) There is automatic cover on all contracts, i.e. without specific notification to insurers.
(d) The policy can be extended to cover all constructional plant and equipment used on all sites throughout the year.
(e) While a blanket policy is usually issued to a contractor to insure him, and normally, his employer and subcontractors also, in respect of accidents occurring in connection with the work described in the policy, it may also be issued to an industrial, commercial or other type of employer who frequently employs contractors for construction, extension or alteration of his own property.

Premium calculation

The sum insured should be the total value ultimately at risk and is made up from:

(a) the contract price (or annual turnover under blanket policies);
(b) temporary buildings and contents, constructional plant tools and equipment etc.;
(c) debris clearance after the operation of an 'insured peril', usually insured as a separate item;
(d) engineers' and architects' fees for reinstatement of damage (excluding fees for the preparation of a claim), usually insured as a separate item;
(e) materials and goods supplied by the employer (principal), thus outside the contract price.

Under both individual and blanket (annual) policies an initial premium is paid. In the case of individual policies it is basically rated on the estimated contract price and for blanket policies as mentioned under the previous heading. The premiums in both policies are retrospectively adjusted when the actual values are known.

The annual policy will incorporate a maximum value for any one contract which will normally be the limit of the insurer's liability for any one contract site. Occasionally the policy indemnity limit may be less than the contract value. Where a total loss is virtually impossible insurers may be prepared to rate the risk on an estimated maximum loss basis.

Clauses 20 and 21 and the contractors' all risks policy: summary of the main risks remaining uninsured

In order to indicate the main risks for which the parties to the ICE Conditions may not have insurance cover the positions of the contractor and the employer are considered in turn. However, it is emphasised that basic policy cover which has been described in this chapter is considered here, not the cover an insurance broker by negotiation with an insurer could produce for one particular insured or tailor-made for one particular contract. Also there is no attempt to indicate some special wordings which some brokers catering for the construction industry have persuaded an insurer or insurers to adopt for the former's clients.

The contractor

(1) *Sums insured for works and constructional plant* The sum to be insured for works and plant must be 'to their full value' by clause 21. See chapter 3 under the heading 'The sum insured' (p. 45), and this chapter on p. 60. It is a question of choosing an adequate method of catering for inflation and deciding what consequential loss cover is required. The aim must be to insure the works and constructional plant brought on to the site for their replacement value.

(2) *Fault, defect error or omission in the design provided by the contractor pursuant to his obligations under the contract* Clause 20(3) indicates this is not an excepted risk, therefore the contractor must insure this risk. The CAR policy normally excludes damage for design (see this chapter) and as a professional indemnity policy will only provide cover against **negligent** design it becomes extremely important to appreciate that the contractor probably cannot obtain cover for non-negligent design for which he is apparently contractually liable. It is arguable that the phrase 'fault defect error or omission in the design' means only negligent design but the words 'faulty design' used in a CAR policy (not in a construction contract) have been decided by the highest appeal court in Australia to include non-negligent design. See the Queensland Goverment Railway case earlier in this chapter. In the case of the alternative wording in the contractors' all risks policy (mentioned in this chapter) covering the damage to the insured property resulting from the defective design it is sometimes difficult to decide what part is faulty design and what is resultant damage. Finally, note the remarks concerning a products guarantee policy at the end of the information under the next heading.

(3) *Faulty workmanship and defective materials* There is a contractual responsibility for this work and materials (see the first proviso in clause 21) but no requirement to insure. Insurance cover is available in respect of loss or damage to other parts of the insured work resulting from such defects but it is unlikely that insurers will cover the faulty work itself. A products guarantee policy would cover the cost of rectifying defective work and financial loss but is probably unobtainable for civil engineering risks.

Assuming the CAR policy cover comes to an end when the contract is completed, attention is drawn to the gap between the CAR policy and the public liability policy. See the 'Summary of risks remaining uninsured' at the end of chapter 8.

The employer

(1) *The excepted risks* Unless the employer arranges insurance to protect himself in respect of these risks in the case of such a loss he will have to pay for it, with

the exception of war and kindred risks and the nuclear risk which in the UK are normally Government responsibilities. 'Riot' outside Northern Ireland so far as the UK area is concerned is usually covered by the CAR policy to the benefit of the employer who is a joint insured with the contractor (see clause 21). No doubt cover could be obtained for loss or damage by pressure waves caused by supersonic aircraft in the UK. However, see a possible danger in relying on the joint named cover in the CAR policy in (3) below.

(2) *Fault defect error or omission in the design of the works* If the engineer's design is found to be the cause of the damage the employer should be able to recover his expenses in this respect from the engineer. However, there may be limits to the engineer's responsibility under his agreement with the employer, or under his professional indemnity policy and his financial resources may be small. Finally, the defence of the 'state of the art' is a factor to be reckoned with in recovering from the engineer, bearing in mind that it has to be proved that the engineer had not reached the recognised professional standard.

(3) *Contractor's inoperative or inadequate insurance* While it is misleading to place too much emphasis on the possibility of the contractor's insurance becoming inadequate the following are the more likely to arise if the situation arises at all:

(a) *Non-disclosure of a material fact.* If the policy is void from inception then, subject to the wording of the policy, the employer might find himself having to pay for the cost of repair.

(b) *The sum insured under the policy being inadequate.* This has been mentioned above under the heading 'The contractor' together with indications of the problems to be overcome.

(4) *Consequential losses* The CAR policy does not cover consequential losses and as indicated earlier in this chapter under this heading concerning policy exception (f) when considering both the contractor's and the employer's consequential losses it is the employer who is likely to suffer the greatest loss in this respect. Delay in completion of the works resulting in loss of anticipated income is probably the more likely source of and resulting consequential loss. Such cover can be provided by an advance profits policy. However, there are other consequential losses from delayed completion (see page 66); such items can be insured if they arise from physical loss or damage to the contract works. All such items are also recoverable if a successful claim can be made against the engineer coming under his professional indemnity policy but such items only add to the possibility of exceeding the limit of indemnity under that policy.

Incidentally, in excluding consequential losses insurers have to be careful that they use the right wording. Merely to exclude consequential loss may not be sufficient as there is authority (*Croudace Construction Ltd* v *Cawoods Concrete Products (1978)*) to interpret 'consequential' as meaning something which is not the direct and natural result of a breach of duty and loss of profit is not indirect and therefore not 'consequential' in the sense just explained, whereas insurers tend to consider loss of profit as a consequential loss.

Chapter 6

Damage to persons and property: clause 22(1)
Indemnity by employer: clause 22(2)

Damage to Persons and Property

22. (1) The Contractor shall (except if and so far as the Contract otherwise provides) indemnify and keep indemnified the Employer against all losses and claims for injuries or damage to any person or property whatsoever (other than the Works for which insurance is required under Clause 21 but including surface or other damage to land being the Site suffered by any persons in beneficial occupation of such land) which may arise out of or in consequence of the construction and maintenance of the Works and against all claims demands proceedings damages costs charges and expenses whatsoever in respect thereof or in relation thereto. Provided always that:

(a) the Contractor's liability to indemnify the Employer as aforesaid shall be reduced proportionately to the extent that the act or neglect of the Employer his servants or agents may have contributed to the said loss injury or damage;

(b) nothing herein contained shall be deemed to render the Contractor liable for or in respect of or to indemnify the Employer against any compensation or damages for or with respect to:

(i) damage to crops being on the Site (save in so far as possession has not been given to the Contractor);

(ii) the use or occupation of land (which has been provided by the Employer) by the Works or any part thereof or for the purpose of constructing completing and maintaining the Works (including consequent losses of crops) or interference whether temporary or permanent with any right of way light air or water or other easement or quasi-easement which are the unavoidable result of the construction of the Works in accordance with the Contract;

(iii) the right of the Employer to construct the Works or any part thereof on over under in or through any land;

(iv) damage which is the unavoidable result of the construction of the Works in accordance with the Contract;

(v) injuries or damage to persons or property resulting from any act or neglect or breach of statutory duty done or committed by the Engineer or the Employer his agents servants or other contractors (not being employed by the Contractor) or for or in respect of any claims demands proceedings damages costs charges and expenses in respect thereof or in relation thereto.

Indemnity by Employer

(2) The Employer will save harmless and indemnify the Contractor from and against all claims demands proceedings damages costs charges and expenses in respect of the matters referred to in the proviso to sub-clause (1) of this Clause. Provided always that the Employer's liability to indemnify the Contractor under paragraph (v) of proviso (b) to sub-clause (1) of this Clause shall be reduced proportionately to the extent that the act or neglect of the Contractor or his sub-contractors servants or agents may have contributed to the said injury or damage.

This is not an insurance clause as it does not require the contractor to effect any insurance cover. It deals with the liability of the contractor to indemnify the employer (and vice versa) against third-party claims but the indemnity to the employer could also include losses of the employer (other than the works). It is long and rather complicated. In the previous edition this clause was referred to as one of the most tortuous and baffling clauses in any standard form in the UK, which is an unenviable reputation.

The pattern of this clause is:

(1)   a contractor's indemnity to the employer in the first paragraph of sub-clause (1);

(2)   five exceptions to that indemnity in sub-clause (1) proviso (b)(i) to (b)(v), as a result of which the employer gives the contractor an indemnity under sub-clause (2); and

(3)   two provisos to reduce both the contractor's and the employer's indemnities in the event of a contributory 'act or neglect of the employer his servants or agents' or 'act or neglect of the contractor or of his subcontractors servants or agents', respectively in sub-clauses 22(1)(a) and 22(2).

Clause 22(1)

Comments will be made on each section.

'except if and so far as the Contract otherwise provides'
A possible situation is given under the next quotation. Where this phrase provides for physical alterations in the wording it is only restating the common law position, i.e. that which is written or typed, as showing the special intention of the parties, will override that which is printed and therefore of non-specific application.

'The Contractor shall . . . indemnify . . . the Employer against all losses'
This requirement to indemnify the employer so far as damage to property is concerned does not include the works (see next heading) but may have a wide application. *Abrahamson on Engineering Law*, fourth edition, page 104, suggests that:

If the contractor becomes entitled to extra payment under clause 12* for artificial obstruction by property which the contractor damages and has to make good in order to complete the contract works, the contractor may be bound under this clause to pay back to the employer any payment recovered under clause 12 for the cost of making good the damage to the property (but not cost received under clause 12 for consequential delay to the works). It is proper that this clause should override clause 12 because the contractor should be insured against this liability.

* It should be noted that the wording of clause 12 gives the contractor who encounters unforeseen physical conditions (other than weather) or artificial obstruction which will give rise to additional cost the opportunity to claim additional payment.

This statement assumes that the existence of insurance is sufficient justification for making the insured liable, which may be the attitude of some law makers (Law Commissions and EEC Directives) but is a doubtful argument as far as existing law is concerned. In fact, if the phrase 'except if and so far as the contract otherwise provides' mentioned earlier includes the situation envisaged by clause 12 then the contractor would not be liable to pay back to the employer the amount recovered under clause 12.

While there is some doubt whether the phrase just quoted includes a clause 12 situation, as clause 12 does not assume damage to property, consider the following position. The contractor damages an underground cable belonging to the Post Office. However, it is not the contractor's fault because he was misled as to the whereabouts of the cable. Nevertheless under section 8 of the Telegraph Act 1878 there is an absolute liability. Assuming this is accepted as a clause 12 claim by the engineer and that action is taken against the employer as being vicariously responsible for his contractor's act (as 'agents' under the Act), would the employer be entitled to claim back the amount the contractor had recovered under clause 12 by virtue of clause 22(1)?

> '*and claims for injuries or damage to ANY person or property WHATSO-EVER (other than the Works for which insurance is required under Clause 21...)*'

The authors' capitalised words in this quotation indicate the extent of this liability although there is some consolation for the contractor in that the indemnity does not include damage to the works. It does, however, relate to damage to any other property and injury to any person. Incidentally, it does not exclude the contractor's employees, therefore conflicting with the insurers' division of such risks into employers' liability to employees and the liability to the public (exclusive of employees).

More important and confusing is the fact that clause 24 deals with accidents to any workman or other person in the employment of the contractor whereas such liability is already included in the phrase quoted in the previous paragraph from clause 22(1). Clause 24 must override clause 22(1) on this point. The phrase 'any . . . property whatsoever' is very wide unlike the usual cover provided by the public liability policy. Most insurers either intend or specifically state that they only cover damage to material or tangible property. See chapter 21 for further details.

The contractor's liability to indemnify the employer for damage to property has the clear exception

> '*(other than the works for which insurance is required under Clause 21 but including surface or other damage to land being the Site suffered by any persons in beneficial occupation of such land)*'

This exception will be considered in two sections.

Unlike clause 20.2 of the Standard Form of Building Contract, it is clear that the works are excluded. Furthermore, it is only the physical destruction of or damage to the works which is excluded, not the consequential loss. See the commentary on clause 21. Moreover, the exclusion does not apply to the employer's property other than the works, e.g. damage to property of the employer on the site other than the works, but it would include unfixed materials, as by clause 21 the insurance must cover such materials.

> *'including surface or other damage to land being the Site suffered by any persons in beneficial occupation of such land'*

This section means the indemnity includes claims by all persons in beneficial occupation of land, which is part of the site but not actually used for the permanent works (see proviso 22(1)(b)(ii) above), for damage to the site land suffered by them in their capacity as occupiers. However, the proviso 22(1)(b)(iii) must be borne in mind; this implies that the contractor will not be liable for any compensation or damages if the employer has no legal right to construct the works over or through the land concerned, or such right is abused. By proviso 22(1)(b)(i) the contractor does not have to indemnify the employer for damage to crops on the site to the extent that the contractor has been given possession.

Finally, the introductory paragraph states that the losses and claims must

> *'arise out of or in consequence of the construction and maintenance of the Works and (the indemnity is) against all claims demands proceedings damages costs and expenses whatsoever'*

Presumably the word 'losses' would include the recovery of the employer's own losses apart from the insured works.

The latter quotation is almost as wide as it can be, and it includes all consequential losses. However, the provisos which follow raise the question of a division of liability if both the contractor and the employer are negligent or in breach of a statutory duty. Furthermore, the quotation will not enable a defendant employer, who has been unable to recover his costs from an unsuccessful plaintiff who has suffered injury on or near the work, to recover his loss from the contractor – see the Court of Appeal decision in *Richardson* v *Buckinghamshire County Council and Others (1971)*.

The Court of Appeal decided that the plaintiff's claim was neither a claim which arose out of or in consequence of the construction of the works or a claim in respect of injuries which arose out of or in consequence of the construction of the works and, therefore, the first defendants were not entitled to be indemnified by the second defendants.

The provisos to clause 22(1)

A claimant who suffers injury or damage in consequence of the construction of the works can take action against either or both the employer and the contractor, but under the introductory paragraph the employer has an indemnity from the contractor except in the cases set out in sub-clause 22(1)(b). If the employer is proceeded against in cases falling within this sub-clause he must pay the damages *prima facie* without any right of indemnity from the contractor, but see later, clause 22(2).

It is most unfortunate that the situation where liability resulting partly from the 'act or neglect' (which phrase in itself — see chapter 1 — raises doubts) of the contractor and partly from an 'act or neglect' of the employer is stated in the contract in a confused way.

Clause 22(1)(a) states that the contractor's liability to indemnify the employer shall be reduced proportionately to the extent that the act or neglect of the employer shall have contributed to the loss injury or damage. On the other hand, clause 22(1)(b)(v) states that the contractor is not liable to indemnify the employer **at all** for injuries or damage resulting from any act or neglect or breach of statutory duty by the employer. Thus the initial conclusion is that the contractor is proportionately liable on the first basis but not liable at all on the second. However, later when dealing with clause 22(2) it will be seen that the intention according to the proviso of this clause is to limit the employer's liability to indemnify the contractor in the same way that the contractor's liability to indemnify the employer is limited under clause 22(1)(a). That is proportionately subject to each party's 'act or neglect' or that of those for whom they are responsible. However, while it is true that clause 22(2) refers to sub-clause 22(1)(b)(v) when indicating that a proportionate basis is to be adopted, it goes on to state 'to the extent that the **act or neglect** of the contractor, etc. may have contributed to the said injury or damage'. Therefore the implication is that there is only a complete relief to the contractor from liability where there is a breach of statutory duty by the employer and, of course, under clauses 22(1)(b)(i) to (iv) or a sole 'act or neglect' of the employer.

In *A.M.F. International Ltd* v *Magnet Bowling Ltd and G. Percy Trentham Co Ltd (1968)*, the building owners Magnet and main contractors Trentham were held liable under the Occupier's Liability Act in the proportion 40% and 60% respectively to another contractor (of Magnet) A.M.F., whose timber was damaged during an exceptionally heavy rainstorm when water entered the building through an unprotected doorway. Magnet were liable because they failed to ensure that Trentham had got the building into a fit state to receive A.M.F.'s material on the delivery date, and had given no instructions regarding anti-flood precautions. The contract between Magnet and Trentham was in the JCT (1957) revised form with quantities. By clause 14(b) of the contract, Trentham indemnified Magnet against liability for damage to property arising out of the execution of the works provided that it was due to the negligence of Trentham.

Furthermore, by items in the bills of quantities, Trentham contracted with Magnet to provide for the diversion of storm water to allow for protection of all work and materials from damage by weather and to prevent water accumulating on the site.

Magnet sought to recover from Trentham in respect of their share of their (Magnet's) liability to A.M.F.: (a) under the indemnity clause; and (b) as damages for breach of the provisions in the bills. It was held regarding (a), following the rule in *Alderslade* v *Hendon Laundry (1945)*, that as Magnet had also been negligent, and as there were other claims than negligence on which the clause could operate, they (Magnet) could not recover under the terms of the indemnity clause. However, regarding (b) Magnet were entitled to recover from Trentham in full for damages for breach of the provisions in the bill of quantities. The general principle established in Alderslade's case is that an exception clause does not protect against negligence unless it expressly so provides on the true construction of the words used, or, unless no content can be given to the words used unless they protect against negligence. The judge accepted that the principle applies as much to claims by persons suing on indemnity clauses as to defendants seeking to escape liability under exception clauses. He felt that here the terms of the indemnity in sub-clause 14(b) (now sub-clause 20.2) were not so clear as to permit the employer to take advantage of that indemnity in respect of damage to the third party for which their own negligence was in part responsible. It has been suggested that, as the ultimate decision was that the employer was entitled to recover in full from the builder as damages for breach of contract and also bearing in mind that the decision on the indemnity sub-clause is contrary to the probable intention of the parties, the decision on the indemnity wording should be treated with reserve. However, neither of these arguments carry much weight legally.

The author of *Hudson's Building and Engineering Contracts*, tenth edition, considers the decision concerning the indemnity clause means that in the standard forms and the vast majority of building contracts such clauses are largely worthless, presumably because liability is often apportioned between the contractor and the employer. In referring to the rule in Alderslade's case, he then says:

But to seek to operate the rule in cases where a partial liability in negligence, however small, arises from failure to oversee and detect the primary and effective negligence or fault of the party giving the indemnity seems to be carrying canons of construction (derived originally from exclusion clauses, and not indemnity clauses, in a totally different field of commerce) to a point where they defeat not only the intentions of the parties but also, quite obviously, of the legal draftsmen they have employed. Not only is the result inconsistent with the obvious intention of the RIBA clause (now clause 18(b))* but it is also, it is submitted, inconsistent with the general principle of building contracts that the architect owes no duty **to the contractor** when supervising the contract to advise him on methods of working, or to detect his breaches of contract, as was indeed stated very clearly by Mocatta J. in the above case. It is submitted that this subject needs reconsideration by the courts.

While it is true that this subject needs reconsideration by the courts, it is doubtful whether it is correct to assume that the negligence of the employer (building owner) will be proved in the majority of claims arising under this clause of the standard form. It follows that the conclusion that the indemnity clause is largely worthless is also doubtful. In any event in the case of the ICE Conditions, the proportionate reduction in the indemnity in the two provisions in clause 22(1)(a) and the proviso 22(2) are obviously intended and may well be sufficient to keep the contractual indemnities alive and avoid the decision in the A.M.F. case.

While discussing this question of apportionment, attention is drawn to the position under clause 22(2) where the employer gives the contractor an indemnity in respect of the matters referred to in the proviso to clause 22(1)(b)(v). In *Kenny v Cooper Type Services Co (1968)* the fourth edition wording of the ICE Conditions was incorporated into a subcontract between the main contractor and the subcontractor. Clause 22(2) was held not to assist the subcontractor in a case where

* The author must mean clause 18(2).

the main contractor has been held liable to his own workman by reasons of his failure to note and remedy a dangerous condition created by the subcontractor. Clause 22(2) provided that the main contractor should indemnify the subcontractor for any 'act or neglect' done or committed by the main contractor his agents or servants. It was held that clause 22 was intended to apply only to cases where one party or the other was solely liable, and this gave the subcontractor no defence. The fifth edition of the ICE Conditions contains a proviso which incorporates a proportionate solution and prevents the effect of the case just mentioned, but this does not alter the attitude of the courts to override these indemnity clauses if they are not worded very carefully.

The obvious intention of clause 22(2) is to incorporate a proportionable liability between the parties which is compatible with sub-clause 22(1)(a) but could have been incorporated into sub-clause 22(1)(b)(v) to prevent the clash mentioned earlier between the last two sub-clauses.

It should be appreciated that where **in law** two parties are liable for the same damage, the claimant can recover from one or both of them. While the very general intention of clauses 22(1) and 24 seems to be for the contractor to indemnify the employer against liabilities which the employer incurs to third parties (contractor's employees in the case of clause 24) except to the extent due to the employer's negligence, in the former case the intention has not been achieved very easily whereas in the latter case it has. Consequently if the contractor is sued by the claimant in cases under paragraph 22(1)(b)(v) the employer must indemnify him subject to the proviso to paragraph 22(2).

The Law Reform (Married Women and Tortfeasers) Act 1935 section 6 for years governed the position at law* and if the rather cumbersome method used in clause 22 to achieve the above intention had been omitted and it merely referred to current law, a party who paid more than his just and reasonable share of the damages, bearing in mind the extent of his responsibility for the damage done, would by current law recover an equitable amount from the other party to the contract. The claimant's position would be unaffected. A reference to the wording of clause 22 of the FIDIC International Contract (third edition) in chapter 18 shows one way of achieving an improvement.

> *Proviso 22(1)(b) commences with the words 'nothing herein contained shall be deemed to render the Contractor liable . . . to indemnify the Employer against . . .'*

This statement is followed by a list of five circumstances where the contractor does not indemnify the employer against any compensation or damages.

The first of the five circumstances where the contractor does not indemnify the employer under paragraph 22(1)(b) reads:

> '(i) damage to crops being on the Site (save in so far as possession has not been given to the Contractor);'

---

* Section 6 was repealed by the Civil Liability (Contribution) Act 1978. The 1935 Act only applied to actions in tort whereas the 1978 Act applies contribution to tort, breach of contract, breach of trust or otherwise or in a combination of these actions.

This is a type of inevitable damage which occurs if the contract is to be carried out and consequently for which the employer cannot expect the contractor to indemnify him. The exception in brackets leaves the contractor supplying an indemnity where he is given possession of only part of the site as regards crops damaged in that part not in his possession.

The next exception is a lengthy one and reads:

> *'(ii) the use or occupation of land (which has been provided by the Employer) by the Works or any part thereof or for the purpose of constructing completing and maintaining the Works (including consequent losses of crops) or interference whether temporary or permanent with any right of way light air or water or other easement or quasi-easement which are the unavoidable result of the construction of the Works in accordance with the Contract;'*

This is really an extension of exception (i) because the contractor has possession of the part of the site occupied by the works (if not the whole site) and it includes any liability arising out of the **use or occupation** of land by the works (including damage to crops). Presumably this could be land not occupied by the works but still within the site and in this respect exception (i) could be similar to exception (ii) but only as regards damage to crops. The remainder of this exception must be wider than exception (i) as it concerns a different subject-matter because it refers later to interference with any easement (after naming some) or quasi-easement which is 'the unavoidable result of the construction of the works'.

An easement is defined as a servitude or a right enjoyed by the owner of the land over the lands of another. An easement must exist for the accommodation and better enjoyment of the land to which it is annexed; otherwise it may amount to a mere licence (*Hill* v *Tupper (1863)*).

Easements are created by express, implied or presumed grants. Taking the easement of support to buildings as a practical example, it may be created as follows:

(a) By express grant or reservation: here the deed creating the grant will specify its exact nature. On the sale of a building, for example, the owner may expressly grant an easement of support in favour of the grantee over the land or building retained or he may expressly reserve for himself a right of support over the building sold.

(b) By implied grant or reservation: in these cases there is a mutual easement of support between two adjacent buildings where the construction is such that each building must give some support to the other, e.g. semi-detached houses and terraced houses.

(c) By presumed grant: where a right to support has been enjoyed for a long time it is the law that it shall not be defeated merely because the person exercising the right cannot produce the grant concerned. In these cases the law presumes that a valid grant was made. The acquisition of easements in this way is known as 'prescription' and is governed mainly by the Prescription Act 1832. As a generalisation this provides that where a building has benefited from the support of another for the period stated (normally twenty years) without interruption, then the building will acquire a right to continued support.

It has already been explained that an easement is the right to use another person's land. Where **one** person owns two pieces of land and habitually uses a path across one of them to reach the other, this does not confer a valid easement called a right of way. The landowner is exercising the right merely in his capacity as the owner of both plots of land. Such rights are known as quasi-easements, as important results may ensue from the two pieces of land later becoming separately owned.

Thus it was stated in the important case of *Wheeldon* v *Burrows (1879)*, that on the grant of part of the land of the grantor, the grantee impliedly acquires all easements and quasi-easements which:

(a)   are continuous and apparent;
(b)   are reasonably necessary to the enjoyment of the land granted; and
(c)   were previously enjoyed as quasi-easements.

By way of illustration, it has been decided that where land and buildings are sold, and the buildings overlook the land retained, an implied easement of light will be created, which will prevent the grantor from building on the land retained so as to obstruct the light coming to the buildings sold.

Clearly it is only right that the employer should be responsible for the preservation of easements and quasi-easements and cannot and should not expect an indemnity from the contractor for any damage thereto or interference therewith. Note that both halves of this second exception must be the unavoidable result of constructing the works in accordance with the contract.

The third exception reads:

'(iii) the right of the Employer to construct the Works or any part thereof on over under in or through any land;'

This exception seems to concern claims to title of land and it is obviously the employer's job to ensure that he has a right to carry out the activity 'on over under in or through' the land concerned. Presumably this exception could also involve the absence of planning permission and/or failure to build in accordance with the permission granted.

The fourth exception is the last of the four exceptions which are of a similar type in that they are clearly matters for which the employer should be responsible. It reads as follows:

'(iv) damage which is the unavoidable result of the construction of the Works in accordance with the Contract;'

The word 'unavoidable' in both this exception and exception (ii) would be difficult to prove and as this burden is on the contractor he is likely to be unsuccessful in discharging it. See also in chapter 21 a discussion on a similar exclusion in the public liability policy. Note also that this exception applies only to the liability for damage and therefore does not include injury to persons — there is a further commentary on this point under the next heading.

Exceptions (i) to (iv) and insurance

In the case of all these exceptions, insurance would probably be impossible to obtain as the public liability policy, which is the usual policy to provide cover for liability for damage to third-party property, will only protect the insured provided (the operative clause says) **accidental** injury or damage occurs or (the policy exception says) no inevitable or deliberate injury or damage takes place. Therefore, assuming 'non-accidental' and 'inevitable' are synonymous with 'unavoidable', as the contractor could not obtain insurance cover this is another reason why the employer should not obtain an indemnity from the contractor. Otherwise the contractor would be personally responsible to pay for these risks. In any event the employer should have resolved any difficulties involving the operation of these exceptions before the contract was entered into, e.g. between the employer and the owner of crops, land or easements as the case may be, so that the work can proceed.

As will be seen in the chapters concerning the employers' and public liability policies it is not the practice to restrict the cover given by the former policy to liability for **accidental** bodily injury but it is (as just mentioned) the practice to do so both for injury and damage (or incorporate a deliberate or inevitable injury or damage exception) in the latter policy.

The effect is that if compensation or damages become payable as a result of unavoidable bodily injury to the contractor's own employees the contractor will be responsible unless some exception in clause 22 or clause 24 applies, and the contractor's employers' liability policy will cover the risk. However, in the case of unavoidable injury to third parties unless a contract exception operates the contractor will have to bear this risk himself as his public liability policy will not apply, and exception (iv) of the contract only operates when there is unavoidable **damage, not injury**. Nevertheless, in the vast majority of cases exception (v) of the contract considered under the next heading will operate in cases of injury to third parties or damage to their property, but consider the following situation.

Without negligence a considerable but inevitable onset of dust from a contract site brings on a disabling allergic attack to a neighbouring trader who is forced to leave his premises throughout the period of the contract. An action in trespass is brought by the trader against the employer. Now, whether this is legally a valid claim or not, who will handle it? Under the contract exception (iv) does not apply as it is an injury not a damage claim, and exception (v) does not apply as assuming 'act or neglect' means negligence (see next heading) there was no negligence by the employer. Consequently the contractor will have to handle the claim, but will his public liability policy operate as the injury was not accidental, it was inevitable? However, the policy decision would probably be that as far as the insured was concerned it was accidental in that the insured could not be expected to realise he was causing an injury, and the insurers would probably handle the claim.

> '(v) injuries or damage to persons or property resulting from any act or neglect or breach of statutory duty done or committed by the Engineer or the Employer his agents servants or other contractors (not being employed by the contractor) or for or in respect of any claims demands proceedings

damages costs charges and expenses in respect thereof or in relation thereto.'
Certain aspects of this exception have already been discussed in this chapter and the
difficulty of interpretation of the words 'act or neglect' was mentioned in chapter 1
but this latter point now requires further and more detailed discussion.

The phrase 'act or neglect' appears in the proviso (a) to clause 22(1) reducing
the contractor's liability to indemnify the employer to a proportionate amount to
allow for the employer's 'act or neglect' or that of his servant or agents. This phrase
appears again in proviso (b)(v) to clause 22(1) with the addition of the words 'or
breach of statutory duty' committed by the engineer or the employer his agents
servants or other contractors (not employed by the main contractors) as an exception to the indemnity given by the main contractor to the employer. Finally, this
phrase appears in the proviso to clause 22(2) reducing the employer's liability
to indemnify the contractor to a proportionate amount to allow for the 'act or
neglect' of the contractor or that of his subcontractors servants or agents.

The exception 'act or neglect of the employer . . .' in the RIBA Contract is apparently limited in its application to common law negligence (*Hosking* v *De Havilland
Ltd (1949), Murfin* v *United Steel Companies Ltd (1957)*) and is not applicable to
breach of statutory duty. Thus the contractor would still have to indemnify the
employer where the claim by the injured person is based on a breach of statutory
duty by the employer.

Hosking's case, which dealt with sub-clause 14(a) of the 1939 RIBA Contract
(containing a similar wording to sub-clause 20.1 of the 1980 contract), decided
that the words 'act or neglect' do not include a breach of statutory duty, as there
would be very little for the indemnity (to the employer) to apply to if these words
in the exception included such a breach by the employer as well as his common law
negligence. The Court of Appeal in Murfin's case when interpreting a similar clause
drew a distinction between statutory liability and negligence and approved of the
decision in *Hosking* v *De Havilland Ltd*.

In Hosking's case the plaintiff (an employee of the defendants) was injured when
a plank placed over a duct by a contractor broke. The plaintiff stepped on to the
plank to proceed to another part of the works. The court held the contractor
liable at common law and De Havillands in breach of the Factories Act current
at that time. However, the contractor was held liable for the total damages and
costs because of the contract conditions, otherwise damages would have been
apportioned between the contractor and De Havillands.

The word 'act' could include an act of nuisance or trespass and this would widen
the effect of the phrase. Obviously in the second use of the phrase in proviso (b)(v)
to clause 22(1) the decision in Hosking's case could not apply as the words 'or
breach of statutory duty' are included in the exception. However, the decision
would apply where the employer committed one of the other two torts mentioned
above where the phrase 'act or neglect' is used. Possibly the difference is not
intended. See also *Strathclyde Regional Council* v *James Waddell* on page 19.

For example, take the situation which occurred in *Andreae* v *Selfridge & Co Ltd
(1938)* mentioned on page 6, but for the sake of argument assume nuisance by
the employer but no nuisance or negligence by the contractor (yet no possibility
to argue the application of the exception in clause 22(1)(b)(v)). To make matters

easy it will also be assumed that damage was caused by dust and the economic loss flowed from this. Does the contractor have to indemnify the employer under clause 22(1)(b)(v) in respect of this third-party claim? It is submitted that if 'act or neglect' means merely negligence and the employer is only guilty of nuisance, the contractor will have to indemnify the employer because the exception does not apply when the employer commits a nuisance. See also Gold's case on page 6.

In view of the fact that the use of this phrase is not only applicable to one of the parties to the contract but also to the servants, agents, and in one case subcontractors of the party and in another the engineer is specifically mentioned, it is appropriate to state the common law position. Incidentally, if the engineer is the agent of the employer it is difficult to understand why it was necessary to mention him. Probably because it depends on his activity*.

While at law a principal is vicariously responsible for his servants' and agents' acts which arise out of and in the course of their employment, in the case of the employment of contractors the position is not so simple.

At common law the general rule is that the principal is not vicariously responsible for the negligence of his contractors. However, there are certain exceptions (for want of a better word) to this rule which are briefly summarised below, and although the examples given from case law mainly concern the relationship of principal and independent contractor the position between main contractor and subcontractor is the same. See *Maxwell* v *British Thomson-Houston Co Ltd (1902)*. Usually these exceptions apply when a personal duty rests on the employer and it is no defence that the duty was delegated to the contractor.

(i)   Where the contractor is employed to do illegal work. In *Ellis* v *Sheffield Gas Consumers Co (1853)* the defendants without obtaining the necessary special powers, employed a contractor to open trenches in the streets of Sheffield. The plaintiff sustained injuries by falling over a heap of stones which had been left by the contractors and the Gas Company was held liable for the contractor's negligence.

(ii)   Where the contractor is employed to do work which involves strict liability for:

(a)   *Unusually hazardous work.* In *Honeywell and Stein* v *Larkin (1934)* the plaintiffs employed the defendants as independent contractors to take photographs by flash light. The defendants set fire to a theatre and the plaintiffs paid for the damage and attempted to recover from the defendants. The Court of Appeal held that the plaintiffs were liable as they had arranged for work to be done which involved some degree of special danger. However, they obtained an indemnity.

* In *Wallis* v *Robinson (1862)* an architect was the agent of the building-owner for the purposes of ordering materials, and in *Clayton* v *Woodman and Son (Builders) Ltd (1962)* the Court of Appeal referred to the architect as the agent of the employer to produce a properly constructed building but the judge in the court of first instance regarded him as an independent contractor. See also *Brighton* v *Cleghorn & Co Ltd (1967)* on page 89 which concerned an engineer.

(b)   *Withdrawal of support to land or buildings.* In *Bower* v *Peate (1876)* the defendant employed a contractor to pull down and rebuild his house. The contractor expressly undertook to support the plaintiff's adjoining house but this was damaged. The defendant was held liable for the damage done. The decision in this case could have been justified on the narrow grounds of nuisance or interference with an easement (see the law on this aspect earlier in this chapter) and it has been included under the above heading for this reason but in fact the court based their decision on the wider principle that there is 'good ground for holding him [the party authorising the work] liable for injury caused by an act certain to be attended with injurious consequences, if such consequences are not in fact prevented, no matter through whose default the omission to take the necessary measures for such prevention may arise'. However, this principle has been criticised.

(c)   *Operations on or adjoining a highway creating dangers thereon.* In *Tarry* v *Ashton (1876)* the defendant employed a contractor to repair a lamp fixed to his house and overhanging the highway. It was not securely fixed and fell on the plaintiff, a passer-by. The defendant was liable because it was the defendant's duty to make the lamp reasonably safe. The contractor's failure to do this meant that the defendant had not performed his duty.

(iii)   Where the principal retains a measure of control over the contractor by the provision of machinery or men or both. In *Levering and Doe* v *The London & St Catherine's Dock Co (1887)* subcontractors undertook to discharge ships with their own gangs, the members of which were on the books of the dock company as permanent workmen. The company provided the gear and when two members of the gangs were injured, the company were held liable as having retained a sufficient degree of control over the gangs although insufficient to make them liable as employers. With the tendency to use 'labour-only' subcontractors this situation may arise more often between main and subcontractors.

(iv)   Where the principal is under a statutory duty to perform work in a particular manner. In *Hole* v *Sittingbourne Railway (1861)* the defendants were authorised by statute to construct a bridge across a navigable river with a provision that the bridge should not detain any vessel navigating the river longer than was necessary. Faulty construction by independent contractors prevented the opening of the bridge for several days but the defendants were held liable.

(v)   Where liability attaches due to the principal's own negligence, such as delegating the work to be carried out to an inexperienced or incompetent contractor or giving insufficient instructions to him. In *Robinson* v *Beaconsfield Rural Council (1911)* the respondent council were under a statutory duty to cleanse cesspools. It delegated the cleansing of certain pools to a contractor but failed to give to him any directions as to the disposal of the filth. The contractor deposited the filth on the appellant's land. It was held that the respondents were liable for their failure to take proper precautions to dispose of the sewage.

While the above list is not completely exhaustive there are numerous situations outside the above list where the employer is not liable for the acts of his contractor because no breach of any duty is imposed on the employer by the law of torts. In addition, even if one or other of the circumstances in the above list is involved the employer is not liable for the collateral or casual negligence of the contractor, that is, negligence which is not created by the type of work being done. For example, in the circumstances of *Tarry* v *Ashton* mentioned earlier, if the contractor while repairing the lamp had allowed a hammer to drop on the passer-by it would have been collateral negligence for which the defendant employer would not be liable.

Various definitions of collateral negligence other than the one given above, have been suggested, as it has caused the courts some difficulty. Two examples are that it is negligence not within the course of the employment of the independent contractor or that it means negligence not arising from the risk which the employer creates when he engages the contractor to do the work in question. However, the definition given in the previous paragraph is concise and conveys a similar meaning to those just mentioned.

For the employer to be liable for the contractor's acts or the main contractor liable for the subcontractor's acts the danger must be inherent in the work or as Fletcher Moulton L.J. said in *Padbury* v *Holliday and Greenwood Ltd (1912)*:

Before a superior employer could be liable for the negligent act . . . of a subcontractor it must be shown that the work which the subcontractor was employed to do was work the nature of which, and not merely the performance of which, cast on the superior employer the duty of taking precautions.

Consequently in allowing for a proportionate reduction for the negligence of his subcontractors the main contractor assumes some measure of liability which he would not otherwise be under; see the proviso to clause 22(2).

Similarly the employer does the same thing in the case of proviso (b)(v) to clause 22(1) although here the employer's responsibility is limited to other contractors (not being employed by the main contractor). Even so, in the case of these other contractors there is an assumption of a greater liability than the common law imposes.

In view of the various insurance interests involved combined with the uncertain meaning of the phrase 'act or neglect', to say nothing of the inconsistency of wording between clause 22(1)(a) ('act or neglect') and clause 22(1)(b)(v) ('act or neglect or breach of statutory duty'), it is surprising that more major third-party claims do not result in disputes. Could it be that the complexity is such that the parties or their insurers instead of being encouraged to fight are scared of a confrontation and a compromise results?

In exception (v) of sub-clause 22(1)(b) the inclusion of the engineer specifically as well as the employer his agents servants or other contractors (not being employed by the main contractor) implies doubt as to whether the engineer would normally be an agent of the employer in designing and supervising the construction of the works. In *Brighton* v *Cleghorn and Co Ltd (1967)*, where a similarly worded exception was considered under the fourth edition of the ICE Conditions of Contract which did not mention the engineer specifically, it was held that the employer was liable to indemnify the contractor. The damage was subsidence due to a change by variation order in the line of a sewer. The contractor had carried out the engineer's instructions which did not include the shoring of the neighbouring building which subsided. The employer would have a right of recovery from the engineer who should have protected himself by a professional indemnity policy.

Clause 22(2)

Clause 22(2) looks at the position of the contractor when he is sued whereas up to this point in clause 22 the situation has been viewed from the position of the employer who is sued. By clause 22(2) the employer undertakes to indemnify the contractor against all claims etc., (the same phrase as in the introductory paragraph of clause 22(1) in respect of the matters referred to in the proviso to clause 22(1) which is set out in sub-clauses 22(1)(a) and 22(1)(b)).

The proviso to clause 22(2) which reads as follows:

> *'Provided always that the Employer's liability to indemnify the Contractor under paragraph (v) of proviso (b) to sub-clause (1) of this Clause shall be reduced proportionately to the extent that the act or neglect of the contractor or his subcontractors servants or agents may have contributed to the said injury or damage.'*

appears to resolve the clash between sub-clause 22(1)(a) and sub-clause 22(1)(b)(v). It will be recollected that the first sub-clause dealt with the indemnity to the employer according to the employer's proportionate liability while the second one, once the employer was liable, gave the employer no indemnity at all if the employer (etc.) was negligent.

Now the proviso to sub-clause 22(2) is actually importing a proportionate liability into sub-clause 22(1)(b)(v) which would resolve the matter by bringing sub-clause 22(1)(b)(v) into line with sub-clause 22(1)(a). However, at the risk of being pedantic, it must be pointed out that in spite of what is said by the proviso to clause 22(2), in fact under sub-clause 22(1)(b)(v) the employer does **not** by the introductory paragraph to clause 22(1)(b) give an indemnity to the contactor but the contractor does not have to give an indemnity to the employer. Nevertheless the intention of clause 22 overall is clearly that each party should indemnify the other in accordance with their respective liabilities, i.e. proportionately.

According to the decision in *City of Manchester* v *Fram Gerrard Ltd (1974)* the phrase 'the contractor or his subcontractors, servants or agents' used here does not include sub-subcontractors. However, the Fram Gerrard case was a decision on the JCT Form and *Petrofina (UK) Ltd* v *Magnaload Ltd (1983)* came to the opposite conclusion (but not concerning the JCT or ICE Forms), that the word 'subcontractor' mentioned in the CAR policy includes the word 'sub-subcontractor'. In view of these contradictory cases any implication is difficult to assess.

Other indemnity clauses

Clause 22 is the main indemnity clause in the ICE Conditions but other indemnity clauses overlap with this clause, e.g. clauses 26(2) (conformance with statutes), 27(7) (compliance with the Public Utilities Street Works Act 1950), 29 (interferance with traffic etc.), 30(2) and 30(3) (claims arising from transport) and 49(5)(a) (reinstatement of highways). However, the public liability policy should have contractual liability and indemnity to principals extensions for use when required so these legal liabilities would be covered provided there is injury, loss or damage to third parties or their property subject otherwise to the terms and conditions of the policy (see chapter 8).

Probably the more important of these overlapping clauses are clauses 26(2) and 29. The former reads as follows:

Contractor to Conform with Statutes, etc.
(2)   The Contractor shall ascertain and conform in all respects with the provisions of any general or local Act of Parliament and the Regulations and Bye-laws of any local or other statutory authority which may be applicable to the Works and with such rules and regulations of public bodies and companies as aforesaid and shall keep the Employer indemnified against all penalties and liability of every kind for breach of any such Act Regulation or Bye-law. Provided always that:
(a)   the Contractor shall not be required to indemnify the Employer against the consequences of any such breach which is the unavoidable result of complying with the Drawings Specification or instructions of the Engineer;
(b)   if the Drawings Specification or instructions of the Engineer shall at any time be found not to be in conformity with any such Act Regulation or Bye-law the Engineer shall issue such instructions including the ordering of a variation under Clause 51 as may be necessary to ensure conformity with such Act Regulation or Bye-law;
(c)   the Contractor shall not be responsible for obtaining any planning permission which may be necessary in respect of the Permanent Works or any Temporary Works specified or designed by the Engineer and the Employer hereby warrants that all the said permissions have been or will in due time be obtained.

As this clause includes any statute and the regulations and bye-laws of any local or other statutory authority it will include the Building Regulations. Thus the contractor is required to indemnify the employer against penalties and liabilities for breach of such statutes etc. subject to the exceptions (a), (b) and (c) above. It is worth noting that in *Street and Street* v *Sibbasbridge and Stratford-on-Avon DC (1980)* the implied term in a building contract to comply with Building Regulations was held to override an express term in the contract. Thus the contractor who had complied with the latter was still held liable for not complying with the former, which required foundations of a greater depth than that specified in the contract documents. The exceptions do not help the contractor against a third party.

Interference with Traffic and Adjoining Properties
29.   (1)   All operations necessary for the execution of the Works shall so far as compliance with the requirements of the Contract permits be carried on so as not to interfere unnecessarily or improperly with the public convenience or the access to or use or occupation of public or private roads and foot-paths or to or of properties whether in the possession of the Employer

or of any other person and the Contractor shall save harmless and indemnify the Employer in respect of all claims demands proceedings damages costs charges and expenses whatsoever arising out of or in relation to any such matters.

Noise and Disturbance

(2)   All work shall be carried out without unreasonable noise and disturbance. The Contractor shall indemnify the Employer from and against any liability for damages on account of noise or other disturbance created while or in carrying out the work and from and against all claims demands proceedings damages costs charges and expenses whatsoever in regard or in relation to such liability.

This clause has been included in this chapter because it overlaps in various ways with the contractor's indemnity to the employer in clause 22(1) bearing in mind provisos (b)(ii) and (b)(iv) of that sub-clause. Also the contractor may have to depend on the employer's indemnity given to the contractor under sub-clause 22(2) because clause 29(1) makes no reference to any such indemnity given by the employer to the contractor.

Incidentally, clause 29(1) is modified to the extent that it only requires the contractor to behave reasonably in the sense that damage must not be caused **unnecessarily or improperly** by interfering with the public convenience or access to or use or occupation of public or private roads and footpaths or to or of properties.

On the other hand, while clause 29(2) refers to all work being carried out without **unreasonable** noise and disturbance the actual indemnity does not tie up with any part of clause 22(1) which is concerned with injury to persons and damage to property. Consequently, no advantage can be taken of clause 22(2) containing the employer's counter-indemnity, and the contractor may be strictly liable for noise and disturbance claims even if, for example, they were unavoidable.

While insurance is not mentioned in clause 29, it is necessary to consider the extent to which the contractor's public liability policy will assist him in covering his liabilities here. Thus, this policy will cover obstruction and trespass (see the operative clause of section 2 in the policy in appendix 1) but not for noise if there is no physical injury or damage, which is very unsatisfactory from the contractor's viewpoint.

There is an environmental impairment liability insurance which in part might cater for some of the risks but this is really a pollution insurance. It covers impairment or diminution of or other interference with any right or amenity protected by law and the definitions clause mentions 'environmental impairment as including the generation of smell **noises** vibrations light electricity radiation changes in tempera ture or any other sensory phenomena'. On the other hand there is no reason why a public liability policy should not be amended to provide the cover required to protect the contractor against a clause 29 claim for noise.

Chapter 7

Insurance against damage to persons and property: clause 23

Insurance against Damage to Persons and Property
23. (1) Throughout the execution of the Works the Contractor (but without limiting his obligations and responsibilities under Clause 22) shall insure against any damage loss or injury which may occur to any property or to any person by or arising out of the execution of the Works or in the carrying out of the Contract otherwise than due to the matters referred to in proviso (b) to Clause 22(1).

Amount and Terms of Insurance
(2) Such insurance shall be effected with an insurer and in terms approved by the Employer (which approval shall not be unreasonably withheld) and for at least the amount stated in the Appendix to the Form of Tender. The terms shall include a provision whereby in the event of any claim in respect of which the Contractor would be entitled to receive indemnity under the policy being brought or made against the Employer the insurer will indemnify the Employer against such claims and any costs charges and expenses in respect thereof. The Contractor shall whenever required produce to the Employer the policy or policies of insurance and the receipts for payment of the current premiums.

Clause 23(1) is an insurance clause as it requires insurance(s) to be effected.

It has already been pointed out in the previous chapter that because clause 22 includes injuries to **any** person it overlaps clause 24 which states that the contractor shall indemnify the employer for injury to any workman or other person in the employment of the contractor or any subcontractor except to the extent that the injury results from any act or default of the employer, etc. In practice this means that a public liability policy is really sufficient to meet the requirements of clause 23 although on the actual wording an employers' liability policy is required as well, but the latter policy is used in practice to meet the liability of the contractor under clause 24. Full details of both these policies will be provided in the chapters concerning policies which provide the cover required by clauses 23 and 24.

The opening phrase reads:

'Throughout the execution of the Works'
This phrase implies that the requirement to maintain the liability insurance (a public liability policy in effect) ends with the completion of the works. In fact, as will be seen from the chapter dealing with the policy cover for this clause the contractor is advised to keep his policy in force because the wording gives policy indemnity for a negligent act performed before or after the inception of the

insurance provided the injury or damage occurs during the currency of the policy. Therefore an annual policy should be kept in force as the negligent act and the consequential injury or damage are not always simultaneous. Furthermore the statute-barred period for bringing an action for damage or injury caused by the contractor's negligence does not start to run until the damage or injury occurs. See later the case of *Sharpe* v *Sweeting & Son Ltd (1963)*.

The obvious intention of clause 23(1) is to see that the contractor is protected by insurance thus ensuring there is adequate backing for the indemnity the employer requires and, incidentally, no financial failure because of the lack of this cover. The question of inadequacy of insurance cover will be dealt with later.

An important aspect of this insurance clause is highlighted by the remark in clause 23(1) in brackets and the phrase which introduces clause 23(2). They read as follows:

> '*the Contractor (but without limiting his obligations and responsibilities under Clause 22) shall insure against any damage loss or injury*' and '*Such insurance shall be effected with an insurer and in terms approved by the Employer . . . for at least the amount stated in the Appendix to the Form of Tender.*'

The result of these two statements is that while the contractor is permitted (and required by insurance practice) to limit his public liability policy to the amount stated in the appendix to the form of tender, as regards any one occurrence or series of occurrences arising out of one event (it is normally unlimited for any one year of insurance) his contractual liability to indemnify the employer is **unlimited**. Therefore he should see that this policy indemnity is as high as reasonably possible. This will be discussed in more detail later in the chapter concerning policy cover.

Employers' liability insurance is compulsory by virtue of the Employers' Liability (Compulsory Insurance) Act 1969 and the limit of indemnity specified by the statute is £2 million but most insurers issue unlimited policies.

The insurance under this clause 23 is **not** required to cover

> '*matters referred to in proviso (b) to Clause 22(1).*'

It should be noted as already mentioned in the first chapter of this book that many limitations and restrictions appear in the public liability policies issued by insurers and the contractual requirement to obtain a policy merely excluding the risks listed in sub-clause 22(1)(b) cannot be met. For example, the exception concerning nuclear radioactive contamination and atomic explosion has already been given as standard in all public liability policies but is still not allowed by the ICE Conditions of Contract.

On the other hand some risks listed in sub-clause 22(1)(b) are excluded by the policy, e.g. 'damage which is the unavoidable result of the construction of the works in accordance with the contract;' at least words are used which have a similar intention.

An important aspect is that this insurance does not have to apply if the claim resulted from the act or neglect, etc. of the employer, etc. under sub-clause 22(1)(b)(v) even though under sub-clause 23(2) the insurance is required to provide

an indemnity to the employer. In practice it depends on the wording of the extension of the policy used to cover the employer as to whether in fact the employer
gets a wider cover than the contract requires. For example, if the policy uses a
wording which only operates an indemnity to principals **to the extent that the
contract so requires** then the cover would conform with the contract, but without
such a phrase the principal would get the cover provided by the policy subject only
to its normal terms and conditions, i.e. just as if the policy had been issued to the
principal as if he were a separate insured.

A further complication of this situation is that the phrase in sub clause 23(1)
under discussion only excludes from this insurance matters referred to in proviso
(b) to clause 22(1) but, in the absence of reference to proviso (a), the insurance
would appear to be **unaffected** in cases of act or neglect, etc. of the employer, etc.
contributing to the damage, although strictly speaking the insurance should not
operate at all if the risk **resulted from** the act or neglect, etc. of the employer, etc.
under sub-clause 22(1)(b)(v). Consequently, if the contractor and employer were
held equally responsible for a negligent act and the claimant chose to take action
against the employer only, in practice it is submitted that the employer would
successfully claim a contribution under sub-clause 22(1)(a), and the difficulty
caused by sub-clause 22(1)(b)(v) would not be raised or if raised it is hoped that
the court would ignore it if the action got that far. In these circumstances the
insurers are unlikely to raise queries whatever their wording.

Amount and terms of insurance: clause 23(2)

Under clause 23(2) the employer requires to approve the insurer although such
approval is not to be unreasonably withheld. Whether in practice employers, or
contractors for that matter, adopt any procedure to ensure that this approval is
carried out formally is doubtful. Presumably it gives the employer a veto if he
should ever need it.

> 'The terms shall include a provision whereby in the event of any claim in
> respect of which the Contractor would be entitled to receive indemnity
> under the policy being brought or made against the Employer the insurer
> will indemnify the Employer against such claims and any costs charges and
> expenses in respect thereof.'

The contractor's public liability policy is required by clause 23(2) to include a
proviso whereby the employer is indemnified in like manner to the insured contractor. Apparently this requirement has been made in place of an insurance in the
joint names of the contractor and the employer as required by the fourth edition of
the ICE Contract Conditions, which created difficulties.

While those adopting the procedure of calling for joint names cover no doubt
intend to obtain for the employer public liability coverage under the contractor's
policy in fact the effect is to reduce the cover in some respects. A public liability
policy does not cover damage to property owned by the insured and often excludes
damage to property in his custody or control, nor does it cover injury to persons

under a contract of service or apprenticeship with the insured. Once the contractor and the employer are both named as the insured the following situation arises.

The contractor has no cover in respect of his legal liability for damage to the employer's property, to property in the custody or control of the employer, and for injury to persons under a contract of service or apprenticeship with the employer because the employer is an insured.

Similarly the employer has no cover in respect of his legal liability for damage to property belonging to the contractor, in the contractor's custody or control, nor for injury to employees under a contract of service or apprenticeship with the contractor. This results in a reduction in the cover given by the contractor's policy but can be overcome by a cross liabilities clause which construes the policy as though separate policies had been issued to each of the joint insureds. While this is the cover now required by the statement in the quotation (without giving the employer the cover mentioned in the first sentence in this paragraph), an easier method of achieving this object is to incorporate an 'indemnity to principals' clause, thus giving the employer a direct right under the policy against the insurers where a third party sues the employer instead of proceeding against the contractor. The main point is that the 'indemnity to principals' clause must have one of the following stipulations (among others) or some wording to similar effect:

(1)   The claim is such that if made upon the insured the insured would be entitled to indemnity under this policy.
(2)   Claims made by the principal shall be treated as though the principal were not insured by the policy.
(3)   It is agreed to indemnify the principal in like manner to the insured.

The intention therefore is to get the employer insurance protection under the contractor's public liability policy, which is reasonable, but this intention has not been included in clause 24 because there is no insurance clause within clause 24. Presumably this is because the contractor has to arrange an employers' liability policy under the Employers' Liability (Compulsory Insurance) Act 1969, but this does not alter the fact that there is no proviso in clause 24 requiring this employers' liability policy of the contractor to indemnify the employer in like manner to the contractor. The result is that the employer may have no cover under the contractor's employers' liability policy in respect of claims made against the employer by employees of the contractor.

Probably this does not matter, as clause 23 (which does include the insurance policy requirement to indemnify the employer in like manner to the contractor), does not exclude claims by employees of the contractor. Furthermore, the contractor's employers' liability policy will probably by its basic wording indemnify the employer (principal) in like manner to the insured so far as claims by the contractor's employees are concerned, but it may be worded to operate only when the contract so requires. This brings us back to the position that clause 23 so requires but clause 24 does not.

As in the case of an insurance effected under clause 21, the contractor must produce for the inspection of the employer the policy (or policies) of insurance and the receipts for current premium payments whenever requested to do so. It is not intended that the employer should retain these documents.

Clause 23 makes no mention of subcontractors, whether the contractor's own or nominated, and insurance is wholly for the contractor and the employer.

Chapter 8

The public liability policy

The operative clause

In the case of the employers' liability policy, if the insured employer is liable to the employee at law, the policy covers that liability in 99% of cases because the policy exceptions are very few and, furthermore, the Employers' Liability (Compulsory Insurance) Act 1969 ensures protection for the employee in certain respects as explained in chapter 10. Under the public liability policy the cover is not as wide as the legal liability of the insured at law because of the restrictions imposed by the policy limitations and conditions.

The public liability policy provides an indemnity against legal liability for bodily injury claims by the public (other than employees) and property damage claims by any third party. The phrases 'legal liability' and 'liability at law' are synonymous. Sometimes the words 'bodily injury' and 'property damage' are qualified by the word 'accident' or 'accidental' in the operative clause. Alternatively, if the word 'accident' or 'accidental' does not appear in the operative clause, it is almost certain that an exception concerning inevitable or deliberate injury, loss or damage will take its place. Consequential loss arising from this injury or damage is covered.

A commentary on this policy is difficult because insurers vary considerably in the subject-matter of the exclusions and in the actual wording when the meaning is similar.

Accident has been legally defined as 'an unlooked-for mishap or an untoward event which is not expected or designed'. Thus, the phrase 'accidental injury or damage' includes injury or damage happening by chance, unexpectedly or without design, and is not dependent upon one specific event, whereas the phrase 'injury or damage caused by accident', used by some insurers, is considered to be limited in this way.

By 'bodily injury' most insurers intend to include death, disease and illness.

The significance of the words 'accidental' or 'accident' appearing in the operative clause on the one hand and an exception concerning inevitable or deliberate injury or damage on the other, lies in the fact that in the former case the onus of proving an accident or an accidental occurrence is upon the insured, whereas in the latter case it rests upon the insurer to prove the exception applies. The authorities for this statement are *Munro Brice & Co* v *War Risks Association (1918)*, which states that the burden of proof is upon the insured to prove an accident when these words appear in the operative clause of the policy, while the case of *Bond Air Services* v

Hill (1955) states that the onus is upon the insurer to prove the operation of an exclusion. The words 'exclusion' and 'exception' are synonymous.

Consequently, the exception as described provides slightly wider cover than the operative clause with the accidental qualification on those few occasions when the circumstances are equally consistent with accident or no accident, because the party upon whom the onus of proof falls will fail since it is impossible to discharge the onus. (See *Wakelin* v *London Western Railway (1886).*

More important than the point just made is the interpretation given to the words 'inevitable' or 'deliberate' damage when these are used in an exclusion. However, it is probably sufficient to say that the word 'accidental' has been legally defined as already mentioned, whereas the exclusions of 'inevitable' and 'deliberate' damage have no legal definition and the insured is to some extent in the insurer's hands concerning interpretation.

One situation which arises out of the phrase 'accidental loss of or damage to property' occurs in the case of deliberate theft by a builder's employee of, say, the building owner's property. Ignoring legal liability, the question is whether the policy wording covers such a claim, bearing in mind that deliberate theft is not accidental or caused by accident, although from the insured's personal viewpoint, it is an accident, and there is legal authority for looking at this situation from the insured's point of view. (See *Trim Joint District School* v *Kelly (1914)*, detailed in chapter 21.) Some insurers make the point clear by a specific exclusion in the policy of loss or damage due to theft with a proviso covering theft caused by the negligence of the insured or by an employee of the insured. This proviso covers the circumstances in *Stansbie* v *Troman (1948)* where the plaintiff was a painter and decorator engaged under contract to do work at the defendant's house. He left the house unoccupied whilst he went to obtain material and, in order that he might be able to gain re-entry, he pulled back the catch of the Yale lock of the front door. He was away from the house for two hours and, during his absence, a thief entered the premises by the front door and stole a quantity of jewellery. It was found that because of his contractual relationship with the defendant, the plaintiff owed a duty to the defendant to take care and that duty had been breached.*

The word 'property' in the policy is capable of a very wide definition. J. Crossley Vaines in his book *Personal Property* states that: 'In English law therefore, "property" comprehends tangibles and intangibles, movables and immovables; it means a tangible thing (land or a chattel) itself, or rights in respect of that thing, or rights, such as a debt, in relation to which no tangible thing exists.'

In the Law of Property Act 1925, the Settled Land Act 1925, and Administration of Estates Act 1925, the definition of 'property' is that it 'includes any ['a' in the Administration of Estates Act] thing in action, and any interest in real or personal property'.

The Master of the Rolls in *Jones* v *Skinner (1835)* said that 'property' is the most comprehensive of all terms which can be used, inasmuch as it is indicative and descriptive of every possible interest which the party can have.

In the Trustee Act 1925 it is defined thus:

* *Petrovitch* v *Callinghams (1969)* is on similar lines.

Property includes real and personal property, and any estate share and interest in any property, real or personal, and any debt and anything in action, and any other right or interest, whether in possession or not.

A person is said to have an interest in a thing when he has rights, titles, advantages, duties, liabilities connected with it, whether present or future, ascertained or potential, provided they are not too remote. So, an easement, e.g. a right of way on or over land, is proprietary interest therein and comes within the meaning of 'property'.

Until an insurer decides to define 'property' in the policy he must accept the wide interpretation of the word by virtue of the *contra proferentem* rule, mentioned in chapter 1 under 'The insurance contract'. If insurers intend to restrict the cover to physical damage to property or damage to material property they should say so in the policy, otherwise it will be interpreted against them.

The wording of section 2 of the policy in appendix 1 reads: 'The company will indemnify the insured against all sums for which the insured shall be legally liable to pay in respect of . . . (b) accidental loss or damage to property'. Consequently, when an electric cable has been severed by the negligence of the insured and a factory a short distance off has been deprived of its electricity and suffers a loss of production, as the words 'in respect of' are in themselves very wide and the policy does not identify 'whose' property, it will give cover if there is legal liability on the insured, i.e., one party's damaged cable supports the consequential loss of another.

Perhaps it should be emphasised that while the insurers will handle the consequential loss claim made, if the cable is not owned by the factory, according to the case of *Electrochrome Ltd* v *Welsh Plastics Ltd (1968)*, there is no duty of care owed by the insured to the factory owner and thus there is no legal liability. An ability to foresee indirect or economic loss to another person as the result of a defendant's conduct does not automatically impose on the defendant a duty to take care to avoid that loss (although see Batty's and the Junior Books cases mentioned under the heading 'Negligence' in chapter 1).

While the policy does not say so, it should be appreciated that consequential loss following injury or damage to property is covered by the policy, as already mentioned.

The cover for accidental obstruction or trespass mentioned in the policy and contained in the operative clause of section 2 of the policy in appendix 1 is not present in all public liability policies but makes it clear that irrespective of loss or damage to property, a legal liability for such obstruction or trespass is covered. This can arise by the blockage of a roadway or trespass on to another's land by the activities of those carrying out the works, and thus consequential loss can occur by the interference with the operation of a third party's business causing, for example, financial loss. Without this extension an insurer would argue this cover does not exist under the policy.

Whether there is bodily injury or illness to, or accidental loss of or damage to the property of, a third party, or a claim arises out of obstruction or trespass, these occurrences must all happen or be caused in connection with the business during the period of insurance, but not exceeding the single accident indemnity limit mentioned in the schedule in respect of any one accident or series of accidents arising out of one event. Reference should be made to chapter 21, page 245, for further information.

The policy exceptions

References under this heading are to appendix 1, section 2: Public liability, unless otherwise indicated.

These exceptions can be classified into risks:

(1)   more properly insured under other policies;
(2)   needing individual consideration and usually, if acceptable, justifying an additional premium;
(3)   undesirable for insurance.

(1) *Risks more properly insured under other policies usually include the following:*

(a)   Claims against the insured by employees of the insured in respect of injury or illness which is a matter for the employers' liability policy. (See exception 1 to section 2.)

(b)   Liability arising as indicated in exception 2. This is liability arising out of the ownership, possession or use by or on behalf of the insured of any mechanically propelled vehicle, locomotive, aircraft, watercraft exceeding twenty feet in length, or hovercraft. Insurance is available in the motor engineering, marine and aviation insurance markets for these types of risks. However, it is usual for the public liability policy to cover unlicensed vehicles (other than those used in agriculture or forestry or by haulage contractors) not requiring a certificate of insurance where the vehicle is to be used only on the insured's private premises or sites to which the public has no right of access.

This policy may be extended to cover the use of such plant as excavators, bull-dozers and the like as a tool of trade where this is excluded from the motor policy. The compulsory insurance required by the Road Traffic Act 1972 is still a matter for the motor policy whether at the time of the incident for which liability is alleged the plant was being used as a tool of trade or not.

(c)   Exception 3 concerns liability for loss of or damage to the contract works, materials, tools, plant appliances and temporary buildings for which the insured is responsible, and exception 5 refers to the cost of making good, replacement or reinstatement of work, materials or goods supplied, installed or erected all of which are or can be the subject-matter of a material damage policy (thus exceptions 3 and 5 overlap). Usually the material damage policy concerned is the contractors' all risks policy, details of which are given in chapter 5 of this book.

(d)   Exception 4 is similar to exception 3 just mentioned. However, the exception under discussion concerns property belonging to the insured. In the first place an insured cannot be legally liable to himself and secondly it is the subject of a separate policy (see (c) above). The second part of this exception concerns property or part of any property worked upon where the loss or damage arises out of such work.

If a workman breaks a window while putting the glass into the frame, any resultant claim for such damage would not be covered, but if he dropped his hammer into a wash basin underneath the window, this exclusion would not affect a claim for such damage. Insurers are reluctant to grant cover against defective workmanship.

Some insurers exclude property in the charge or control of the insured. The words 'charge or control' are generally taken to indicate ownership or custody. Now ownership of property has already been stated to be excluded by the policy. However, property in the builder's legal possession which is not actually being worked upon but merely being handled incidental to the work concerned does not come within the exclusion. For example, where decorators move furniture to get at their work, the furniture is not in their charge or control. On the other hand, if joiners are employed to repair furniture it is in their charge or control once they begin work upon it.

Similarly, neither a building nor the whole of a particular part of a building is in the insured's charge or control when he is merely executing some work of repair on a particular room. The test is the intention to take possession of or work upon something. Nevertheless, there could be a doubt when the insured builder takes over the complete building, or the owner is not in occupation. So far as property in the charge or control of the insured is concerned, the policy will often provide cover for buildings which are in the insured's temporary occupation for the purposes of reconstruction, alteration or repair.* Even wording of this nature may not completely resolve the position concerning the contents of the buildings as to whether these are considered to be in the insured's charge or control. It is usual, however, to include cover for loss or damage to employees' property, as such property is not normally regarded as being in the insured's custody or control. In exception 4(b) in the policy in appendix 1, damage to property being that part of any property worked upon where the loss or damage arises out of such work is not excluded if it is caused by fire, explosion or by removal or weakening of support to any land, property or building. Thus the catastrophe risks are covered. However, this extension is by no means universal and exception 3 still applies.

(e)   Exclusion 5 applies only to the cost of **making good** the defective work or materials which are excluded, not consequential losses flowing from such defects. The usual example here is where a plumber installs a central heating system with a leaking radiator due to faulty workmanship. The cost of making good the fault would not be covered by the policy, but any liability for damage to property due to the leaking of the water would be covered. Similarly, the canopy repair

* See a rather surprising decision on the 'custody or control' exclusion in *Oei* v *Crawford and The Eagle Star (1982)* details of which are given in an article in *Insurance Week Incorporating Policy Holder* of 8 June 1984 by F.N. Eaglestone.

in Sharpe's case would be excluded (see later). To some extent there is in this exclusion an overlap with the previous one concerning property being worked upon. However, the exclusion under discussion does make it clear that work produced by the builder is excluded so far as making good is concerned as well as actual goods supplied. There is a doubt whether goods supplied would include an actual building but in view of the way this exclusion is worded there can be no doubt. The words 'work erected' would include a building.

In *Wayne Tank & Pump Co Ltd* v *Employers Liability Assurance Corporation Ltd (1973)*, the Court of Appeal decided that the equipment supplied by the plaintiffs was 'goods supplied' within the meaning of this exception. In this case Harbutts' Plasticine factory was burnt to the ground when a pipeline installed by the plaintiffs disintegrated. However, the exceptions clause excluded from cover liability consequent on 'damage caused by the nature or condition of any goods . . . supplied by' the plaintiffs which is much wider than the exception under discussion, and this case does not go so far as to say that 'goods supplied' includes a building.

It is convenient at this point to consider the lapse of time during which the defect may remain latent as this is no bar in law. This is because the period of limitation concerned (three years in which to bring a claim consisting of or including personal injuries and six years for property damage alone) only starts to run when the right of action accrues and this does not take place until the injury or damage occurs. Consequently, in *Sharpe* v *Sweeting & Son Ltd (1963)*, builders were held responsible for injuries when an eight-year-old canopy, negligently reinforced, fell on the wife of the tenant, and in *O'Connor* v *Swan and Edgar & Another (1963)* when a five-year-old ceiling fell on an employee the contractors were held responsible for the injuries. This raises the incidental question of cover when the negligent act and the consequential injury or damage are not contemporaneous. It will be recollected that the wording of the operative clause of the public liability policy refers to an incident happening in connection with the business and **occurring during the period of insurance**.

This wording gives a policy indemnity for a negligent act performed before or after the inception of the insurance provided the injury or damage occurs during the currency of the policy.

(f)   Some policies exclude liability in respect of professional negligence, advice or design. These risks are properly insured in the professional negligence market. Those insurers who do not include such an exclusion do not intend to give full professional negligence cover, but they assume that the average contractor will not be involved in the preparation of plans, designs, drawings and specification unless he has his own specialised department dealing with these aspects. Therefore, to ensure that the insurer is not giving more than the cover he intends to give, a question is usually asked on the proposal form on this point unless, of course, an exclusion exists in the policy as already mentioned.

(g)   This final exception, number 6, already discussed, refers to liability arising from loss or damage due to theft unless caused by negligence of the insured or an employee of the insured (see the case of *Stansbie* v *Troman (1948)* above).

(2) *Risks needing individual consideration include:*

(a)   Liability in respect of loss or damage to property caused by vibration or by the removal or weakening of support. Contractors might undertake work which might affect the support of other property. Most insurers catering for the building and allied trades do not include this exception in their policies but may have a lower limit of indemnity on this type of peril, although they are quite prepared to cover this risk after satisfactory survey up to the limit of indemnity required on payment of an additional premium. Some insurers give automatic cover for the subsidence perils under their basic policy limiting to the same indemnity figure any one occurrence as applies to the general cover, i.e. a figure expressed as a limit applicable to any single occurrence (or in the aggregate during the policy year). The aggregate figure is unsatisfactory to a contractor in the event of several claims in the policy year as subsequent claims may be completely or partly excluded if the earlier accident has consumed the aggregate figure or most of it. However, this is a minor point compared with the necessity for the contractor to have some automatic cover for the subsidence perils. It is usual to apply an excess to this type of cover to avoid small claims.

(b)   The geographical limitation. The combination of general exception 1 and the 'work abroad' clause of the policy in appendix 1 means that protection is given to the insured against his legal liability for bodily injury or damage claims to third parties occurring within Great Britain, the Isle of Man or the Channel Islands (most insurers include Northern Ireland) but for commercial visits abroad by the insured or his employees the policy gives worldwide cover provided there is no manual work. A minority of insurers will also cover manual work performed in Europe (with certain excluded territories) but no cover if injury or damage is caused by a contractor or subcontractor to the insured.

A few insurers cover actions brought in the European Economic Community under their employers' and public liability policies. There have been cases of American nationals proceeding against English insureds in the American courts. In some American states the courts seem to ignore the geographical limitation under the policy. Thus it can be a matter of some concern to insurers who receive premiums for risks in this country which contemplate the English rate of damages not the much higher American rate of damages. However, even if the policy jurisdiction clause is overridden by an overseas court (possibly on the grounds that it is unconstitutional), normally a judgment for the plaintiff would be of no value to him unless the defendant insured had assets in the country concerned, and to involve the insurer the latter would have to have assets in the country concerned.

(c)   Liability assumed by the insured by agreement and which would not have attached in the absence of such agreement. The object of the policy is to protect the insured against his legal liability in tort (see chapter 1), but as the operative clause is not worded in this way it may be necessary to exclude any additional liabilities the insured may have acquired in his agreements with other parties, namely contractual liabilities.

There is a tendency on the part of insurers to include liability assumed under agreements. In fact it is essential in the case of the construction industry. This is a large enough subject to justify a separate heading and the matter will be discussed in more detail below under 'Separate sections of the policy' and 'Contractual liability including indemnity to principals'.

(3)   *Risks undesirable for insurance*

(a)   War and kindred risks (see appendix 1, general exception 3). War risks for property on land are such that no insurer could undertake them. They are a matter for the government, as happened in the First and Second World Wars.

(b)   Liability in respect of radioactive contamination. This is necessary because legislation directs liability to the nuclear reactor owner. Similar responsibility for loss or liability arising from explosive nuclear assemblies would probably fall upon the UK government department concerned or the Atomic Energy Authority.

This exception does not include radio-isotopes or equipment such as X-ray machines as they are considered capable of being covered by the insurance market. These articles can be used to test welding and in this respect the construction industry may be involved. The proposal form usually includes an appropriate question.

In these circumstances it is clear that insurance policies were never intended to cover the effects of nuclear explosives and insurers decided to make this clear by the clause or a similar one which appears as general exception 2 in appendix 1. The same exception includes the radioactive contamination wording.

Indemnity limit

This limit usually appears in the policy schedule under two headings. Firstly, it is in respect of loss or damage to land property or buildings by the removal or weakening of support to such property arising in consequence of such loss or damage except bodily injury claims. The schedule also provides for an excess in respect of this particular indemnity limit.

The other indemnity limit concerns a single accident limit in respect of all other claims. By a limit of indemnity in respect of a single accident, the insurers mean in respect of any one accident or series of accidents arising out of any one event. By negotiation this can be raised but never removed altogether. A property damage excess is now quite common.

The Finance (No 2) Act 1975 when the relevant part was in force gave a statutory indication of what may be a reasonable indemnity, namely £250,000 for any one accident. However, this only applied to personal injury claims. It is in the public liability property damage field that contractors and, consequently, insurers face the catastrophe. There is no limit to the amount of the plaintiff's claim, either financially or in law. A catastrophe claim such as a spreading fire affecting adjacent property caused by the negligent use of a blow lamp could force into liquidation a company which was not insured or not adequately insured. It has happened. A contractor should think in terms of damage to property rather than in terms of bodily injury claims in assessing his limit of indemnity under his public liability policy. A spreading fire can result in a consequential loss claim that is as large as the cost of repairing or replacing the property damaged, e.g. loss of profits, and loss of use of the premises, which is covered by the public liability policy.

It has been suggested that inflation in recent years has overtaken the old ideas of indemnity limits and that insurers must now think in terms of limits of millions of pounds, and not hundreds of thousands.

Separate sections of the policy

See appendix 1 for details of the following.

(1) *Motor contingent liability*

This section covers the insured's legal liability for injury or illness, loss or damage caused by or through or in connection with any motor vehicle (not belonging to or provided by the insured) being used in the course of the business.

At law an employer is vicariously liable for the acts of his employees arising out of and in the course of their employment. Thus, when an employee uses his own car on his employer's business, it is possible for the employer to find himself a target for an action because of the employee's negligence while acting for the insured employer. This applies even though the employee has his own motor insurance covering his negligent acts. Sometimes the employee's insurance does not operate for some reason or does not exist. In any event, there is nothing to stop a plaintiff proceeding directly against the employer at law. The loss or damage to the motor vehicle concerned is not covered as the employee is expected to have his own motor insurance. The other exclusions are equally self-explanatory. The limit of indemnity for loss or damage to property is £100,000 which is often less than the overall policy limit of indemnity.

(2) *Defective Premises Act 1972*

Until 1 January 1974, three rules of the common law operated concerning defective premises:

   (a)   In the sale of a new house the vendor could rely upon the doctrine of *caveat emptor*, let the buyer beware.
   (b)   There was little law against letting a tumbledown house.
         Note that where the builder of a house was also the vendor or lessor of it, he had exemption from liability for defects in the house at the time of the sale or lease (*Otto* v *Bolton and Norris (1936)*). Where the builder of the house was neither the vendor nor the lessor, he was liable on the ordinary rules of negligence (as in *Sharpe* v *Sweeting and Son (1963)*). This last rule still applies.
   (c)   The owner of the premises ceased to be liable in respect of pre-disposal negligence after disposing of the premises. Thus at common law, if a builder (or any other person) negligently constructed or negligently

carried out work on a house and then sold or let it he was not liable if the purchaser or tenant suffered bodily injury or damage to property although according to *obiter dicta* in *Dutton* v *Bognor Regis UDC (1972)*, the rule was doubted in the Court of Appeal. Consequently, in this situation it was advisable for the builder to build and then sell, not sell the land and then build for the purchaser.

The above Act came into force on the 1 January 1974 and not at the time it received the royal assent because the Law Commission wished to give insurers the time to alter their rates and, possibly, the wordings of their policies in respect of those concerned with property to take account of increased liabilities.

From this date, taking the situation in the same order as above:

(a) The builder, the developer and the professional expert involved come under a non-excludable statutory duty of care in respect of new houses (which includes the conversion or enlargement of a building) towards the purchaser and his successors in title, unless the National House-Builders' Registration Council or similar scheme applies. While the usual three- or six-year periods of limitation apply for injury or damage to property, they begin to run when the dwelling is completed because the Act so dictates (sections 1 and 2).

(b) The landlord came under a wider duty to third parties suffering injury or damage because of breach of an obligation to repair. The obligation may arise under the tenancy agreement, by statute, e.g. the Housing Act 1957 section 6 or the Housing Act 1961 section 32, or by virtue of a right to enter. However, a right of entry does not impose an obligation on the landlord to the tenant where the injury or damage arises from a breach of obligation by the tenant (section 4).

(c) The owner of premises is liable for pre-disposal negligence even after he disposes of the premises. The usual periods of limitation already mentioned apply but they run in negligence (as in the case of all torts) from the date the injury or damage was caused (section 3).

For further details of this Act see *The Defective Premises Act 1972 and Implications for Insurers* by F.N. Eaglestone published by PH Press. However, it is sufficient here to say that apart from the considerable widening of a vendor's and a lessor's liability on the sale or letting of premises and for the results, after their disposal, of negligence occurring before disposal, all persons who undertake work for or in connection with the provision of a dwelling owe a statutory duty to see that their work is carried out in a workmanlike or professional manner, with proper materials so that the dwelling is fit for human habitation.

Turning to policy liability, sections 1 and 2 of the Act are excluded from the public liability policy as it is a professional indemnity risk, and regarding section 4 the policy can cover the builder who is also a landlord. It is section 3 which requires an extension to the public liability policy and the wording concerned is given in appendix 1.

(3) *Property in the ground precautions*

The three requirements under this section are not demanded by the majority of insurers. They are clear with the exception of no. 3 which refers to compliance with section 26 of the Public Utilities Street Works Act 1950.

This Act (which does not operate in Northern Ireland) does not apply to road works in the general sense of making or repairing roads, but applies to the placing, etc. of equipment such as pipes and cables in streets or roads or controlled land. It should be noted that undertakers proposing to carry out works must supply plans and give notice, etc., and the owning undertakers must agree plans and disclose the location of their equipment. If an owning undertaker is given the opportunity to disclose the existence of its equipment or requires the new equipment to be sited elsewhere and fails to do so, and such failure leads to its equipment being damaged then the proviso to section 26(6) exonerates the operating undertaker and, consequently, its contractors from liability to pay for repairs. Possibly the greatest number of claims arise from this source.

(4) *Fire precautions*

This section probably indicates the source of the largest claims under this type of policy. In fact it is a constant worry to public liability insurers that they are in effect covering a proportion of the field which should fall to the lot of the fire insurer. Unfortunately the construction industry is notoriously careless in handling oxy-acetylene or similar welding or cutting plant, and blowlamps or torches. Consequently, a fire insurer having paid a claim in the name of his insured naturally recovers from the negligent contractor's public liability insurer. The situation has reached such proportions that the public liability insurer has been forced to make common and reasonable fire precautions a requirement under the policy. *In fact it cannot be emphasised too strongly that non-compliance with this section could very easily put the largest contractor into liquidation as not many can afford to pay several million pounds of fire damage out of their own pocket because of failure to comply with this section.* Even if an insurer does not incorporate this requirement into his policy there is a 'reasonable precautions' condition (no. 3 in appendix 1) in most policies which the insurer could invoke in the circumstances outlined above.

In *British Food Freezers & Ors* v *Industrial Estates Management for Scotland (1977)* the public liability insurers of a firm of contractors, who were engaged to strip out old heaters and install new ones at premises occupied by the plaintiffs, were held not liable to indemnify their insured because of a breach of the condition of the policy providing that 'the insured shall exercise reasonable care in the selection of competent employees and shall take all reasonable precautions to prevent accidents'.

The work carried out by the contractors necessarily involved some use of oxy-acetylene burning equipment on plant attached to roof members of the factory. It was known that there was a high risk of fire in the storage area as highly inflammable materials were stored there.

There was a condition in the contract that the contractors should take all reasonable precautions. Lord Dunpark found that the contractors took no fire precautions whatsoever. There was a duty on the contractors to plan their work so as to prevent damage arising from a risk which any reasonable heating contractors, in their position, would have realised was glaringly obvious and substantial.

The failure to take precautions was the responsibility of the contracts manager who was responsible to the board of directors for performance of the duty. They had left it to their contracts manager to make such arrangements as he thought appropriate and thereby delegated to him the discharge of their duty of care. In the circumstances, Lord Dunpark held that the insurers were not bound to make any payment under the policy in respect of the damage caused by the fire.

(5) *Costs*

The remarks under this heading are self-explanatory but see also chapter 10. See appendix 1 under heading 'Clauses applicable to sections 1 or 2 of the policy'.

(6) *Contractual liability including indemnity to principals*

The expression 'contractual liability' can be used in two senses. It can mean all contractual terms the insured enters into which would include an indemnity given to principals. On the other hand it can exclude indemnity to principals and in this sense it is used in both this policy and the employers' liability policy, otherwise there would be no point in incorporating clauses under 'Contractual liability' and also 'Indemnity to principals' in the policy (see appendix 1).

The difference between contractual liability and indemnity to principals is that in the case of indemnity to principals the principal is treated as an insured and can claim directly against the insurer under the policy, whereas under the contractual liability heading (sometimes called an assumed liability) the principal can only claim against the insured under the contract and the insured contractor can in turn obtain an indemnity under the policy. Points to note about these two paragraphs in the policy are:

(a) The extensions are subject to the terms and conditions of the policy. This is particularly significant in the case of indemnity to a principal who does not get a greater benefit under the policy than the insured has already.

(b) The insurer has sole conduct and control of all claims covered by these extensions and in the case of indemnity to principals the insured contractor is required to have arranged with the principal for such conduct and control of all claims by the insurers.

(c) In the case of the contractual liability extension there is a proviso that the insurer shall not be liable for loss or damage to property hired, leased or rented when liability would not have attached in the absence of a hire lease or rent agreement.

Note that other separate clauses common to both the public and employers' liability policies, i.e. Labour Masters, Legal Defence, Business, Indemnity to Personal Representatives and Indemnity to Other Persons are dealt with in chapter 10 concerning the employers' liability policy. For 'Work Abroad' see page 104.

Policy conditions

These are the same as indicated in chapter 5 concerning the contractors' all risks policy; see appendix 1. However, there is one exception as condition 6, which does not apply to the contractors' all risks policy, applies to the public liability policy. The wording used in appendix 1 for condition 6 is not universal in that once the condition is invoked it sometimes states that the insurers will pay the limit of indemnity for any one accident.

Single or annual policy (employers' and public liability)

There are advantages and disadvantages in both.

Some contractors compromise by taking annual policies covering the contractors' basic work but excluding hazardous work. For example a pipe-layer could have an employers' liability policy covering all jobs where excavation did not exceed five metres, or a welder with a public liability policy which excluded work in oil refineries or gas works. This leaves single policies for hazardous contracts.

(1) *The annual policy – the advantages:*
   (a) To the insurer
      (i) Gives better details of the claims experienced over the years as it covers all work
      (ii) There are less administrative costs than in the case of a single policy – policy issue, adjustment of premium on individual contract prices, possible extension of contract period
   (b) To the insured
      (i) He knows the rate of premium to include in his tender price
      (ii) He does not have to watch for necessity to extend the policy because of extension of contract period

(2) *The annual policy – the disadvantages:*
   (a) To the insurer
      (i) Underwriting tends to be less scientific being a mathematical calculation
      (ii) He can lose touch with the insured's activities
   (b) To the insured
      The rate may be higher than would be available on a single contract cover which may make just that difference in losing the contract when tendering.

Clauses 22 and 23 and the public liability policy; summary of the main risks remaining uninsured

As indicated at the end of chapter 5 when introducing the same type of summary in respect of the contractors' all risks policy only the basic cover is being considered for the public liability policy not the cover an insurance broker might produce by special negotiation or a special policy wording insurers might be persuaded to adopt for that broker's clients in the construction industry.

The contractor

(1) *Limit of indemnity* The contract may specify a minimum limit (in the appendix to the tender) for any one accident or series of accidents arising out of one event up to which the contractor is required to insure his liability to third parties (other than employees). Under the ICE Conditions and at common law the liability to third parties is unlimited. Rarely if ever will insurers cover the liability in respect of any one accident or series of accidents arising out of one event up to an unlimited amount although very high limits can be obtained.

(2) *The gap between the contractors' all risks policy and the public liability policy concerning the contract works and future structural damage* Usually the cover under the CAR policy ceases when the contract is completed including any liability under the maintenance period plus fourteen days. On the other hand the public liability policy excludes loss or damage to the contract works. Yet it is clear that there is a liability for many years in tort and contract after the contract is completed for defective work and materials and defective design. Neither the final certificate (clause 60(3)) nor the maintenance certificate (clause 61(2)) contain any permanent binding force. The freedom to recover damages for defective work for the full period of limitation is generally speaking satisfactory to employers (for their entitlement to sue in tort see *Dutton* v *Bognor Regis UDC (1972)* which was upheld and explained by the House of Lords in *Anns* v *Merton BC (1977)*, and the period of limitation runs from the date the damage **comes into existence**, *Pirelli General Cable Works Ltd* v *Oscar Faber & Partners (1982)*, which might be longer than the period of limitation in contract which runs from the date of breach of contract). However, the gap is a worry for contractors.

The usual solution is to say that what is required is a professional indemnity policy but it has already been explained that this policy has its limitations (see the summary at the end of chapter 5 – only professional negligence is covered which does not include non-negligent (contractual) liability), and it is not very practical to say that a products guarantee policy is also required to cover replacement of the defective product as it is probably impossible or very difficult to obtain this cover. The National House-Building Council scheme and Decennial (or Building Guarantee) Insurance have limited application. The former is limited to private dwellings and the latter is usually confined to commercial buildings. Both are usually offered to the owner not the contractor and both are limited to ten years and major structural damage. It is true that in the case of decennial insurance the owner may be able to pay an extra premium to avoid the insurer using his subrogation right against the contractor but why should he do this? The whole situation really leaves civil engineering works without any cover for major structural damage. Insurers ask why members of the construction industry should expect their work to be guaranteed as the insurers regard it as a contract hazard which that industry's members should bear themselves. Probably the short answer is that the insurance industry generally regard this risk, particularly in the case of civil engineering works, as too hazardous to insure.

The employer

(1) *The employers' responsibility to third parties because he commissioned the works* Clause 22, which requires that the contractor should indemnify the employer in respect of third-party claims, also has certain exceptions, notably when injury to persons or damage to property occurs for which the employer, etc. is responsible. The five exceptions in proviso (b) are all liabilities for which the contractor at common law would not have to indemnify the employer so there is no legal imposition on the employer and the public liability policy follows the contract requirements (see clause 23), although the contractor's policy is required to provide a limited indemnity to the employer. Obviously the employer's liability can be serious if third-party buildings collapse due to the activities on the works which may be due to ground structural faults or the taking of a calculated risk. In any event there would be a liability on the employer for nuisance and normally no liability on the contractor assuming no negligence on his part. Now if the public liability policy follows the contract and the employer only gets the cover for which the contract requires the contractor to indemnify the employer, then the employer would not be able to rely on the contractor's policy. It all depends on the policy wording but it should not be assumed that the employer is in effect being given a separate public liability policy in his own name by clause 23. Assuming the worst, i.e. that the employer gets no cover unless it is a liability for which the policy would indemnify the contractor, which is one way of reading the contract wording, then the employer is short of public liability cover for his own liabilities such as that described above, i.e. a nuisance liability. Now in the JCT Form of Contract clause 21.2.1 (in the 1980 edition; 19(2)(a) in the previous edition), requires insurance cover for a similar collapse risk to third-party property, to protect the employer. However, while no doubt this type of cover could be given to the employer in the ICE Conditions it has its limitations. To name the more important ones:

(a)   It only covers damage to property.
(b)   It only applies to collapse and kindred risks.
(c)   Design risks are excluded.
(d)   Inevitable damage is excluded.
(e)   There is usually a substantial excess.

Alternatively the employer could take out his own public liability policy but this to some extent would be duplicating the cover given by the contractor's public liability policy. In any event the employer should not expect the insurance market to provide a complete indemnity.

(2) *Consequential losses* The public liability policy covers any consequential losses flowing from the injury to persons or damage to property covered by the

operative clause, although the policy does not specifically say so. The problem is that while there is an exclusion the purpose of which is to exempt loss or damage to the contract works, insurers claim that this means consequential losses arising from such loss or damage must also be excluded. It really depends on how the exclusion is worded. If the wording is that 'liability in respect of loss or damage to the contract works' is excluded then it is probably also sufficient to exclude the consequential loss. There is an Australian authority that the 'words "in respect of" are difficult of definition but have the widest possible meaning of any expression intended to convey some connection or relation between the two subject-matters to which the words refer.' (*Trustees Executors and Agency Co Ltd* v *Reilly (1941)*.)

Chapter 9

Accident or injury to workmen: clause 24

Accident or Injury to Workmen

24. The Employer shall not be liable for or in respect of any damages or compensation payable at law in respect or in consequence of any accident or injury to any workman or other person in the employment of the Contractor or any sub-contractor save and except to the extent that such accident or injury results from or is contributed to by any act or default of the Employer his agents or servants and the Contractor shall indemnify and keep indemnified the Employer against all such damages and compensation (save and except as aforesaid) and against all claims demands proceedings costs charges and expenses whatsoever in respect thereof or in relation thereto.

It has already been explained when discussing clause 22(1) that there is an overlap between these two clauses in that clause 24 seems to repeat, in respect of claims by the contractor's employees and those of his subcontractor for bodily injuries the indemnity given by the contractor to the employer in the introductory paragraph to clause 22(1), and the exception in proviso (b)(v) to that clause.

However, there are differences as follows:

(1)   The phrase 'act or default' is used instead of 'act or neglect' in setting out the exception where the employer does not get an indemnity because of his own actions or those of his servants or agents. It has been suggested that the phrase 'act or default' is narrower than 'act or neglect' as it indicates some positive act rather than an act or error of omission. In fact the word 'default' in dictionaries is often defined as 'failure or omission to do any duty or carry out any liability' or 'neglect to do what duty or law requires'. Therefore the negative aspect is present in the phrase 'act or default'. Probably more significant is that in common parlance the word has a very wide interpretation and the phrase 'act or default' has no direct legal connections as the phrase 'act or neglect' has with negligence. This is an example of bad drafting as there is probably no intention to differ in the extent of the indemnity by the change of phrase, so why change it or better still why not use legal terms, as suggested in chapter 1?

The point is that where this type of phrase operates an exception to an indemnity the wider the phrase is interpreted the less room there is for the indemnity to apply. Thus if the employer says to the contractor in the contract you will indemnify me in respect of injury or damage to third parties except when that injury or damage is caused by my 'act or neglect' (or 'act or default') and you both interpret the phrase to mean negligence, nuisance, trespass, breach of statutory duty or breach of contract, the contractor is not really giving the employer an indemnity as there is little left in law for which the contractor must indemnify the employer. Clearly from *Hosking* v *De Havilland* and *Murfin* v *Royal* mentioned in chapter 1 the courts are reluctant to cut down an indemnity to this extent where they have a choice.

Therefore, possibly the phrase 'act or default' intends, like the phrase 'act or neglect', only to impart the common law concept of a reasonable duty of care (negligence).

(2)   There is no specific reference to the engineer, or other contractors (not being employed by the contractor), as in clause 22(1)(b)(v) (the exception to the contractor's indemnity to the employer), although the word 'agents' may include the employer's engineer.

(3)   While the phraseology used in the introductory paragraph to clause 22(1) is different from that used in clause 24 there is probably no significance in this.

There is no complementary insurance clause with clause 24, presumably because the insurance of liability to employees is now compulsory in this country by virtue of the Employers' Liability (Compulsory Insurance) Act 1969. Details of this Act are given in the next chapter. As to whether clause 24 overrides clause 22, it seems reasonable to argue that being more specific in its wording it should take precedence where the injured claimant is an employee of the contractor or of his subcontractors.

Chapter 10

Employers' liability insurance

During most of the first half of this century insurers transacting this class of business had separate departments dealing with the employers' liability policy which not only provided cover for employers' liability at common law and under statutes but, in addition, for the then **alternative** benefits which were payable under the Workmen's Compensation Acts. When the State took over payment of the benefits which replaced those under the Workmen's Compensation Acts by the National Insurance (Industrial Injury) Act 1946 and made these new benefits payable **in addition** to any damages at common law and under statutes, the insurers continued to keep a separate employers' liability policy, although later this risk was incorporated by some insurers in one policy document with other policies, e.g. the public liability policy (see chapter 8) and the contractors' all risks policy (see chapter 5 and appendix 1).

Vicarious liability

In chapter 1 under 'Absolute or strict liability', the rule of vicarious liability was mentioned. This is an extension or elaboration of negligence. An employer is vicariously liable for the negligence of his employees acting within the scope of their employment, even if he was wholly free from blame himself. An employer may be held liable in respect of an injury sustained by one of his employees through the negligence of a fellow employee. The situation often occurs in claims handled by insurers under employers' liability policies, although public liability policies apply if the negligent employee injures or damages the property of a third party who is not a fellow employee, or damages the property of a fellow employee.

This vicarious liability rule is an application of the principle of agency, namely that he who acts through another acts himself. It has the practical advantage that the necessity for insurance rests with the employer who usually can insure more economically than individual employees. Furthermore, it is argued that it is reasonable to expect the employer to insure this risk because, by instructing his employees to engage in activities on his behalf, he creates the risk that they may negligently injure others.

Employers' Liability (Compulsory Insurance) Act 1969

This Act came into force on 1 January 1972. From that date all employers are required to insure against their liability for personal injury or disease sustained by their employees in the course of their employment, other than the employers of domestic employees, in Great Britain.

The following statutory instruments deal with the commencement of this Act and the general regulations concerning its operation:

SI 1971 No. 1116, The Employers' Liability (Compulsory Insurance) Act 1969 (Commencement) Order, 1971
SI 1971 No. 1117, The Employers' Liability (Compulsory Insurance) General Regulations, 1971

Insurers are required from 1 January 1972 to issue approved policies of insurance and certificates confirming the contract of insurance whenever a policy of insurance is entered into or renewed. From 1 January 1973, employers are required to display the certificate of insurance (or a copy of it) at each of their premises for the information of their employees, and produce it as required.

The general regulations, *inter alia*, prohibit certain conditions in policies of insurance which would entitle insurers to deny liability under the policy, and limit the amount for which an employer is required by the Act to insure to £2 million in respect of any one occurrence giving rise to liability.

Regulation 2 of the General Regulations deals with the prohibition of certain conditions in policies of insurance and reads as follows:

(1) Any condition in a policy of insurance issued or renewed in accordance with the requirements of the Act after the coming into operation of this Regulation which provides (in whatever terms) that no liability (either generally or in respect of a particular claim) shall arise under the policy, or that any such liability so arising shall cease

    (a) in the event of some specified things being done or omitted to be done after the happening of the event giving rise to a claim under the policy;

    (b) unless the policyholder takes reasonable care to protect his employees against the risk of bodily injury or disease in the course of their employment;

    (c) unless the policyholder complies with the requirements of any enactment for the protection of employees against the risk of bodily injury or disease in the course of their employment; and

    (d) unless the policyholder keeps specified records or provides the insurer with or makes available to him information therefrom,

is hereby prohibited for the purposes of the Act.

(2) Nothing in this Regulation shall be taken as prejudicing any provision in a policy requiring the policyholder to pay to the insurer any sums which the latter may have become liable to pay under the policy and which have been applied to the satisfaction of claims in respect of employees or any costs and expenses incurred in relation to such claims.

It should be noted that the prohibition is only for the purposes of the Act and that paragraph 2 of this regulation allows the insurer, having paid the claim of an employee to whom the employer was legally liable, to recover from the insured employer if he would have had a right to repudiate under those policy conditions. In other words, in the insurer's relationship with the insured, the conditions operate. In order to take advantage of the right contained in regulation 2, a similar section to that contained in the present motor policies, entitled 'Avoidance of certain terms and right of recovery', is now included in the majority of employers' liability policies. This states:

The indemnity granted by this policy is deemed to be in accordance with the provisions of any law relating to compulsory insurance of liability to employees in Great Britain, Northern Ireland, the Isle of Man, or the Channel Islands. But the insured/policyholder shall repay to the Company all sums paid by the Company which the Company would not have been liable to pay but for the provisions of such law.

It should be noted that similar legislation is operative in Northern Ireland, the Channel Islands and the Isle of Man.

Nevertheless, a repudiation under the conditions mentioned in paragraphs (a)–(d) was rare enough prior to the passing of the 1969 Act and therefore, in practice, the position is not likely to be very different now the Act is in operation. The obvious purpose of this legislation is to avoid the situation in which employees or their dependants suffer financially in spite of the employer's legal liability for the employee's injury or death. This arises when damages cannot be paid either because the employer has no insurance policy or because his employer's liability insurers reject the claim because of breach of certain policy conditions combined in either case with the fact that the employer has no financial means to meet the claim.

It will be seen that the general regulations under the statutory instrument dealing with the prohibition of certain conditions in employers' liability policies do not interfere with the freedom to apply effective underwriting conditions so far as hazardous work is concerned, e.g. in relation to depths of excavations below ground or height of buildings above ground. While it is not normally the practice to apply excesses in underwriting the employers' liability policy, it is apparent that the regulations would not allow an excess to be deducted from an employee's claim for damages. However, there is no reason why the policy terms should not require an employer to reimburse the insurer to a stated extent in respect of any particular claim. (For further details concerning this legislation see *The Employers' Liability (Compulsory Insurance) Act 1969: Implications for insurers*, F.N. Eaglestone, published by PH Press.)

Employers' liability policy

As stated in chapter 4 this policy covers the legal liability of an employer (master) to his employees (those persons under a contract of service or apprenticeship with the employer) for bodily injury or disease arising out of and in the course of his employment.

The modern policy comes nearer than any other considered in this book, to justifying being called 'exclusionless'. The extent to which this is justified will be seen from the operative clause etc appearing in the policy set out in appendix 1. Sometimes this policy is endorsed excluding hazardous work but not usually with civil engineering contractors. However see under 'Business' on page 123.

The indemnity given in respect of employees employed outside Great Britain, etc. only applies when the action for damages is brought against the insured in a court of law in Great Britain, etc. Most insurers include Northern Ireland in this list. At least one insurer includes 'any member country of the EEC' excluding in that case social security compensation.

It will be noted from appendix 1 that cover is restricted to employment within or 'visits abroad' outside the listed territorial limitations. While there may be various interpretations of 'visits abroad', the intention is that cover shall not include risks which are more properly the province of a foreign insurance, e.g. workmen's compensation insurance where it still applies abroad. If an employer engages local labour for work abroad, he should arrange a policy to meet local requirements.

The policy contains no war risk exclusion. Insurers seem to rely on the fact that the Government undertook to deal with occupational war injuries by a special scheme during the Second World War and presume similar arrangements will be made in any future war. Also it is not easy to visualise how an employer could be legally liable to his employees for injury caused by a war.

The bodily injury or disease must be sustained during the period of insurance. 'Sustain' is probably used to indicate that injuries or diseases which do not manifest themselves immediately may not be diagnosed until several years have elapsed when the policy may no longer be in force. The insured will then have to show that the policy was in force when the injury or disease was caused. Even if the word 'sustain' is not used, it is made clear that the bodily injury or disease must be caused during the period of insurance.

Legal actions concerning personal injury must usually be commenced within three years of the date of accident. However, the law allows exceptions because some diseases take a time to develop and may not become obvious for some years, possibly even after the policy has lapsed, e.g. the insured changes his insurer. The Limitation Act 1980 caters for this situation to a considerable extent by allowing an employee to bring an action although more than three years have passed since the accident date in certain circumstances. See pages 12 and 250.

The phrase 'arising out of and in the course of his employment' has given rise to litigation, particularly in deciding whether employees travelling to and from work as passengers along a public road are being carried in circumstances arising out of and in the course of their employment, if injured while travelling. The question is whether the contractor's employers' liability policy or his motor policy applies. The employer (principal) is not likely to be involved in such an accident unless he supplied the vehicle. However, the old test which applied in the days of the Workmen's Compensation Acts was recently confirmed in *Vandyke* v *Fender (1970)* as still applying. The employees were being carried in circumstances arising out of and in the course of their employment, and thus the contractor's employers' liability policy operates if the employees were obliged to travel in the vehicle, i.e. they had no option as they were under the control of their employer (master) who is the contractor. If the employees could have taken alternative transport rather than the contractor's vehicle then the employers' liability policy would not operate. The fact that they would be foolish to do so is quite irrelevant as it is a question of whether they were under their employer's (master's) control at the time of travel or whether he was merely granting a facility which the employees could take or leave as they chose.

The employers' liability policy does not cover damage to employees' property, since it is limited to bodily injury to employees only. Thus, damage to an employee's clothing is dealt with by the public liability policy, although as a small item of damage in a personal injury claim it might be paid under the employers' liability policy purely as a matter of administrative convenience in spite of the fact that there is no cover for this item in the employers' liability policy.

Separate clauses of the policy

These clauses state:

Costs
The Company will also
(1)   Pay all costs and expenses incurred with its written consent.
(2)   Pay the Solicitor's fee incurred with its written consent for representation of the insured at any Coroner's inquest or fatal injury enquiry or proceedings in any Court (or Court of Summary Jurisdiction) arising out of any alleged breach of a statutory duty resulting in bodily injury or disease which may be the subject of indemnity under this Policy.

Indemnity to personal representatives
In the event of the death of the Insured the Company will in respect of the liability incurred by the insured indemnify the insured's personal representatives under the terms of this Policy, provided that such personal representatives shall as though they were the insured, observe, fulfil and be subject to the terms, exceptions and conditions of this Policy in so far as they can apply.

Indemnity to other persons
Subject to the terms exceptions and conditions of this Policy at the request of the insured, the Company will indemnify

(1)   Any director, partner or employee of the Insured in respect of any claims for which the Insured would be entitled to indemnity under the Policy if claims were made against the Insured.
(2)   Any director, partner or employee of the Insured, for whom with the consent of the Insured an employee is undertaking private work.
(3)   Any officer or member of the Insured's canteens, clubs, sports, athletic, social or welfare organisations and first aid, fire and ambulance services

Provided that such persons shall as though they were the insured observe, fulfil and be subject to the terms exceptions and conditions of this Policy in so far as they can apply.

These separate clauses of the policy define the extra cover the insurer will give.

(a)   All costs and expenses include, for example, examination of defective machinery by engineers after an accident, or expenses of expert witnesses. These, it will be noted, must be incurred with the written consent of the insurer.

This section referring to the solicitor's fee and representation in any Court refers to criminal proceedings arising out of breaches of the Factories Acts or similar legislation. The indemnity under the policy only applies to a solicitor's fee not to any fine or other penalty imposed by the court.

A successful criminal prosecution in respect of a breach of statutory duty may be given in evidence in a civil action by an injured employee. Thus if an employee is injured by failure to fence a machine in accordance with section 14 of the Factories Act 1961 and the employer is successfully prosecuted by the Inspector of Factories, this conviction may be given in evidence when the employee brings his action in the civil court for damages. Consequently the insurer and insured have a common interest in any such criminal conviction. Hence the insurer's interest in ensuring that the insured employer is properly defended at the criminal proceedings.

Similarly this section applies to fatal inquiries and coroner's court proceedings.

(b)   The paragraph concerning indemnity to personal representatives deals with the automatic continuation of policy cover in the event of the death of the insured. Without such cover the policy would be inoperative since it is a personal contract between the insured and the insurers.

It should be borne in mind that a personal action at law does not cease with the death of either party to it. The Law Reform (Miscellaneous Provisions) Act 1934 provides that if either party to an action dies the action survives for the benefit of or against the deceased's estate.

This cover can only apply where the insured is an individual as a corporate body's life is unaffected by the death of its members.

(c)   This clause gives an indemnity to any director or partner of the insured firm as well as the officers, committees and members of the contractor's canteen, social, sports and welfare organisations including the first aid, fire and ambulance services.

Exclusions and contractual liabilities

The modern policy does not exclude liability which attaches by virtue of a contract or agreement which would not have attached in the absence of such contract or agreement, which is an exclusion common to other liability policies. Special sections of the policy provide for:

Indemnity to Principals   Where any contract or agreement entered into with any Principal so requires the benefits of this Policy shall apply jointly to the Insured and the Principal (named in the Schedule hereto) and the Company will indemnify the Principal within the terms of this Policy for any claim resulting from injury, illness, loss or damage (as herein defined) where such injury, illness, loss or damage occurs during the currency of the Policy and arises out of, in the course of or by reason of the carrying out by the Insured and/or his Subcontractors of work for which an indemnity is provided by this Policy.

Contractual liability   Under Sections 1 and 2 of this Policy (subject to the terms and conditions of these Sections and the general exceptions of the Policy) the Company will indemnify the Insured in respect of liability for bodily injury to or illness of any person or loss or damage to property (as defined herein) under indemnity clauses forming part of any contract to which this Policy applies.

With regard to both these clauses it should be noted that under the Unfair Contract Terms Act 1977 it is no longer possible for a person to contract out of liability for negligence causing personal injury; however, this cover should not be affected for the future. See chapter 1 to appreciate the problem, although the ICE Conditions in clause 24 make each party to the contract responsible for his own negligence. In any event, the employer can under contract pass liability to a non-consumer. Note the difference between an exclusion clause and an indemnity clause under the Act.

It should also be noted that sections 1 and 2 referred to under contractual liability involve the employers' liability and public liability sections of the combined policy concerned. Also the 'Indemnity to principals' paragraph caters for both sections of the policy. The provisos to the 'Indemnity to principals' clause in appendix 1 should be noted.

It will be seen that the direct indemnity to the principal himself which is given by the insurers is in addition to the liability assumed by the insured under his contract with the principal and both this direct indemnity and the assumed liability operate only where the contract or agreement entered into so requires. Only one or other of these covers is required by contract at any one time and the difference is that in the case of the assumed or contractual liability (as it is sometimes called) it is only the insured contractor who can enforce the policy, whereas in the case of the indemnity to the principal he has the advantage of being able under the wording of the policy to bypass the insured contractor completely and claim indemnity directly from the contractor's insurers since the principal is in effect a joint insured under the policy. While these clauses provide automatic cover in respect of any indemnities given by the insured, they only do so as far as employees of the insured are concerned. They do not, for example, apply to employees of subcontractors as such liability is covered by the public liability policy.

There is a proviso to these extensions of cover which reads as follows:

Provided that in respect of any indemnity to any principal or liability assumed under contract the Company shall not be liable in respect of any legal liability of whatsoever nature directly or indirectly caused by or contributed to by or arising from

    (a)   ionising radiations or contamination by radioactivity from any nuclear fuel or from any nuclear waste from the combustion of nuclear fuel

    (b)   the radioactive, toxic, explosive or other hazardous properties of any explosive nuclear assembly or nuclear component thereof.

This proviso excludes liability from ionising radiations, contamination by radioactivity or explosive nuclear assembly. It should also be noted that the insured is covered in respect of his normal legal liabilities so far as these risks are concerned; this proviso only operates in respect of indemnities given by the insured under contract. This is because the contractual liability and indemnity to principals extensions result in the principal being given cover for which he would normally be insured under his own public liability policy, as the insured's employees are third parties to him and thus are a matter for his public liability policy to deal with. Now it is not intended in giving this extension of cover to a principal that protection should be given to a principal for any nuclear risk under a general public liability policy, which is covered by arrangement with the government under statute.

Labour masters

For the policy wording see appendix 1. The relationship between a main contractor and members of labour gangs including self-employed persons is frequently difficult to decide. See the legal position on page 247. In these circumstances the employers' liability policy usually handles claims for injury to them and the public liability policy deals with accidents caused by those so-called employeees. While these remarks apply to paragraphs (a) to (c) in this clause in appendix 1 the most practical example of the operation of paragraph (d) concerns the Contractors' Plant Association Model Conditions for the Hiring of Plant. See pages 237 to 239 for the legal position. Thus drivers of such plant are deemed to be employed by the insured by this policy clause.

Legal defence

For the policy wording see appendix 1. This clause will pay the legal costs of those senior employees for defending a prosecution under the Health and Safety at Work etc Act 1974. In case it is thought that this cover is unnecessary as the senior official's firm would pay for his defence, this is incorrect as the Companies Act 1948 forbids this unless the defence is successful. The legal costs and expenses cover goes beyond the cover of the employers' liability policy which is strictly based on a claim for bodily injury or disease and some insurers may charge for this cover. For some information about the 1974 Act see pages 9 and 189. A separate legal costs and expenses policy with wider cover is obtainable.

Business

For the policy wording see appendix 1. The business, i.e. the specific activity of the insured, is a vital rating factor with this class of policy and the insurer tends to confine cover to the particular trade or activity for which a premium has been paid. Thus any change, particularly to a more hazardous type of work, must be notified to the insurers.

Indemnity limit

It is not the practice of the insurance market to impose a limit of indemnity under an employers' liability policy; that is, the policies provide an indemnity for an unlimited amount. In fact the Employers' Liability (Compulsory Insurance) Act 1969 provides that the employer is to insure and maintain insurance for an amount of not less than £2 million in respect of claims relating to any one or more of his employees arising out of any one occurrence. The result is that the employers' liability policies provide more than the minimum cover required by this Act.

Policy conditions

These are the same as those detailed for the public liability policy in chapter 8 except that condition 6 only applies to the public liability policy.

Single or annual policy

The same remarks apply under this heading as were made under the public liability policy under the same heading.

Chapter 11

Remedy on contractor's failure to insure: clause 25

Remedy on Contractor's Failure to insure

25. If the Contractor shall fail upon request to produce to the Employer satisfactory evidence that there is in force the insurance referred to in Clauses 21 and 23 or any other insurance which he may be required to effect under the terms of the Contract then and in any such case the Employer may effect and keep in force any such insurance and pay such premium or premiums as may be necessary for that purpose and from time to time deduct the amount so paid by the Employer as aforesaid from any monies due or which may become due to the Contractor or recover the same as a debt due from the Contractor.

This clause states that the employer may effect and keep in force any insurance which the contractor may be required to arrange under the terms of the contract if the contractor fails upon request to produce to the employer satisfactory evidence that there is in force the insurance referred to in clauses 21 and 23.

The opportunity is taken to remind the reader of the insurances required.

Material damage insurance

Firstly, under clause 21 the contractor must insure in the joint names of the employer and the contractor against all loss or damage, from whatever cause arising (other than the excepted risks) for which the contractor is responsible under the terms of the contract, the permanent works, the temporary works and the constructional plant each to their full value. In the case of the permanent and temporary works, unfixed materials or other things delivered to the site for incorporation therein, are to be included for their full value.

Secondly, the contractor must also insure in such manner that the employer and contractor are covered for all loss or damage arising during the period of maintenance, from whatever cause arising for which the contractor is responsible under the terms of the contract, occurring prior to the commencement of the period of maintenance.

Thirdly, the contractor must also insure in such manner that the employer and contractor are covered for any loss or damage occasioned by the contractor in the course of any operation carried out by the contractor for the purpose of complying with his obligations under clauses 49 (dealing with the period of maintenance and defects) and 50 (dealing with the contractor's responsibility to carry out searches and tests required by the engineer in writing).

Liability insurances

Under clause 23 the contractor must insure throughout the execution of the works against any damage loss or injury to any property or person which may arise out of the execution of the works or occur in carrying out the contract, otherwise than due to matters referred to in proviso (b) to clause 22(1). These insurances must be effected with an insurer and in terms approved by the employer and for at least the limit of indemnity stated in the appendix to the form of tender.

The terms of such insurance must include a provision whereby the insurer will indemnify the employer against any claims, costs, charges and expenses in respect of any indemnity which the contractor would be entitled to under the policy. This requires insurances giving an indemnity to the principal.

Insurances generally

Both clauses 21 and 23 conclude by stating that the contractor shall wherever required produce to the employer the policy or policies of insurance and the receipts for payment of the current premiums, which ties up with the operative part of clause 25, namely that the employer may effect and keep in force any such insurance and pay such premium or premiums as may be necessary for that purpose if the contractor fails upon request to produce satisfactory evidence that there is in force the insurance referred to in clauses 21 and 23.

What is satisfactory evidence? This has already been indicated in the final part of the clauses concerned. The evidence may be the original policies and premium receipts or preferably in the contractor's interest photostatic copies. It is sometimes suggested that a certificate summarising the cover will be sufficient but this should be avoided as it is impossible to summarise satisfactorily all the terms exceptions and conditions of a policy, thus the usual certificate is useless. This type of certificate should not be confused with the motor and employers' liability insurance certificates required by statutes; these serve another purpose.

If the employer pays such premium(s) as may be necessary under clause 25 to effect the insurances required under the terms of this contract he may from time to time deduct the amount so paid from any monies due or which may become due to the contractor. This would include any monies due on other contracts. Alternatively the employer may recover the amount so paid as a debt due from the contractor.

On the contractor failing to insure under clauses 21 and 23 if the employer effects the necessary insurances in his own name (as he would have to do) in the event of a loss occurring for which the contractor is liable under the contract, the insurer, having paid the employer would be entitled to sue the contractor by virtue of subrogation. However, if it is uneconomical for the contractor to eliminate all excesses or to obtain proper cover, for example, against faults in design, while technically clause 63(1)(d) could be invoked (expulsion from site and works after seven days notice), it may be impracticable to do so especially as insurances against certain risks may be unobtainable. See the summaries at the ends of chapters 5, 8, 15 and 18.

The employer should check that the insurance policies which have been produced are extended or renewed at the end of each policy period as required.

Chapter 12

Vesting of goods and materials not on site and insurance requirements: sub-clause 54(3)

Vesting in Employer

(3)   Upon the Engineer approving in writing the said goods and materials for the purposes of this Clause the same shall vest in and become the absolute property of the Employer and thereafter shall be in the possession of the Contractor for the sole purpose of delivering them to the Employer and incorporating them in the Works and shall not be within the ownership control or disposition of the Contractor.

Provided always that:

(a)   approval by the Engineer for the purposes of this Clause or any payment certified by him in respect of goods and materials pursuant to Clause 60 shall be without prejudice to the exercise of any power of the Engineer contained in this Contract to reject any goods or materials which are not in accordance with the provisions of the Contract and upon any such rejection the property in the rejected goods or materials shall immediately revest in the Contractor;

(b)   the Contractor shall be responsible for any loss or damage to such goods and materials and for the cost of storing handling and transporting the same and shall effect such additional insurance as may be necessary to cover the risk of such loss or damage from any cause.

The purpose of clause 54 is to give the engineer authority to certify payment for goods and materials which have not yet reached the site subject to the following conditions.

The provisions of the whole of clause 54 apply to those goods and materials, before they are delivered to the site, which are in the appendix to the form of tender (prepared by the engineer; see footnote (e) of that appendix) and which at the time of the contractor's application for payment are:

'substantially ready for incorporation in the Works; and . . . are the property of the Contractor or the contract for the supply of same expressly provides that the property therein shall pass unconditionally to the Contractor upon the Contractor taking the action referred to in sub-clause (2) of this Clause.'

Sub-clause 54(2) states that the evidence of the intention of the contractor to transfer the property in the goods or materials to the employer shall be given by the contractor or the supplier taking the following action:

(a)   providing the engineer with documentary evidence that the property in the goods or materials has vested in the contractor;

(b)   identifying the goods and materials so as to show that their destination is the site, that they are the property of the employer, and (where they are not stored at the premises of the contractor) to whose order they are billed;

(c) setting aside and storing the said goods and materials so marked or identified to the satisfaction of the engineer; and

(d) sending to the engineer a schedule which lists and gives the value of every item of the goods and materials so set aside and stored and inviting him to inspect the same.

Turning to sub-clause 54(3), and bearing in mind the requirements of sub-clause 54(2) just mentioned, the vesting of the goods and materials in the employer is subject to the engineer's approval in writing. In spite of such vesting in the employer the contractor is responsible for any loss or damage, cost of storage, handling and transporting of such goods and materials. Furthermore **the contractor is required to effect any additional insurance which is necessary** to cover the risk of such loss or damage arising from **any** cause. It will be recollected that at the beginning of this chapter the following requirements (concerning goods and materials) were mentioned before the contractor could apply for payment. The goods and materials must:

(i) appear in the appendix to the form of tender;

(ii) be substantially ready for incorporation in the works;

(iii) be the property of the contractor, or

(iv) unconditionally become the property of the contractor if he takes the action mentioned above in sub-clause 54(2).

It is item (iv) that is governed by the Sale of Goods Act 1979. When ownership of goods passes from seller to buyer is a little complicated. It depends on the nature of the goods. 'Specific goods' are defined in the 1979 Act as 'goods identified and agreed upon at the time a contract of sale is made'; see section 61(1) of this Act. Now section 18, rule 1 states that:

Where there is an unconditioned contract for the sale of specific goods, in a deliverable state, the property in the goods passes to the buyer when the contract is made, and it is immaterial whether the time of payment or the time of delivery, or both, be postponed.

Non-specific or 'unascertained' goods are defined by description only and section 18, rule 5 states that:

Where there is a contract for the sale of unascertained or future goods by description and goods of that description and in a deliverable state are unconditionally appropriated to the contract . . . the property thereupon passes to the buyer.

While the words 'appropriated to the contract' are rather vague it appears to be accepted that ownership passes upon delivery to the buyer. In fact rule 5(2) says so and indicates an exception where a seller delivers the goods to a carrier or other bailee or custodier for transmission to the buyer, when ownership seems to pass on delivery to the carrier, etc.

In general building work most contracts for sale will be for 'unascertained goods', e.g. tiles from stock. If the builders merchant enters into a contract of sale with the builder for the sale of tiles from stock as there is no appropriation to the contract until delivery it is only at this stage that the property in the goods passes. If the sale had been for 'the 1000 Marley tiles stocked in my yard' this would probably be a sale of specific goods and the property would pass on the making of the contract.

In the case of engineering construction a number of purchases will be for specific goods when, as just explained, the ownership of the property passes at an early stage, i.e. on the making of the contract.

A comparatively recent case known as the 'Romalpa case'* was heard by the Court of Appeal and it has an effect on the types of transaction just mentioned, and also on goods actually on site, i.e. delivered to the site.

Dutch manufacturers of aluminium foil contracted to sell foil to a British firm Romalpa. The contract of sale provided that the suppliers of the foil retained ownership of the foil supplied until they had been paid under **all** transactions the full amount owing. Romalpa went into liquidation having used some of the foil. Some foil still remained in stock and the liquidator sold part of that foil but kept the proceeds in a separate account.

The Dutch firm claimed ownership of the foil still in stock, and the money in the liquidator's separate account. Now section 19(1) of the 1893 Act states as follows:

Where there is a contract for the sale of specific goods or where goods are subsequently appropriated to the contract, the seller may, by the terms of the contract or appropriation, reserve the right of disposal of the goods until certain conditions are fulfilled. In such case . . . the property in the goods does not pass to the buyer until the conditions are fulfilled.

As a condition of the contract of sale had not been fulfilled the foil still belonged to the sellers and thus they succeeded in their claim. Probably they would also have succeeded if they had claimed the balance of the purchase price of the foil through the payments received by Romalpa for the foil used in manufacturing.

In these circumstances it appears that suppliers of goods can incorporate the type of clause used in the Romalpa case in their contract of sale, thus reserving ownership of the goods until they have received payment.

Both contractors and engineers on behalf of their employers should be on the look-out for a section 19(1) stipulation which prevents the contractor assuming that he can comply with the circumstances listed in clause 54(2) of the ICE Conditions, believing the goods are his when they are not. Equally it behoves the engineer to examine the contract of sale in some detail. It was stated earlier that the purpose of clause 54 is to give the engineer authority to certify payment for goods and materials which have not yet reached the site, and when this occurs such goods and materials become the absolute property of the employer although in the possession of the contractor. However, the contractor in accordance with sub-clause 54(3) proviso (b) is responsible for all loss or damage to such goods and materials, *inter alia*, and must 'effect such additional insurance as may be necessary to cover the risk of loss or damage from any cause'.

* *Aluminium Industrie Vassen* v *Romalpa (1976)*. Some doubt has recently been expressed by the courts on the validity of a Romalpa type clause.

It has been suggested in respect of a similar clause in the Standard Form of Building Contract that architects should ensure that the contractor has confirmed continuation of cover by the supplier's policy and that any premium for continued cover should be paid by the suppliers.

There is no difficulty where the materials and goods are in the custody of the contractor as the insurance covering the contents of his own premises will protect such off-site materials and goods (at least up to fire and special perils cover – see later), but where they are in the custody of a manufacturer, supplier, or sub-contractor, the more satisfactory solution is for the contractor to make his own insurance arrangements in respect of these goods and materials.

In the first place, the proviso (b) in sub-clause 54(3) states that the contractor 'shall effect such additional insurance as may be necessary', not that somebody else should effect such insurance.

Secondly this proviso goes on to require the contractor 'to cover the risk of such loss or damage from **any** cause', so clearly a contractors' all risks policy is required not the fire and special perils cover, which is probably the widest cover the person holding the goods and materials will have.

The fire and special perils policy covers loss and damage by fire, lightning, explosion, storm, tempest, flood, bursting or overflowing of water tanks, apparatus or pipes, earthquake, aircraft and other aerial devices or articles dropped therefrom, riot and civil commotion.

Thirdly, apart from the legal reason (the strict wording of the clause) for the contractor to arrange his own insurance cover for off-site goods in the hands of a third party, there are other reasons. A supplier, for example, may say that he has insured the goods which are the contractor's property but still in the supplier's possession. In fact he may have only covered his personal interest (e.g. his lien upon the goods) which may only be for balance of payment, and goods held in trust may otherwise be excluded. Alternatively, while goods in trust may be covered this item in the policy of the supplier could be limited to such goods for which the insured is responsible, meaning legally responsible, and as just explained the supplier would not be legally responsible in the circumstances envisaged unless the supplier gave the contractor an indemnity or was guilty of a tort, e.g. negligence, causing loss or damage to the goods.

What is required is cover for the contractor's goods irrespective of the supplier's legal responsibility. If the words 'for which the insured is responsible' were omitted from this item in the supplier's policy covering goods in trust, then the supplier could recover the full value of the goods. However, he would hold any balance in excess of his interest as trustee on behalf of the buyer (the contractor). See *Waters v Monarch Fire and Life Assurance Co (1856)*, and *Hepburn v A. Tomlinson (Hauliers) Ltd (1966)*. The inclusion in the policy of the words just mentioned in inverted commas, if strictly interpreted, limits the amount that the insured (the supplier) can recover to the extent of his legal responsibility to the owner (the contractor) in addition to his personal interest in the goods. This is not sufficient to cover the contractor's interest and indirectly (as explained earlier) the employer's interest. In Hepburn's case there was apparently no item of goods 'in trust', in the policy concerned. However, it was a goods in transit policy taken out by a carrier endorsed as follows: 'tobacco . . . the property of the Imperial Tobacco Co (of Great Britain and Ireland), Ltd'.

The Court held that the identification of the tobacco as that of the tobacco company was not sufficient to show that their proprietary interest was insured, but the conditions which followed in the policy showed that this must be so. Enough has been said to show that there are dangers in leaving third parties to insure off-site goods. It is a matter of the construction of the supplier's policy as to whether the contractor's interest in the goods is covered and as Hepburn's case went to the House of Lords for a decision the legal expense must have been extremely high and this is to be avoided. Therefore, the contractor should make his own insurance arrangements for off-site goods.

The normal wording of the 'all risks' policy should include 'materials and goods intended for' but not yet 'delivered to and placed on the works', in accordance with sub-clause 54(3). This policy should also cover (as sub-clause 54(3) will include) the transit risks and the materials and goods while on the contractor's premises from which they will be sent to the building site, and where this is granted risks which are more specifically insured under fire and burglary policies must be excluded from the latter policies (usually insurers give cover for off-site goods and transit cover in their basic wording of the CAR policy).

Chapter 13

Subcontractors under the ICE Conditions: clauses 58 and 59; and the form of subcontract

Possibly most of the difficult problems in the law concerning contract wording in the construction industry are produced by the system of nominated subcontracting.

The doctrine of privity of contract must be understood in order to appreciate the implications of clauses 58 and 59 concerning subcontractors. Privity of contract is the relation which exists between the actual parties to a contract, as where A agrees with B to pay him a sum of money. Privity of contract is necessary to enable B to sue A on this contract. A contract cannot confer rights on a third party as only a party to a contract can sue on it. There are exceptions to this rule which do not concern the aspect under discussion. Thus if a building owner A in the terms of his contract with the main contractor B confers any benefit on a subcontractor C, then C cannot enforce the promise for his benefit as he is not a party to the contract between A and B. Similarly C is not bound in any way to conform to the requirements concerning him which are set out in the contract between A and B until he (C) enters into a separate contract either with A or B, and then of course C is only bound by the contract to which he is a party. If the terms of that subcontract, as it is called, do not conform with the requirements of the contract between A and B, that is a matter betweeen A and B only, and C cannot be involved. There is a strong line of legal authorities to support these statements from *Tweddle* v *Atkinson* in 1861 to *Beswick* v *Beswick* in 1967.

Clauses 58 and 59 are part of the contract between 'the employer' (the building owner) on the one hand and 'the contractor' (the main contractor as distinct from a subcontractor) on the other, and the following requirements of clauses 58 and 59 concerning subcontractors only apply between the employer and the contractor in accordance with the doctrine of privity of contract.

Before considering clauses 58 and 59 it is necessary to mention:

(a) Clauses 3 and 4 which govern the assignment and subletting of the contract.

(b) Clause 3 makes it very clear that the written consent of the employer is required before assignment of the contract or any part thereof can take place. In effect clause 3 refers to an assignment of liability by the contractor so that another contractor takes over liability from the contractor and is an expression of the common law position.

(c) Clause 4 does not allow the subletting of the whole of the works and only allows subletting of part of the works with the written consent of the engineer. It is legally possible to sublet part of the contract **work** (unless there is a term to the contrary in the contract), but as this clause says, this leaves the liability of the contractor to the employer unaffected.

Provisional and prime cost sums: clause 58

Clause 58 consists of the following sub-clauses:

(1) provisional sum;
(2) prime cost item;
(3) design requirements to be expressly stated;
(4) use of prime cost items;
(5) nominated subcontractors – definition;
(6) production of vouchers, etc.;
(7) use of provisional sums.

Provisional Sum
58. (1) 'Provisional Sum' means a sum included in the Contract and so designated for the execution of work or the supply of goods materials or services or for contingencies which sum may be used in whole or in part or not at all at the direction and discretion of the Engineer.

Prime Cost Item
(2) 'Prime Cost (PC) Item' means an item in the Contract which contains (either wholly or in part) a sum referred to as Prime Cost (PC) which will be used for the execution of work or for the supply of goods materials or services for the Works.

Design Requirements to be Expressly Stated
(3) If in connection with any Provisional Sum or Prime Cost Item the services to be provided include any matter of design or specification of any part of the Permanent Works or of any equipment or plant to be incorporated therein such requirement shall be expressly stated in the Contract and shall be included in any Nominated Sub-contract. The obligation of the Contractor in respect thereof shall be only that which has been expressly stated in accordance with this sub-clause.

Use of Prime Cost Items
(4) In respect of every Prime Cost Item the Engineer shall have power to order the Contractor to employ a sub-contractor nominated by the Engineer for the execution of any work or the supply of any goods materials or services included therein. The Engineer shall also have power with the consent of the Contractor to order the Contractor to execute any such work or to supply any such goods materials or services in which event the Contractor shall be paid in accordance with the terms of a quotation submitted by him and accepted by the Engineer or in the absence thereof the value shall be determined in accordance with Clause 52.

Nominated Sub-Contractors – Definition
(5) All specialists merchants tradesmen and others nominated in the Contract for a Prime Cost Item or ordered by the Engineer to be employed by the Contractor in accordance with sub-clause (4) or sub-clause (7) of this Clause for the execution of any work or the

supply of any goods materials or services are referred to in this Contract as 'Nominated Sub-contractors'.

Production of Vouchers, etc.
(6) The Contractor shall when required by the Engineer produce all quotations invoices vouchers sub-contract documents accounts and receipts in connection with expenditure in respect of work carried out by all Nominated Sub-contractors.

Use of Provisional Sums
(7) In respect of every Provisional Sum the Engineer shall have power to order either or both of the following
(a) work to be executed or goods materials or services to be supplied by the Contractor the value of such work executed or goods materials or services supplied being determined in accordance with Clause 52 and included in the Contract Price;
(b) work to be executed or goods materials or services to be supplied by a Nominated Sub-contractor in accordance with Clause 59A.

The sub-clauses here are mainly self-explanatory but the following points should be noted:

Sub-clause 58(1) defines 'provisional sum'. The engineer at his direction and discretion, **may use**, in whole or in part or not at all, a provisional sum. This includes not only work executed by the main contractor but the execution of work and the supply of goods and services by a nominated subcontractor.

Sub-clause 58(2) defines 'prime cost (PC) item'. A prime cost item **will be used** for the execution of work as well as the supply of goods materials or services. The distinction between this sub-clause and the previous one should be particularly noted. These items must be the subject of a nomination or the contractor must do the work himself – see sub-clause 4 below.

Sub-clause 58(3). If any provisional sum or prime cost item includes the provision of design or specification services concerning the permanent works or any equipment or plant to be incorporated therein, then this requirement must be expressly stated in the contract and included in any nominated subcontract. Presumably the contractor's obligation in respect of such services is limited to that which is expressly stated in the sub-clause because under clause 8 the contractor is not responsible for design of any permanent works. This is a commendable way to deal with the difficult question of design services performed by nominated subcontractors.

Sub-clause 58(4). The engineer has power to order the contractor to employ a subcontractor nominated by the engineer in respect of every prime cost item subject to the contractor's right of objection – see clause 59A. Subject to the contractor's consent, the engineer also has power to order the contractor to execute work, supply goods, materials or services in respect of any prime cost item. In this event the engineer is required to pay for such work, etc. in accordance with the terms of an agreed quotation or in the absence of this by determining the value in accordance with clause 52. The effect of this sub-clause is that express provision is made for the nomination of subcontractors, and also for the contractor to be nominated for specialist work if he is able to do such work.

Sub-clause 58(5). The definition of nominated subcontractors includes nominated suppliers.

Sub-clause 58(6). This wording has been criticised as literally it only requires production by the contractor as required by the engineer of the documents concerned in respect of nominated subcontract work which has been carried out and not for work to be carried out, and not in relation to the supply of goods materials or services (as distinct from work).

Sub-clause 58(7). It should be remembered that in the case of provisional sums the contractor has no power to refuse to carry out work or supply goods materials or services as he has in the case of prime cost items — see sub-clause 58(4) above. Sub-clause 7(a) deals with payment for work done by the contractor and 7(b) in conjunction with sub-clause 59(A)(5) with payment for work done by a nominated subcontractor.

Nominated subcontractors: clause 59

There are three main sub-clauses here. The first, 59A, deals with the procedure to nominate, the second, 59B, with the procedure on forfeiture of the subcontract, and 59C with payment of nominated subcontractors.

Clause 59A: the procedure to nominate

Sub-clause (1)   nominated subcontractors — objection to nomination;
　　　　　(2)   engineer's action upon objection;
　　　　　(3)   direction by engineer;
　　　　　(4)   contractor responsible for nominated subcontracts;
　　　　　(5)   payments;
　　　　　(6)   breach of subcontract.

Nominated Sub-contractors — Objection to Nomination
59A. (1)   Subject to sub-clause (2)(c) of this Clause the Contractor shall not be under any obligation to enter into any sub-contract with any Nominated Sub-contractor against whom the Contractor may raise reasonable objection or who shall decline to enter into a sub-contract with the Contractor containing provisions:
　　(a)   that in respect of the work goods materials or services the subject of the sub-contract the Nominated Sub-contractor will undertake towards the Contractor such obligations and liabilities as will enable the Contractor to discharge his own obligations and liabilities towards the Employer under the terms of the Contract;
　　(b)   that the Nominated Sub-contractor will save harmless and indemnify the Contractor against all claims demands and proceedings damages costs charges and expenses whatsoever arising out of or in connection with any failure by the Nominated Sub-contractor to perform such obligations or fulfil such liabilities;
　　(c)   that the Nominated Sub-contractor will save harmless and indemnify the Contractor from and against any negligence by the Nominated Sub-contractor his agents workmen and servants and against any misuse by him or them of any Constructional Plant or Temporary Works provided by the Contractor for the purposes of the Contract and for all claims as aforesaid;
　　(d)   equivalent to those contained in Clause 63.

Engineer's Action upon Objection
　　(2)   If pursuant to sub-clause (1) of this Clause the Contractor shall not be obliged

to enter into a sub-contract with a Nominated Sub-contractor and shall decline to do so the Engineer shall do one or more of the following:

    (a)   nominate an alternative sub-contractor in which case sub-clause (1) of this Clause shall apply;

    (b)   by order under Clause 51 vary the Works or the work goods materials or services the subject of the Provisional Sum or Prime Cost Item including if necessary the omission of any such work goods materials or services so that they may be provided by workmen contractors or suppliers as the case may be employed by the Employer either concurrently with the Works (in which case Clause 31 shall apply) or at some other date. Provided that in respect of the omission of any Prime Cost Item there shall be included in the Contract Price a sum in respect of the Contractor's charges and profit being a percentage of the estimated value of such work goods materials or services omitted at the rate provided in the Bill of Quantities or inserted in the Appendix to the Form of Tender as the case may be;

    (c)   subject to the Employer's consent where the Contractor declines to enter into a contract with the Nominated Sub-contractor only on the grounds of unwillingness of the Nominated Sub-contractor to contract on the basis of the provisions contained in paragraphs (a) (b) (c) or (d) of sub-clause (1) of this Clause direct the Contractor to enter into a contract with the Nominated Sub-contractor on such other terms as the Engineer shall specify in which case sub-clause (3) of this Clause shall apply;

    (d)   in accordance with Clause 58 arrange for the Contractor to execute such work or to supply such goods materials or services.

Direction by Engineer

    (3)   If the Engineer shall direct the Contractor pursuant to sub-clause (2) of this Clause to enter into a sub-contract which does not contain all the provisions referred to in sub-clause (1) of this Clause:

    (a)   the Contractor shall not be bound to discharge his obligations and liabilities under the Contract to the extent that the sub-contract terms so specified by the Engineer are inconsistent with the discharge of the same;

    (b)   in the event of the Contractor incurring loss or expense or suffering damage arising out of the refusal of the Nominated Sub-contractor to accept such provisions the Contractor shall subject to Clause 52(4) be paid in accordance with Clause 60 the amount of such loss expense or damage as the Contractor could not reasonably avoid.

Contractor Responsible for Nominated Sub-contracts

    (4)   Except as otherwise provided in this Clause and in Clause 59B the Contractor shall be as responsible for the work executed or goods materials or services supplied by a Nominated Sub-contractor employed by him as if he had himself executed such work or supplied such goods materials or services or had sub-let the same in accordance with Clause 4.

Payment

    (5)   For all work executed or goods materials or services supplied by Nominated Sub-contractors there shall be included in the Contract Price:

    (a)   the actual price paid or due to be paid by the Contractor in accordance with the terms of the sub-contract (unless and to the extent that any such payment is the result of a default of the Contractor) net of all trade and other discounts rebates and allowances other than any discount obtainable by the Contractor for prompt payment;

    (b)   the sum (if any) provided in the Bill of Quantities for labours in connection therewith or if ordered pursuant to Clause 58(7)(b) as may be determined by the Engineer;

    (c)   in respect of all other charges and profit a sum being a percentage of the actual price paid or due to be paid calculated (where provision has been made in the Bill of Quantities for a rate to be set against the relevant item of prime cost) at the rate inserted by the Contractor against that item or (where no such provision has been made) at the rate inserted by the Contractor in the Appendix to the Form of Tender as the percentage for adjustment of sums set against Prime Cost Items.

Breach of Sub-contract

(6)   In the event that the Nominated Sub-contractor shall be in breach of the sub-contract which breach causes the Contractor to be in breach of contract the Employer shall not enforce any award of any arbitrator or judgment which he may obtain against the Contractor in respect of such breach of contract except to the extent that the Contractor may have been able to recover the amount thereof from the Sub-contractor. Provided always that if the Contractor shall not comply with Clause 59B (6) the Employer may enforce any such award or judgment in full.

Sub-clause 59A(1).   Subject to sub-clause (2)(c) of clause 59A, which provides for alteration of the contract by the engineer with the employer's consent, there are two reasons allowed here for the contractor to avoid entering into a subcontract with a nominated subcontractor. In the first place the objection must be reasonable. Probably the usual type of objection under this heading involves the competency of the subcontractor to carry out the work concerned, which would include the lack of finance as well as the more usual reasons of lack of ability or plant or equipment. It is not unknown where there is a query concerning the subcontractor's financial position for the main contractor to call for a bond from the subcontractor. See chapter 17, 'Contract guarantee and performance bonds'. The alternative valid objection is that the nominated subcontractor refuses to enter into a subcontract incorporating the provisions listed in clause 59A(1)(a) to (d) – see below.

(a)   Contains the common requirement in subcontract forms that as regards the subcontract works goods materials or services the nominated subcontractor will undertake towards the contractor such obligations and liabilities as will enable the contractor to discharge his own obligations and liabilities towards the employer under the terms of the contract. It should be remembered that the subcontract completion date will probably be different from that of the main contract but the contractor must still be able to discharge his own responsibility to complete on time.

(b)   The previous paragraph (a) dealt with contractual liabilities concerning the works, etc.; it is assumed that the paragraph (b) indemnity to the contractor concerns other liabilities under the contract. It follows that the phrase 'such obligations or fulfil such liabilities' here refers to the obligations and liabilities mentioned in the previous paragraph (a).

(c)   The previous paragraphs (a) and (b) dealt with contractual liabilities, therefore it is presumed that this paragraph (c) concerning indemnity to the contractor refers to common law negligence as well as the misuse of any constructional plant or temporary works, by the nominated subcontractor his agents, workmen and servants.

(d)   The provisions here provide conditions under which the subcontract can be forfeited and they must be equivalent to the conditions of forfeiture in the main contract set out in clause 63 of that document. The procedure and circumstances under which a nominated subcontract is forfeit are set out in clause 59B. In fact the blue subcontract form published by the Federation of Civil Engineering Contractors for use with the ICE Conditions of Contract contains a clause similar to clause 63 of this main contract. Both clause 59B and the blue subcontract form will be considered later in this chapter.

Sub-clause 59A(2). If the contractor exercises his right under sub-clause 59A(1) and declines to enter into a subcontract with a nominated subcontractor the engineer under sub-clause 59A(2) has several courses open to him but these do not directly concern the subject-matter of this book and the same is true of the remainder of this clause 59A (with the exception of sub-clauses 59A(4) and 59A(6)), but the full clause is set out above for the sake of completeness.

Sub-clause 59A(4). The responsibility for the work executed or goods materials or services supplied by a nominated subcontractor, except as otherwise provided in clauses 59A and 59B, is upon the contractor as if he has himself executed such work, etc., or sublet the same under clause 4. Nevertheless, see the effect of sub-clause 59A(6) below.

Sub-clause 59A(6). If the contractor becomes liable to compensate the employer for a breach of the main contract caused by a breach by the subcontractor of the subcontract, even though the contractor did not object to the nomination, the employer cannot enforce any arbitration or court award, except to the extent that the contractor recovers the amount from the subcontractor. There is a proviso in this sub-clause that if the contractor does not comply with clause 59B(6) concerning recovery of the employer's loss from the subcontractor the employer may enforce any award or judgment against the contractor in full.

This sub-clause seems to be an avoidance of the privity of contract principle mentioned at the beginning of this chapter, which concerns the insurers of the main contractor and subcontractor respectively, in that the contractor's legal responsibility to the employer for the nominated subcontractor's work is reduced, because the employer seems to be guaranteeing the subcontractor's ability to reimburse the contractor for any claim made against the contractor by the employer for the subcontractor's breaches. However, this sub-clause is limited to the employer's own damage and does not include the contractor's own (which may be called 'private') damage caused by the subcontractor's breaches. See sub-clause 59B(4) for the latter situation. These breaches which might cause damage to the employer are, for example, delay, defective work, implied warranties of fitness for purpose or merchantable quality or even express design or performance obligations.

Insurances which might protect the employer following breaches of subcontract. While the consequences of defective work, such as damage to other parts of the insured property, would be covered by a contractors' all risks policy (which protects both the employer and the contractor — see chapter 5), the defective work itself would not be covered nor would the delay consequent upon the defective work. However, delay following damage to the works (assuming the latter is covered by an all risks policy) could be covered by an advance profits policy which would protect the employer. Implied warranties and design risks although excluded by the contractors' all risks policy could be covered by a professional indemnity policy (provided the breach involves no more than professional negligence), but this policy is issued to contractors and subcontractors (or rather to their design departments) which will protect the employer indirectly in that the subcontractor's insurance would be able to pay his dues and they would not fall on the employer. However, the professional indemnity policy, if the subcontractor has one, which is unlikely, would not cover performance guarantees as it usually only covers professional negligence not a contractual warranty. A products guarantee policy would be necessary to cover this latter risk and such policies are very difficult if not impossible for the building and allied trades to obtain.

Junior Books Ltd v *Veitchi Ltd (1982)*. It is convenient at this point to mention this recent House of Lords case which seems to alter the restricted view taken by the courts hitherto concerning economic loss as a valid claim in tort. In 1969 and 1970 the respondents had a factory built by main contractors. In 1968 the architects nominated the appellants as subcontractors to lay flooring and relied on the fact that the appellants were flooring specialists. The appellants were in breach of their duty of care to mix and lay the flooring with reasonable care and cracks in the flooring developed. As a result the respondents suffered the following items of damage or loss:

Necessary relaying or replacement of flooring,	£50,000
Storage of books during the carrying out of the work,	£ 1,000
Removal of machinery to enable the work to be done,	£ 2,000
Loss of profits due to disturbance of business,	£45,000
Wages paid to employees who cannot work,	£90,000
Fixed overheads producing no return,	£16,000
Investigation of necessary treatment of flooring,	£ 3,000

With the exception of the £50,000 required to replace the floor and possibly the £3,000 necessary to investigate the treatment required to do so, all the items could reasonably be described as items of economic or financial loss.

The majority of the court were influenced by Lord Wilberforce's statement in *Anns* v *Merton London Borough Council (1978)* when he quoted two tests which have to be satisfied. The first is 'sufficient relationship of proximity' and the second any considerations negativing, reducing or limiting the scope of the duty or the class of person to whom it is owed or the damages to which a breach of duty may give rise.

On the first test the relationship was as close as it could be short of actual privity of contract. The appellants must be taken to have known that if they worked negligently the resulting defects would at some time require remedying by the respondents expending money upon the remedial measures as a consequence of which the respondents would suffer financial or economic loss. On the second test there was nothing to restrict the duty of care arising from the proximity just mentioned. The court felt that a relevant exclusion clause might in some circumstances limit the duty of care as the defendant's disclaimer defeated the plaintiff in the Hedley Byrne case and possibly where, for example, the employer in clause 59 of the ICE Conditions guarantees the main contractor against the subcontractor's breaches, but there were no such clauses in the Junior Books case. So the decision was that where the relationship of the parties was close the duty of care was **not** limited to an avoidance by the appellant of causing foreseeable harm to persons or property other than the subject-matter of the work, by negligence, but extended to a duty to avoid causing pure economic loss consequent upon the defects in the work and to avoid defects in the work itself.

This was an appeal from Scotland but it was accepted that there was no difference between the Scots law of delict and the English law of negligence so this appeal raises a question of fundamental importance in the development of both systems of law. Incidentally the court was not told why the respondents proceeded in delict against the appellants rather than against the main contractors in contract, nor why the main contractors had not been joined as parties to the proceedings. The form of contract was not mentioned in the law report as the action was only in tort. One of the judges doubted whether the decision of the majority of the Court of Appeal in *Spartan Steel & Alloys Ltd* v *Martin & Co (Contractors) (1973)* (detailed on page 240), is correct and suggested that the dissenting judgment of Edmund-Davies L.J. is to be preferred. The majority decision in that case to limit a claim for economic loss to part only of such loss, i.e. that said to flow directly from the physical damage, always seemed unnecessarily cumbersome and uncertain. If the wrongdoer could foresee that there would be a loss of profit arising from the material damaged in the furnace when the electricity was negligently cut off, then he should also be able to foresee that the delay to the queue of material waiting for the use of the furnace would also affect production. Furthermore, to admit such claims would not open the floodgates to all and sundry who might be in contract with the owner of the furnace, which is the 'contrary to public policy' argument. There is really no logical objection to admitting all foreseeable loss suffered by the plaintiff who is directly affected by the wrongdoing. The Junior Books decision should make the loss adjuster's and claims official's interpretation of the law easier than it has been in the past because it was found difficult to know, when applying Spartan's decision to a particular claim, when liability for economic loss ceased. Now probably* all economic loss of this nature will be considered as foreseeable and thus a valid legal claim, and the difficulty with the insurance cover in the future will be that only the economic loss flowing from third-party damage will be covered by the normal public liability policy unless it is extended to cover financial loss without the accompanying physical loss or damage.

* For a full discussion of this subject see *Insurance Law Reports* Volume 6 – Theme 'Economic Loss from Careless Acts' published by George Godwin. Note in the type of claim under discussion there is damage to property in the ground thus the policy requirement of physical damage is present to support an economic loss claim.

Edmund-Davies L.J. in his dissenting judgment in Spartan's case said:

I cannot see why the £400 loss of profit [on the ruined 'melt' in the furnace] here sustained should be recoverable and not the £1767 [the further loss of profit caused by the inability to put four more 'melts' through the furnace before power was restored]. It is common ground that both types of loss were equally direct consequences of the defendant's admitted negligence, and the only distinction drawn is that the former figure represents the profit lost as a result of the physical damage done to the material in the furnace at the time when the power was cut off. But what has that purely fortuitous fact to do with legal principle? In my judgment, nothing, and [later] I consider the plaintiffs are entitled to recover the entirety of the financial loss they sustained.

Clause 59B: procedure on forfeiture of the subcontract

This clause deals with the forfeiture of the subcontract and is a long, complex clause. As it does not deal directly with insurance (if the subcontractor is bonded it would affect the insurer standing surety), it is proposed to consider only certain aspects. Financial responsibility is placed as far as possible on the person at fault. Thus either the subcontractor, if his action or omission results in forfeiture, or the contractor, if he terminates the subcontract incorrectly, is responsible. In the first case the contractor recovers from the subcontractor (subject to the latter's solvency) his own and the employer's loss. In the second case the contractor bears any extra expense and is allowed no extension of time.

Sub-clause 59B(1): forfeiture of subcontract. This sub-clause merely repeats clause 59A(1)(d) as a contractual requirement, i.e. incorporation of 'equivalent' forfeiture in subcontracts as set out in clause 63.

Sub-clauses 59B(2), (3) and (4): termination of subcontract. The sub-clause 59B(2) sets out a complicated procedure if the contractor thinks he is entitled to repudiate the subcontract. The contractor may request the employer's consent via the engineer; nevertheless it is clear that with or without consent he may terminate it. A notification to the engineer without a request to terminate may enable the employer to direct the contractor to terminate the subcontract.

In any of the above circumstances the employer:

(i)   must via the engineer renominate a subcontractor, or enter into a direct contract with another or the present contractor, or cancel the work and pay the contractor for his loss of profit (sub-clause 59B(3));

(ii) must make financial adjustments with the parties concerned necessarily due to the termination of the subcontract including the contractor's 'private' loss due to delay as part of the employer's loss (sub clause 59B(4)).

Sub-clause 59B(5): termination without consent. This sub-clause, as mentioned above, makes the contractor financially responsible if he terminates the subcontract without the employer's consent and in circumstances not entitling him to do so.

Sub-clause 59B(6): recovery of employer's loss. This sub-clause which deals with recovery of the employer's loss recognises the basic intention of clause 59B that action against the subcontractor can only be taken by the contractor, because of their privity of contract. However, other parts of this clause recognise that the contractor's freedom of action in the nomination of a subcontractor has been restricted and this brings responsibilities upon the employer. This was made clear in the case of *North West Metropolitan Regional Hospital Board* v *T.A. Bickerton and Son (1970)*. While this case concerned the standard form of building contract the situation is similar and is stated under the ICE Conditions, namely, that it is the employer's responsibility if the need to renominate arises. The intention of clause 59B seems to be that if the engineer on behalf of the employer required nomination in the first place it is assumed that the same circumstances apply if the need to renominate arises.

Clause 59C: payment to nominated subcontractors

As already mentioned this clause does not directly concern the subject of this book as it deals with direct payments to nominated subcontractors if the contractor defaults on such payments.

Chapter 14

Form of subcontract designed for use in conjunction with the ICE General Conditions of Contract Indemnities: clause 12

This form of subcontract does not state whether it is the result of negotiation between the representative bodies of the parties concerned. However, it is published by the Federation of Civil Engineering Contractors, which is one of the three bodies sponsoring the ICE Conditions of Contract. Apparently it is appropriate whether or not the subcontractor has been nominated by the employer under the principal contract. This form includes five schedules for completion by the parties. The fifth schedule concerns 'insurances'. This form also includes 'Notes for the guidance of contractors on the completion of the schedules' and these notes concerning the fifth schedule read as follows:

Reference should be made to clause 13 (Insurances) of the Sub-Contract. In completing the two Parts of the Schedule the parties should take care to ensure that all insurances required by the Main Contract are effected by one or other of them and that there is no unnecessary duplication of insurance.

Part 1 should specify insurances to be effected by the Sub-Contractor.

Part 2 should specify the policy of insurance which the Contractor is effecting in pursuance of clause 21 of the Main Contract Conditions, if it is intended that the Sub-Contractor shall have the benefit thereof. In such case his interest should be noted either generally or specifically on the policy and this Part of the Schedule should so state. If the Sub-Contractor is not to have any benefit under the policy of the Contractor, then that Part can be left blank.

The fifth schedule itself only contains two sub-headings after the main headings of 'Fifth Schedule' and 'Insurances' and they are:

Part 1 Subcontractor's insurances;
Part II Contractor's policy of insurance.

Appendix 4 gives examples of how this fifth schedule might be completed.

The relevant clauses of the subcontract form are clause 12: indemnities; clause 13: insurances, which is dealt with in the next chapter.

Indemnities
12. (1) The Sub-Contractor shall at all times indemnify the Contractor against all liabilities to other persons (including the servants and agents of the Contractor or Sub-Contractor) for bodily injury, damage to property or other loss which may arise out of or in consequence of the execution, completion or maintenance of the Sub-Contract Works and against all costs, charges

and expenses that may be occasioned to the Contractor by the claims of such persons.

Provided always that the Contractor shall not be entitled to the benefit of this indemnity in respect of any liability or claim if he is entitled by the terms of the Main Contract to be indemnified in respect thereof by the Employer.

Provided further that the Sub-Contractor shall not be bound to indemnify the Contractor against any such liability or claim if the injury, damage or loss in question was caused solely by the wrongful acts or omissions of the Contractor, his servants or agents.

(2) The Contractor shall indemnify the Sub-Contractor against all liabilities and claims against which the Employer by the terms of the Main Contract undertakes to indemnify the Contractor and to the like extent, but no further.

This is a liability clause the purpose of which is to distribute the liability for claims by third parties (including employees of the subcontractor and contractor) between the parties to the subcontract and to keep within the terms of clause 22 of the principal contract.

This clause 12 of the subcontract commences with a general indemnity by the subcontractor to the contractor against all such third-party claims mentioned in the previous paragraph. However, a proviso follows to the effect that the contractor is not entitled to this indemnity if he has the right by the terms of the principal contract to an indemnity from the employer. Similarly the contractor is not entitled to this indemnity if the injury loss or damage was caused **solely** by the wrongful acts or omissions of the contractor his servants or agents. The effect of the word 'solely' where third parties sue both the main contractor and the sub-contractor similar to the case of *A.M.F. International Ltd* v *Magnet Bowling Ltd (1968)* (page 80), is to leave the indemnity by the subcontractor to the main contractor unaffected. An exception appears under sub-clause 12(2) whereby the contractor must indemnify the subcontractor against all liabilities and claims in circumstances where the employer has to indemnify the contractor under the main contract.

These indemnities by the employer are usually found in clause 22(2) of the principal contract, which refers back to the proviso to clause 22(1), which generally speaking cover claims which are inherently and/or inevitably resulting from the carrying out of the works.

The result of clause 12 is to bring into effect in the subcontract all the principal contract indemnities so that where the contractor has an indemnity from the employer, the subcontractor if claimed upon in the first place, can pass the claim on to the contractor who in turn can pass it on to the employer. On the other hand, where the employer has the indemnity under the principal contract he can pass it to the contractor who in turn will pass it on to the subcontractor under clause 12. The only qualification is that the claim must arise out of or in consequence of the execution, completion or maintenance of the **subcontract works**. See later clause 14 for details of the maintenance obligation in chapter 15.

Insurances required under the form of subcontract: clause 13

Insurances

13. (1) The Sub-Contractor shall effect insurance against such risks as are specified in Part I of the Fifth Schedule hereto and in such sums and for the benefit of such persons as are specified therein and unless the said Fifth Schedule otherwise provides, shall maintain such insurance from the time that the Sub-Contractor first enters upon the Site for the purpose of executing the Sub-Contract Works until he has finally performed his obligations under Clause 14 (Maintenance and Defects).

(2) The Contractor shall maintain in force until such time as the Main Works have been completed or ceased to be at his risk under the Main Contract, the policy of insurance specified in Part II of the Fifth Schedule hereto. In the event of the Sub-Contract Works, or any Constructional Plant, Temporary Works, materials or other things belonging to the Sub-Contractor being destroyed or damaged during such period in such circumstances that a claim is established in respect thereof under the said policy, then the Sub-Contractor shall be paid the amount of such claim, or the amount of his loss, whichever is the less, and shall apply such sum in replacing or repairing that which was destroyed or damaged. Save as aforesaid the Sub-Contract Works shall be at the risk of the Sub-Contractor until the Main Works have been completed under the Main Contract, or if the Main Works are to be completed by sections, until the last of the sections in which the Sub-Contract Works are comprised has been completed, and the Sub-Contractor shall make good all loss of or damage occurring to the Sub-Contract Works prior thereto at his own expense.'

(3) Where by virtue of this clause either party is required to effect and maintain insurance, then at any time until such obligation has fully been performed, he shall if so required by the other party produce for inspection the appropriate policy of insurance together with receipts for premiums payable thereunder and in the event of his failing to do so, the other party may himself effect such insurance and recover the cost of so doing from the party in default.

The first point which would probably occur to the insurance official is the assumption that the subcontractor will have several insurances but the contractor will apparently have only one policy. The various policies available were described in chapter 4, and as far as construction contracts are concerned they fall into two categories. The material damage policy covers the works, and the policy concerned in the ICE Conditions is the contractors' all risks policy. The other category is the legal liability policies of which there are two main policies: the employers' liability policy; and the public liability policy.

It would therefore be logical to expect both part I and part II of the fifth schedule to refer to these policies of insurance. It is true that a contractors' combined policy, as explained in chapter 4, covers in one document under three different sections risks which are normally covered by several policies, namely material damage and legal liabilities. Perhaps this is what the drafters of part II had in mind in referring to the 'Contractor's policy of insurance'. The only other assumption is that clause 13 only applies to insurance of works. However, this is hardly likely as the preceeding clause clearly envisages legal liabilities and moreover part I of the fifth schedule refers to the subcontractor's insurances in the plural.

Sub-clause 13(1)

As sub-clause 13(1) only refers to part I of the fifth schedule it only applies to those insurances required from the subcontractor and it is arguable that as the contractor under the main contract is responsible for insuring the works (including presumably the subcontract works) subject to the excepted risks in sub-clause 20(3) of the main contract, no insurance of the subcontract works is usually required from the subcontractor. However, even on this argument, he would be obliged to have the usual employers' and public liability policies because of the indemnity mentioned in clause 12 and also because of the statutory requirement mentioned in chapter 10. If the subcontractor had a combined employers' and public liability policy it would be the subcontractor who had only one policy of insurance not the contractor. Nevertheless, when one comes to part II it is clear that the subcontractor may have to cover the subcontract works under his own contractors' all risks policy in addition to the usual liability policies just mentioned.

Apart from indicating that the insurances to be effected by the subcontractor are those specified in part I of the fifth schedule, sub-clause 13(1) also specifies that part I will detail the sums and persons to be insured. Presumably as this part, unlike part II, is certainly dealing with liability policies the intention is to indicate a limit of indemnity for any one occurrence under the subcontractor's public liability policy, bearing in mind that the employers' liability policy is normally unlimited or if the statute concerned is followed a figure of two million pounds applies. Should it be decided that the subcontractor should also take out a contractors' all risks policy on the subcontract works then this must be indicated in part I with the appropriate sum insured. However, unless the subcontract works are not included in the main contractors' all risks policy it is difficult to envisage circumstances where it would be necessary for the subcontractor to take out such a policy on the subcontract works, as normally this would only result in the payment of a double premium for the subcontract works, and according to the guidance notes duplication of insurance is to be avoided.

If a separate policy is not required from the subcontractor(s) as just indicated, it is important, if it is intended that the subcontractor(s) should be indemnified in like manner to the main contractor under the latter's policy, that the subcontractor's name(s) should appear as joint insured for their respective contract works.

Sub-clause 13(1) concludes by indicating the period of the subcontractor's insurances, i.e. from the time the subcontractor first enters upon the site to execute the subcontract works until he has finally performed his obligations under clause 14 (maintenance and defects).

Sub-clause 13(2)

Sub-clause 13(2) deals with the insurance that is required to be maintained by the contractor and the statement reads:

> '*The Contractor shall maintain in force until such time as the Main Works have been completed or ceased to be at his risk under the Main Contract, the policy of insurance specified in Part II of the Fifth Schedule hereto.*'

So again the stipulation appears to be for **one** policy and the policy the drafters have in mind seems to be the contractors' all risks policy covering the works, etc. (no thought is apparently given to the legal liability policies also required by the contractor), as the sub-clause continues as follows:

> '*In the event of the Sub-Contract Works, or any Constructional Plant, Temporary Works, materials or other things belonging to the Sub-Contractor being destroyed or damaged during such period in such circumstances that a claim is established in respect thereof under the said policy, then the Sub-Contractor shall be paid the amount of such claim or the amount of his loss, whichever is the less, and shall apply such sum in replacing or repairing that which was destroyed or damaged.*'

As already stated, it could be that some type of combined policy including the legal liability risks (employers' and public liability) as well as the contractors' all risks policy was in the mind of the drafters but probably not. In any event it would have been as well to have indicated that the liability policies also have to be provided by the contractor.

Sub-clause 13(2) should be read in conjunction with sub-clause 14(1) of this Subcontract Form and clauses 20 and 21 of the principal contract. The last sentence of sub-clause 13(2) explains that apart from the insurance requirements of the first part of sub-clause 13(2),

> '*the Sub-Contract Works shall be at the risk of the Sub-Contractor until the Main Works have been completed under the Main Contract, or if the Main Works are to be completed by sections, until the last of the sections in which the Sub-Contract Works are comprised has been completed, and the Sub-Contractor shall make good all loss of or damage occurring to the Sub-Contract Works prior thereto at his own expense.*'

So, subject to what has just been said, all accidental or other damage, even after completion of his own work is the responsibility of the subcontractor. This statement is echoed in sub-clause 14(1) (concerning maintenance and defects), which adds an exception where defects are caused by the act neglect or default of the employer or contractor or their servants or agents.

Sub-clause 20(3) of the principal contract specified a list of 'excepted risks' in respect of which a contractor is not required to insure the works under clause 21. If these risks occcur and cause damage to the works reinstatement will be at the employer's cost. It should be remembered that the excepted risks are, generally speaking, war and kindred risks, radioactive contamination and that arising from explosive nuclear assemblies, sonic waves, damage done by the employer (and others not employed by the contractor) while in occupation and damage due to the design of the works.

While no specific mention is made of the excepted risks in the subcontract clauses, it is clear that the intention is that the subcontractor should be entitled to payment in the event of the excepted risks causing damage to the subcontract works, as the contractor will be paid by the employer and should pass the payment on to the subcontractor.

As regards risks other than the excepted risks under clause 21 of the principal contract, the contractor has to insure the works against these risks in the joint names of himself and the employer. Nevertheless, the decision given by the contractor in part II of the fifth schedule decides whether the subcontractor is expected to insure his subcontract works. There is one exception which sometimes appears in the contractors' all risks policy and which is in contradiction to the requirements of sub-clause 13(2) already quoted above. This concerns damage to the tools, plant and appliances belonging to subcontractors or other employees. The phrase in sub-clause 13(2) reads: 'In the event of the Sub-Contract Works, or any Constructional Plant Temporary Works, materials or other things belonging to the Sub-Contractor being destroyed or damaged . . . the Sub-Contractor shall be paid the amount of such claim or the amount of such loss . . .'

If by part II of the fifth schedule the contractor is expected to insure the constructional plant of the subcontract works he should ensure that the exclusion (if it appears on his policy) is removed, probably at the cost of an additional premium.

The question of including subcontractors as joint insureds in the contractors' all risks policy has already been mentioned. However, it should be appreciated that if the subcontractors are not included in this way the insurers will have subrogation rights against the subcontractors if damage is caused by the negligent actions of the subcontractors. Furthermore, the public liability policies of the subcontractors will not operate to protect them completely in this respect as damage to subcontract works is excluded in one way or another as are parts of the works on which they are working or which are under their charge or control. See chapter 8, 'The public liability policy'. Therefore, the subcontractor would be forced to take out his own contractors' all risks policy with the duplication of coverage mentioned earlier and possibly difficulties of settlement and even litigation.

Sub-clause 13(3)

Sub-clause 13(3) states that either party to the subcontract who is required to effect insurance shall if so required by the other party produce the appropriate policy, together with receipts for premiums payable thereunder. If he fails to do so the other party may effect such insurance and recover the cost of doing so from the party in default.

Clause 13 of the ICE subcontract form: summary of the main risks remaining uninsured

To avoid repetition attention is drawn to the introductory paragraphs of the summaries at the ends of chapters 5 and 8, which are applicable here as far as basic policy cover is concerned.

Insurance of the subcontract works

The subcontractor's obligation to complete his own works at his own risk (assuming they are not covered, under the fifth schedule of the subcontract form, by the main contractor's CAR policy) and maintain them until substantial completion of the main contract (see clause 13(2) and clause 14(1) of the ICE subcontract), means that it would be prudent to insure these works on an all risks basis. In this connection the problems concerning cover set out on page 74 would also apply to the subcontractor's insurance of the subcontract works, particularly item (2), as a specialist subcontractor may be responsible for his design work. Thus it is clear from clauses 8, 58(3) and 59A(6) of the main contract that it is not the intention to make the contractor responsible for the subcontractor's design services unless expressly so stated. So the employer will suffer if anybody does, from the sub-contractor's faulty work.

However, where there are provisional items or prime cost items which include design then, subject to the contractor being informed at the time of tender and this design responsibility being incorporated into the subcontract, the employer will have a right of recovery against the contractor who will then be able to recover from the nominated subcontractor. Clause 59A(6) provides that if a nominated subcontractor is in breach of the subcontract and that breach results in the contractor being in breach of the main contract, then the employer has no redress from the contractor in excess of any amount that the contractor has recovered from the subcontractor. This is always subject to the contractor under clause 59B(6) taking all necessary steps against the subcontractor. As an alternative to the use of the ICE clauses just mentioned, the employer could enter into a form of warranty with the nominated subcontractor and then the employer would have the benefit of a contract directly with the subcontractor which would provide a right of recovery in the event of, for example, defective design.

As regards insurance protection for the employer, reference should be made to the heading in chapter 13 entitled 'Insurances which might protect the employer following breaches of subcontract' for details (page 137). It may be disappointing to those outside insurance that insurers do not cover the replacement of defective design which is not due to negligence (i.e. a contractual liability, not a professional negligence liability). It is questionable whether the insurance market should take on the design risks which may contain a high possibility of physical damage to the project. Furthermore, wider design cover is unlikely to appear if the risks placed before the insurance market are all of a very hazardous type. Usually underwriters can only accept insurances if the risks presented are of varying degrees of hazard.

Insurance of the subcontractor's liabilities to third parties

If the main contractor's CAR policy includes the subcontract works and the subcontractor's responsibility therefor and if that CAR policy has a third party section, it is possible that the main contractor's third party cover will include the subcontractor's liabilities to third parties and the fifth schedule of the subcontract should indicate this. However, the usual situation is that the subcontractor has to provide his own public liability policy (in fact he should have an annual policy) and then the problems indicated in the summary at the end of chapter 8 concerning the main contractor's public liability cover also apply to the subcontractor's public liability cover, so far as matching the subcontract requirements are concerned. For present purposes the words 'contractor' and 'employer' in the chapter 8 summary will have to be read as 'subcontractor' and 'contractor' respectively.

The insurance of the subcontractor's liability to his employees is compulsory by law in the United Kingdom. See page 116 *et seq*.

Chapter 16

Sureties under the ICE Conditions: clause 10

Sureties

10. If the Tender shall contain an undertaking by the Contractor to provide when required 2 good and sufficient sureties or to obtain the guarantee of an Insurance Company or Bank to be jointly and severally bound with the Contractor in a sum not exceeding 10 per cent of the Tender Total for the due performance of the Contract under the terms of a Bond the said sureties Insurance Company or Bank and the terms of the said Bond shall be such as shall be approved by the Employer and the provision of such sureties or the obtaining of such guarantee and the cost of the Bond to be so entered into shall be at the expense in all respects of the Contractor unless the Contract otherwise provides. Provided always that if the form of Bond approved by the Employer shall contain provisions for the determination by an arbitrator of any dispute or difference concerning the relevant date for the discharge of the Sureties'/Surety's obligations under the said Bond:

> (a)   the Employer shall be deemed to be a party to the said Bond for the purpose of doing all things necessary to carry such provisions into effect;
>
> (b)   any agreement decision award or other determination touching or concerning the relevant date for the discharge of the Sureties'/Surety's obligations under the said Bond shall be wholly without prejudice to the resolution or determination of any dispute or difference between the Employer and the Contractor pursuant to the provisions of Clause 66.

Apart from the ICE Conditions of Contract, the common standard building contract forms do not stipulate the provision of a performance bond by the contractor, but this is a requirement which can be and is requested from time to time under all types of construction contract.

The ICE Conditions make provision for sureties in clause 10 by means of the tender which contains an undertaking by the contractor to provide, when required, a 10% bond for the due performance of the contract. The 'form of tender' in the conditions contains such an undertaking and the appendix to the form of tender provides the amount of the bond as a percentage of the tender total. Clause 10 allows for two sureties (see the next chapter on individual sureties) or the guarantee of an insurance company or bank. While the terms of the bond are to be approved by the employer there is a form of bond at the end of the Conditions which the 'form of tender' requires to be used. The cost of the bond is an expense of the contractor unless the contract otherwise provides. However, as the contractor knows at the time of tender that a bond is required, he is able to include in his tender figure any necessary charge for the provision of the bond. In effect, therefore, the principal defrays the cost and it is not a charge on the contractor.

If a bond, although required, is not provided, it seems from the South African case of *Swartz and Son (Pty) Ltd* v *Wolmaransstadt Town Council (1960)* that the employer would be entitled to repudiate the contract. In practice acceptance of the tender is usually conditional on a bond being provided within a certain time limit which should lapse before work commences.

It is not unknown for contractors to commence or attempt to commence work before the bond has been provided but the employer should not permit this to happen. When a contractor does start work and is then unable to provide a bond, he may be ordered off the site with consequential financial loss.

Performance bond wording

Form of Bond

BY THIS BOND ¹We ....

¹ Is appropriate to an individual, ² to a Limited Company and 3 to a Firm. Strike out whichever two are inappropriate.

of .... in the

County of .... ²We .... Limited

whose registered office is at .... in the

County of .... ³We ....

and .... carrying on business in partnership under

the name or style of ....

at .... in the

⁴ Is appropriate where there are two individual Sureties, 5 where the Surety is a Bank or Insurance Company. Strike out whichever is inappropriate.

County of .... (hereinafter called " the Contractor ") ⁴and ....

.... of ....

in the County of .... and ....

of .... in the County of ....

.... ⁵and .... Limited

whose registered office is at .... in the

County of .... (hereinafter called " the ⁴Sureties/Surety ") are held and firmly

bound unto .... (hereinafter

called " the Employer ") in the sum of .... pounds

(£ ....) for the payment of which sum the Contractor and the ⁴Sureties/Surety bind

themselves their successors and assigns jointly and severally by these presents.

Sealed with our respective seals and dated this .... day of ....

19 ....

WHEREAS the Contractor by an Agreement made between the Employer of the one part and the Contractor of the other part has entered into a Contract (hereinafter called "the said Contract ") for the construction and completion of the Works and maintenance of the Permanent Works as therein mentioned in conformity with the provisions of the said Contract.

NOW THE CONDITIONS of the above-written Bond are such that if:—

(a)   the Contractor shall subject to Condition (c) hereof duly perform and observe all the terms provisions conditions and stipulations of the said Contract on the Contractor's part to be performed and observed according to the true purport intent and meaning thereof or if

(b)   on default by the Contractor the Sureties/Surety shall satisfy and discharge the damages sustained by the Employer thereby up to the amount of the above-written Bond or if

(c)   the Engineer named in Clause 1 of the said Contract shall pursuant to the provisions of Clause 61 thereof issue a Maintenance Certificate then upon the date stated therein (hereinafter called " the Relevant Date ")

this obligation shall be null and void but otherwise shall remain in full force and effect but no alteration in the terms of the said Contract made by agreement between the Employer and the Contractor or in the extent or nature of the Works to be constructed completed and maintained thereunder and no allowance of time by the Employer or the Engineer under the said Contract nor any forbearance or forgiveness in or in respect of any matter or thing concerning the said Contract on the part of the Employer or the said Engineer shall in any way release the Sureties/Surety from any liability under the above-written Bond.

PROVIDED ALWAYS that if any dispute or difference shall arise between the Employer and the Contractor concerning the Relevant Date or otherwise as to the withholding of the Maintenance Certificate then for the purposes of this Bond only and without prejudice to the resolution or determination pursuant to the provisions of the said Contract of any dispute or difference whatsoever between the Employer and Contractor the Relevant Date shall be such as may be:—

(a) agreed in writing between the Employer and the Contractor or

(b) if either the Employer or the Contractor shall be aggrieved at the date stated in the said Maintenance Certificate or otherwise as to the issue or withholding of the said Maintenance Certificate the party so aggrieved shall forthwith by notice in writing to the other refer any such dispute or difference to the arbitration of a person to be agreed upon between the parties or (if the parties fail to appoint an arbitrator within one calendar month of the service of the notice as aforesaid) a person to be appointed on the application of either party by the President for the time being of the Institution of Civil Engineers and such arbitrator shall forthwith and with all due expedition enter upon the reference and make an award thereon which award shall be final and conclusive to determine the Relevant Date for the purposes of this Bond. If the arbitrator declines the appointment or after appointment is removed by order of a competent court or is incapable of acting or dies and the parties do not within one calendar month of the vacancy arising fill the vacancy then the President for the time being of the Institution of Civil Engineers may on the application of either party appoint an arbitrator to fill the vacancy. In any case where the President for the time being of the Institution of Civil Engineers is not able to exercise the aforesaid functions conferred upon him the said functions may be exercised on his behalf by a Vice-President for the time being of the said Institution.

Signed Sealed and Delivered by the said ⎫
   in the presence of:— ⎭

The Common Seal of ⎫
                    LIMITED ⎬
was hereunto affixed in the presence of:— ⎭

(*Similar forms of Attestation Clause for the Sureties or Surety*)

Taking in turn each paragraph of the wording of the form of bond given in the ICE Conditions of Contract as set out above, the following comments should be noted. Note also three amendments numbered 1 to 3 which follow.

(i)   The first paragraph sets out the three parties to the contract and the pecuniary limit of the bond. Provision is made for the contractor being either an individual or a limited company or a firm, and similarly provision is made for two individual sureties (see the next chapter on this aspect) if these are obtained and alternatively for a bank or insurance company. The wording in appendix 2 obviously does not cater for the alternatives to an insurer just mentioned.

(ii)   The second paragraph identifies the contract concerned. Some wordings then say 'which contract with all its terms and conditions is hereby made a part of the agreement', thus making the contract a part of the bond and ensuring that all parties to the bond have the same rights and duties that the contract provides.

(iii)   The third paragraph states that if the contract is completed, as required, by the contractor or if he defaults and the surety satisfies and discharges the damages sustained by the employer, up to the amount of the bond, then the obligation under the bond shall be null and void. Consequently, the contractor and surety are released from all liabilities under the bond. The appendix 2 wording is slightly different here but the effect is the same. There is also an indulgence (or forbearance) clause. This gives the employer the right to alter the terms of the contract by agreement with the contractor or the extent or nature of the work, and even allows time or forbears or forgives any matter or thing concerning the contract without releasing the surety from liability under the bond. Without this clause the common law would operate to protect the surety's right and any variation of the contract without the surety's agreement would probably release the surety from liability. This clause appears in appendix 2.

(iv)   The wording in appendix 2 caters for the release of the surety upon the happening of whichever shall last occur of the following events:
      (a)   the issue of a certificate of practical completion of the works by the architect or engineer; or
      (b)   the subsequent expiration of any defects liability period or periods of maintenance provided by the contract.
Furthermore, any suits at law or proceedings in equity against the surety to recover any claim under the bond must be instituted within six months after the date of the relevant event or be absolutely barred.
   See amendment 1 below concerning release of the surety under the latest ICE wording.

(v)   Finally there is an attestation clause which provides for the seals of both the contractor and the surety.

The fifth edition of the ICE Conditions was revised in January 1979 and the following amendments are applicable to this chapter:

(1)   The amendment to the form of bond provides for its release at the date of expiration of the period of maintenance when the work is to the engineer's satisfaction (see clause 61(1)). This amendment was made because the previous form of bond contained no date for its release. This means that if the contractor becomes insolvent that any defects which arise after the release of the surety will be the employer's responsibility.

(2)   As a result of the amendment to the form of bond, an amendment was made to clause 10 to provide for the employer to be deemed a party to the bond in the event of a dispute arising over the relevant date for the discharge of the surety's obligation under the bond. A general 'without prejudice' provision has been added with the intention of making it clear that a determination of the relevant date or any other decision concerning the grant or withdrawal of the maintenance certificate will in no way affect any subsequent dispute or difference covering the same ground and falling to be decided under clause 66 (arbitration).

(3)   An amendment was made to clause 60(5)(a) (payment of retention money) because as previously drafted, its application could result in a situation whereby, when retentions are released in respect of 'sections', the amount of retention money left in the hands of the employer could be less than one half of the reten-

tion fund as envisaged by clause 60(5)(b). The amendment has been made to ensure that the aggregate sums that may be released under the provisions of clause 60(5)(a) do not exceed one half of the retention money under the contract. As to the connection between this clause and the performance bond, see later.

If the contractor fails to meet his obligations, the bond becomes enforceable and the surety will be called upon to satisfy the damages sustained by the obligee (the employer), which will be the additional cost of completing the contract, up to the amount of the fixed penalty (the bond amount).

Most major works are carried out under contracts providing for payments to be made as the work proceeds. In the ICE Contract clause 60(2) it is clear that the period of interim statements is one month. By clause 60(4), the retention is 5% of the amount of the statement up to 5% of the tender amount not exceeding £1500 or, where the tender exceeds £50,000, 3% of the tender total. This fund which is owed to the contractor, i.e. the retention, will be set off against the cost of any default by the contractor and this reduces the claim against the surety. This right exists at common law but usually is given in any counter indemnity signed by the contractor, details of which are given in the next chapter and in appendix 3.

The wording of the form of bond in the ICE Conditions is enforceable only on failure to perform and proof of damage and **not** on demand as in some bank guarantees which tend to be used in international contracts. For an example of the operation of an on demand guarantee without proof of any default, see the case of *Edward Owen Engineering Ltd* v *Barclays Bank International Ltd (1977)* detailed in the next chapter. The case of *General Surety and Guarantee Co* v *Francis Parker Ltd (1977)* deals with the presumption against such an absolute requirement where the wording is not sufficiently explicit.

Chapter 17

Contract guarantee and performance bonds

Introduction and background

In connection with many contracts awarded to contractors for performance of works (whether a standard form of contract requiring a bond applies or not), it is a normal and necessary protection for the principal (the employer in the ICE Contract) to obtain security from the contractor guaranteeing the satisfactory completion of the work. If a bond or some other form of security is not obtained a default on the part of the contractor could cause a loss to the principal – in the case of public authorities (which includes local authorities), a loss to the ratepayers.

A number of well-known main contractors are now insisting that nominated subcontractors (and even their own subcontractors) provide them with security by the provision of a bond in similar manner to the obligation of the main contractor to the employing authority. In these circumstances, the main contractor is the principal.

There are several different ways of arranging security, for example, by means of a cash deposit, or the provision of a surety.

A surety is a person who binds himself usually by a bond to satisfy the obligation of another person, if the latter fails to do so. A bond is a contract under seal. Historically, a bond is an unqualified promise to pay a sum of money subject to the condition that if a certain event happens the promise shall be void.

The stability of the party providing the bond must be beyond question, but unfortunately private sureties can die, disappear or become bankrupt. It has been argued that there is a doubt as to the validity of private sureties in the case of performance bonds (in spite of the mention of two individual sureties in the footnotes to the 'form of bond' in the ICE Conditions), as the Insurance Companies Act 1981 in section 2 states that **no person** shall carry on any insurance business in the United Kingdom unless authorised to do so under section 3 or 4.

Section 3 gives the Secretary of State power to authorise a body to carry on in the UK such of the classes of insurance business specified in Schedule 1 or 2 of the Act, or such parts of those classes as may be specified in the authorisation. Furthermore, section 7 limits the authorisation to a company defined in section 455 of the Companies Act 1948, a registered society or a body corporate established by royal charter or Act of Parliament and already authorised to carry on insurance business, and there are other limitations. Finally, Schedule 2 Part I item 15, headed 'Suretyship', refers *inter alia* to performance bonds. Presumably banks arranging bonds in the normal course of banking business would not be affected by the 1981 Act. In America federal law precludes banks from being involved in domestic surety business. In Europe banks are generally speaking allowed to transact domestic surety business.

When an insurance company issues a surety bond it does not affect a contractor's banking lines of credit. There are, of course, occasions when only a banker's guarantee will be accepted by the employer, particularly with projects in certain Middle East countries. It is for these occasions that the contractor's bank bond facilities should be retained and utilised. It must also be considered that, in the main, banks in this country will not provide bonds where the employer is in the private sector. If an insurer is prepared to bond the contractor it does not matter whether the employer is in the public or private sector.

For this reason and those mentioned later, public authorities, etc. now prefer the provision of the bond to be made by the insurance market where there are a small number of well-established and experienced insurers prepared to underwrite the business. The principal is relieved from making those searching enquiries which are necessary to decide if the contractor is suitable to carry out the work. It is true, of course, that the most thorough examination cannot exclude the possibility of default (large and well-known firms have been forced into liquidation). Therefore, a performance bond issued by a reputable insurer is the accepted safe method by which a principal can be protected against loss. Thus, performance bonds are a vital safeguard to any principal against losses which inevitably follow the default of a contractor and, unfortunately, the construction industry ranks high in the lists of insolvencies.

While banks often issue a form of financial guarantee (which is an undertaking to pay on demand) by the terms of the insurance bond the insurer accepts responsibility for the performance of the obligations under the contract.

It is often simpler and quicker to obtain a bond from a bank because of their general knowledge of a client's affairs and because of their less complex underwriting approach to bonds, but banks are inclined to regard bonds they issue as part of their customer's credit facilities. Insurers therefore argue that banks only treat the transaction as they would any financial transaction (e.g. an overdraft), i.e. they want adequate security whereas insurers carry out very thorough investigations all of which are of benefit to the employer as well as the insurer, and may at times assist the contractor. It follows therefore that any contractor who uses his bank for bond requirements does so at the expense of possible working capital (advances which the bank could otherwise make had it not got the bond commitment). The modern form of bond normally provides that the sum of money payable by the surety shall be the actual amount of loss suffered by the employer up to a maximum limit of the amount of the bond. The event which defeats the promise in the case of construction contractors is the fulfilment of the contractor's obligations under the contract. Where a bond is given in this country the courts will enforce it only if the person to whom the bond is given can show that he has suffered loss by the default of the contractor.

A bond is usual rather than a simple contract of guarantee as the latter would require proof of legal consideration as between the surety and the obligee, i.e. the employer in the ICE Contract, and the use of a bond has long since become the accepted practice. Being under seal, a bond does not require consideration. A bond wording is given in appendix 2. Another wording is given on the last page of the ICE Conditions of Contract, fifth edition. See previous chapter.

Although the normal amount of a performance bond is 10% of the contract price there is a trend to increase this figure as it is doubtful whether 10% is sufficient. Perhaps 20% may become more usual if inflation continues.

In some overseas countries a 100% bond may be required, although insurers are usually reluctant to agree these.

The cost of performance bonds is relatively low, although it does vary according to circumstances. The stability and experience of the contractor are of prime importance, but other features may affect rating, e.g. bond percentage, the size of retention, and period of contract.

As with all insurance contracts the underwriter assesses the risk and he will charge a higher premium for those risks which he considers to be more hazardous.

Underwriting considerations

The following aspects are investigated:

(a)   the technical ability of the contractor;
(b)   the usual type of contract undertaken by the contractor;
(c)   the contractor's tender amount;
(d)   a progress report on the contracts already bonded by the insurer for the contractor;
(e)   whether the contractor's business is well run and successful;
(f)   the whole financial position will be considered, the extent of the contractor's indebtedness, and whether overdrafts and loans are reasonable in relation to the size of the business. The amount of liquidity is vital.

In the case of a first application a standard questionnaire is completed by the contractor and the last three years' audited accounts are made available to the insurers. If the contractor is a member of a group of companies, the same information is required from the group.

The object of these investigations is to discover whether the contractor satisfies the following criteria, that is, whether the contracting firm:

(1)   is run competently;
(2)   has a record of completing contracts from the technical and profitability viewpoints as well as having adequate resources of manpower, plant and equipment;
(3)   has sufficient financial resources to carry out the operations of the business after considering the possibility of unknown future contingencies.

Many contractors in the construction industry overtrade; their financial resources may be inadequate to meet their probable commitments. With luck they survive but if, for example, they meet with unexpected problems on a contract or they are faced with unanticipated higher interest charges on their loans, they may encounter serious cash flow problems and at the worst be unable to carry on their business.

Counter indemnities

Although there is a common law right of indemnity from the bonded contractor in favour of the surety company, there is a growing practice among sureties to require a specific imdemnity form to be sealed by the bonded contractor. A counter indemnity should always be required as a matter of course, from the ultimate parent company of the contractor. A wording of a form of counter indemnity (or guarantee) is given in appendix 3.

It is important in the case where the contractor is a subsidiary company that the insurer surety should obtain a counter guarantee from the holding company, especially as the subsidiary is usually a limited liability company and consequently a separate legal entity. The main reason is that it cannot be assumed that the holding company will support the subsidiary as it may be economically sound for the parent company to allow the subsidiary to go into liquidation rather than continue to incur losses. Where a proper counter guarantee is taken and the contractor encounters difficulties, the holding company will either have to support the subsidiary financially or reimburse the surety and, in order to keep the reimbursement figure to a minimum, the parent company will no doubt assist in the completion of the contract with the least possible delay. If the parent company will not provide the counter guarantee required, the insurer will have to decide whether the risk is acceptable on any other basis, e.g. a form of collateral security.

Additionally, where the directors have substantial personal shareholdings, or where a substantial part of a contractor's capital comprises directors' loans or undrawn remuneration, the surety should require each director to sign a personal guarantee.

In essence, a counter indemnity is a written undertaking given by the contractor or some other party in consideration of the surety issuing a bond to the principal to repay to the surety all losses, costs, charges and expenses which the surety may be called upon to pay or discharge under or by virtue of the bond to be issued. Where personal guarantors are required, the signatories accept personal responsibility to fulfil the terms of the guarantee.

Most counter indemnities require the reimbursement of monies which the insurer has paid in respect of its obligations under the bond. On the other hand, some insurers may attempt to introduce a form of 'on demand' wording which requires reimbursement before the insurers have actually made any payment. However, the case of *General Surety and Guarantee Company Ltd* v *Francis Parker Ltd (1977)* makes it clear that the wording has to be precise before there is a liability to pay on demand. The court commented that a simple way of dealing with this matter is to provide that as between the contractor and the insurer a demand by a principal should be conclusive evidence of the insurer's liability to the principal.

Acceptability

The insurers, in practice, will set their standards for the acceptance of risks and regrettably many applications will prove to be unacceptable. Sometimes the decision can be changed by the offer of additional security from the contractor. A charge on the business is not usually possible because the bank will already have this as a security for the overdraft. However, the directors or principals (or their wives) often have personal assets, property or land, which are unencumbered and these can be offered as a collateral security. Sometimes the equity in the principal's own private dwelling house may be offered, but this is less than satisfactory. If it is acceptable it is necessary for the principal's wife to join in the guarantee.

Occasionally a substantial cash deposit may be available. However, this has the effect of reducing the contractor's working capital and can only exacerbate his difficulties. If one insurer records a declinature, the contractor is forced to try other underwriters or he may persuade his bank to offer assistance.

If a bond cannot be secured it means that there is a serious doubt about the contractor's ability to complete the contract and the employer must either award the contract to another or take the risky course of allowing the contractor to proceed without a bond being in force.

Bulk schemes

Sometimes insurers come to an agreement with a local authority or other regular employer of contractors to have the first right to provide the bond required from any contractor whose tender is accepted by the employer. In return the insurer quotes a special advantageous rate of premium.

At one time, it was thought that agreements such as these bulk contract guarantee schemes would have to be registered under the Fair Trading Act 1973. However, while it is true that the contractor has no freedom of choice of insurer, as already mentioned, he is not in fact paying for the bond. The real parties to the agreement are the employer and the insurer, and the employer clearly has a choice and need not enter into such an agreement unless it is to his advantage, i.e. he obtains a very satisfactory rate of premium. In return the insurer gets bulk business. It is also arguable that public and local authorities are not strictly speaking 'persons who carry on business' within the meaning of the 1973 Act.

On 2 July 1976 the district and county councils associations issued a directive to their members as follows: if the employer nominates the bond guarantor in the tender documents, provision should be made for alternative sureties to be put forward by the tenderer for the approval of the employer. Such approval should not be unreasonably withheld.

Certain local authorities and some large private sector employers now nominate their own bond consultants in tender documents. The successful contractor must arrange his bond for that contract through the nominated consultants but normally such schemes allow any surety acceptable to the employer to be used.

On demand bonds

Bonds issued in this country usually require proof of the amount of the loss in-curred before any payment can be made. However, in connection with overseas bonds it is common for an 'on demand' wording to be employed. In other words in the event of the contract not proceeding for some reason or the contractor being in default, the principal can demand payment of the full bond amount without giving any proof of the amount of loss or indeed that any loss has occurred at all.

In *Edward Owen Engineering Ltd* v *Barclays Bank International Ltd (1977)* an English company contracted to erect glasshouses in Libya. The contract never started because of difficulties regarding the letter of credit arranged by the Libyan customers for the payment of instalments. However, a guarantee had been given by the bank on behalf of the contractors, the bank had said: 'we confirm our guarantee . . . payable on demand without proof or conditions'. The bank had obtained a counter guarantee from the contractors and it was the contractors who brought the action in an attempt to prevent Barclays Bank from paying under the bond. The Court of Appeal held that in the absence of fraud the bank had to fulfil its duty to honour the guarantee it had issued.

Insurers generally are reluctant to accept these 'on demand' wordings, for it is only right that any claim shall be investigated properly and an amount paid which represents the real loss to the principal.

How claims arise

A contractor may, because of financial problems, be forced to stop work on a site. A receiver may be appointed on behalf of creditors or a liquidator may be appointed to wind up the company. Occasionally, however, difficulties may arise between the contractor and the principal which lead to the withdrawal of the contractor from the site. The contractor may remain in business but in such an event a court action will probably be necessary to decide the liabilities of the respective parties.

When a receiver is appointed his main duty will be to look after the interest of the creditors; there is nothing to prevent him carrying on the business and if a contract is likely to be profitable it will be continued and completed. There is, however, usually little incentive to complete a contract which has lost and will continue to lose money.

The principal will be anxious that the contract should be completed on time with the minimum of additional cost or trouble. When the original contractor has stopped work it is usually necessary for fresh tenders to be obtained for completion of the outstanding work. There are often difficulties in connection with completed work which may not be up to the required standard. It is inevitably much more costly for new contractors to come on to a site to complete and/or repair someone else's work.

Underwriters when assessing the risk pay particular attention to the level of other tenders originally obtained. If there are numerous tenders, fairly close together, it can be assumed there will not be much difficulty in finding other contractors to complete in the event of default. If, however, there are few tenders with wide variations, underwriters will ask many more questions. Often with specialist work, there are just one or two contractors who are interested in tendering for contracts, similarly in isolated areas, the number of potential contractors may be limited. These problems can all increase the likely cost of a claim in the event of a default.

As soon as they are aware of the possibility of a claim sureties will usually appoint engineers or surveyors to protect their interests. The amount and quality of work completed will have to be assessed, together with the work still to be done. The interests of principals and sureties do not necessarily coincide; sometimes principals do not wish to put the work out to tender or to accept the lowest tender available.

Where there are counter guarantors involved (for example, where personal guarantees are obtained from directors) there is a further possible conflict of interest. The counter guarantors may not agree with the actions taken either by the principals or the insurers – they may, for example, disagree about the extent of defects in work already done.

The protection of the site is important during the period between the cessation and recommencement of work. The principals have a duty to ensure the safety of already completed work and the materials on the site. Arrangements have to be made for the CAR policy to be maintained until fresh cover is effected by the new contractor. In practice the liquidator or receiver may arrange cover and accept responsibility for the premium. Uncompleted sites are especially liable to thefts and damage by vandals and it is in the interests of all that effective security arrangements be implemented. The principals are likely to add the cost of these to the claim being made.

It is unusual for the amount of any retention money held to be sufficient to pay the additional costs involved in completing the work. There is an inevitable time-lag and inflation alone tends to increase the costs. It is not unusual for the whole of the bond amount to be swallowed up, especially when the original tender price was low.

After paying a claim the insurers are entitled to recover under any counter indemnity they may have or to realise any collateral security. Unfortunately personal guarantees are often of little value, as these may have been given to all and sundry. There may be no money available or it may be effectively tied up in other names or other countries so that recovery is not possible.

The sureties will rank as ordinary creditors in the winding up of the company, but the contractor's bankers will usually have preference as will the Inland Revenue. The amount available for distribution to other creditors is often very small.

Bonding of subcontractors

This has to be considered both from the point of view of the main contractor in connection with either domestic or nominated subcontractors, and also from that of the employer – but only with regard to nominated subcontractors. The main contractor is responsible for completing the works and if one of his domestic subcontractors fails, the main contractor has to solve the problem at his own expense. It is, however, becoming increasingly common for main contractors to insist on bonds from their own domestic subcontractors, particularly where the subcontract amounts are substantial. The position with regard to nominated subcontractors is far more complex and one has to refer to the main contract document to establish the responsibilities of the various parties in the event of the failure of a nominated subcontractor.

The ICE Conditions do not offer the same protection for the employer as some unilateral contracts do. However, the main contractor is required, under clause 59(B) paragraphs 4 and 6, in the event of forfeiture of the subcontract to endeavour to recover for the employer what is referred to as the 'employer's loss', to the extent of taking legal proceedings against the subcontractor. Included in any claim against the subcontractor who has been dismissed would need to be the loss and expense incurred by the main contractor while new instructions are awaited, since the architect or engineer would normally be entitled to a reasonable period to accommodate the situation during which time the contractor would have no claim against the employer. If, however, the contractor can show that despite having taken the steps required by the contract against the subcontractor he has been unable to recover the whole or part of the employer's loss, then the employer has to repay to the contractor as much as was irrecoverable.

Third edition FIDIC contract

This contract provides that where required, a contractor shall obtain a bond from an insurance company or bank or other approved sureties but does not include specimen bond wordings and does not, now, even specify the bond percentage which is to be the amount stated in the letter of acceptance.

Other types of bonds

There are various types of bonds, other than performance bonds, that can be required at any stage in a contract.

Bid or *tender bonds* are intended to assure the buyer or the employer that the bid is a responsible one and that if the tenderer's bid is accepted they will proceed and effect the form of contract including any necessary subsequent bonding requirements. If the tenderer fails to do so the losses of the buyer would form a claim under the bond to be met by the tenderer or his surety.

Advance payment, progress or *repayment bonds* are similar to performance bonds in that the buyer receives a guarantee that any monies advanced will not be lost through default or poor performance by the contractor.

Maintenance bonds are normally requested in connection with construction contracts and guarantee that once the construction has been completed the contractor will fulfil his obligations throughout the maintenance period and may be in lieu of retention monies during the maintenance period.

Retention bonds are given in lieu of retention monies required throughout the contract period or against the early release of the retention element of monthly progress payments.

All the various classes of bonds mentioned can be worded to be either conditional or on demand. A conditionally worded bond places the onus upon the employer to prove default by the contractor and the bond generally indemnifies the employer up to the value of the bond, but only for any loss that the employer may have incurred. On demand bonds, on the other hand, in their purest form are worded that the surety will pay to the employer the full amount of the bond on the employer's first demand without reference to the contractor. They are, effectively, a blank cheque in favour of the employer which can be cashed at any time without the contractor even being in default on the contract. By their very nature they are basically a banking instrument.

Chapter 18

The International Civil Engineering Contract (FIDIC)

The contract considered in this chapter is the third edition of the Conditions of Contract (International) for Works of Civil Engineering Construction published in March 1977, and known as the 'FIDIC' International Contract. The first edition was prepared in 1957 on behalf of the Fédération Internationale des Ingénieurs-Conseils (FIDIC) and the Fédération Internationale du Bâtiment et des Travaux Publics (FIBTP). There is no difference in the wording of the second edition, published in 1969, it merely arranged for certain additions outside the wording of the actual conditions as did the 1973 reprint. The FIBTP changed its name and three new sponsoring bodies were added. However, the third edition takes into consideration the improvements in the fifth edition of the ICE Conditions of Contract and adds some of its own. This contract in its wording no longer closely follows the current edition of the English ICE Conditions of Contract; nevertheless, the basic requirements are so similar within each clause of the ICE Conditions already considered that it lends itself to a commentary in this book.

The fourth edition of the English ICE Conditions of Contract published in 1955, was replaced in June 1973 by the fifth edition in the United Kingdom. The authors are only qualified to express views on the English insurances and legal aspects of this contract, consequently there will be no discussion of the differences in principles and practice of insurance or the law outside the United Kingdom. However, as the original text of the principal conditions is the English language version (while there is provision for the 'Ruling language', by clause 5 and part II, to be other than English and thus a translation of the English text), English will normally be the language used. Whether English law will apply is another matter altogether but again there is provision in the contract by clause 5 and part II of the conditions, for the law governing the contract by which it is to be interpreted to be identified.

The question of language and interpretation is obviously important but fortunately for the foreign translator the criticisms of the fourth edition clause 22 and even those remaining in the fifth edition have largely been eliminated in the 1977 (third) edition of the International Contract. While the differences in wording are considerable the basic intentions closely follow those of the current (fifth) edition of the ICE Conditions of Contract, as already stated, and while details will be commented upon as each clause is considered the general pattern is given below in comparison with that of the ICE Conditions.

ICE Conditions	*International Contract*
Clause 20 three sub-clauses (1) Care of the works. (2) Responsibility for reinstatement. (3) Excepted risks.	two sub-clauses (1) Care of the works which includes the requirements of the 'Responsibility for reinstatement' clause. (2) Excepted risks.
Clause 21 Insurance of the works etc., including a proviso exonerating contractor from insuring otherwise than as required by the contract, subject to any relevant requirement in the bill of quantities.	Insurance of the works etc., but no proviso.
Clause 22 two sub-clauses (1) Damage to persons and property with two provisos as follows: (a) proportionate reduction contractor's liability to indemnify employer to extent of employer's contribution; (b) no liability of contractor to indemnify employer in five cases. (2) Indemnity by employer.	two sub-clauses (1) Damage to persons and property with four exceptions to contractor's liability to indemnify employer. (2) Indemnity by employer.
Clause 23 two sub-clauses (1) Insurance against damage to persons and property. (2) Amount and terms of insurances including provisions for an indemnity to employer and production of policies.	three sub-clauses (1) Third party insurance. (2) Minimum amount of third party insurance. Very similar wording to sub-clause (2) of ICE Conditions except that provision for indemnity to employer does not appear here. (3) Provision to indemnify employer.
Clause 24 Accident or injury to workmen. No requirement to insure as compulsory in the UK in respect of employers' (masters') liability.	two sub-clauses (1) Accident or injury to workmen. This wording is almost identical to clause 24 of the

ICE Conditions.

(2) Insurance against accident etc., to workmen. Necessary as not always compulsory to insure abroad.

Clause 25   Remedy on contractor's failure to insure.

Remedy on contractor's failure to insure. Almost identical in wording but includes insurance required under clause 24 as well as 21 and 23.

Care of works and excepted risks: clause 20

Care of Works

20. (1) From the commencement of the Works until the date stated in the Certificate of Completion for the whole of the Works pursuant to Clause 48 hereof the Contractor shall take full responsibility for the care thereof. Provided that if the Engineer shall issue a Certificate of Completion in respect of any part of the Permanent Works the Contractor shall cease to be liable for the care of that part of the Permanent Works from the date stated in the Certificate of Completion in respect of that part and the responsibility for the care of that part shall pass to the Employer. Provided further that the Contractor shall take full responsibility for the care of any outstanding work which he shall have undertaken to finish during the Period of Maintenance until such outstanding work is completed. In case any damage, loss or injury shall happen to the Works, or to any part thereof, from any cause whatsoever, save and except the excepted risks as defined in sub-clause (2) of this Clause, while the Contractor shall be responsible for the care thereof the Contractor shall, at his own cost, repair and make good the same, so that at completion the Permanent Works shall be in good order and condition and in conformity in every respect with the requirements of the Contract and the Engineer's instructions. In the event of any such damage, loss or injury happening from any of the excepted risks, the Contractor shall, if and to the extent required by the Engineer and subject always to the provisions of Clause 65 hereof, repair and make good the same as aforesaid at the cost of the Employer. The Contractor shall also be liable for any damage to the Works occasioned by him in the course of any operations carried out by him for the purpose of completing any outstanding work or complying with his obligations under Clauses 49 or 50 hereof.

Excepted Risks

(2) The 'excepted risks' are war, hostilities (whether war be declared or not), invasion, act of foreign enemies, rebellion, revolution, insurrection or military or usurped power, civil war, or unless solely restricted to employees of the Contractor or of his sub-contractors and arising from the conduct of the Works, riot, commotion or disorder, or use or occupation by the Employer of any part of the Permanent Works, or a cause solely due to the Engineer's design of the Works, or ionising radiations or contamination by radio-activity from any nuclear fuel or from any nuclear waste from the combustion of nuclear fuel, radio-active toxic explosive, or other hazardous properties of any explosive, nuclear assembly or nuclear component thereof, pressure waves caused by aircraft or other aerial devices travelling at sonic or supersonic speeds, or any such operation of the forces of nature as an experienced contractor could not foresee, or reasonably make provision for or insure against all of which are herein collectively referred to as 'the excepted risks'.

A comparison between this clause and clause 20 of the fifth edition of the ICE

Conditions of Contract discloses the following differences.

(i)   The contractor's full responsibility for the care of the works dates in both contracts from the commencement of the works but whereas the ICE Conditions add fourteen days to the date of the issue of the certificate of completion, the FIDIC Contract uses the date stated in the certificate of completion as the date the contractor's responsibility ceases. Similarly, the fourteen-day period applies to any part of the works issued with a certificate of completion in the case of the ICE Conditions but not in respect of the FIDIC Contract.

(ii)   Both contracts make it clear that in the event of damage loss or injury happening from any of the excepted risks the contractor need only make it good to the extent required by the engineer and then only at the cost of the employer. However, in the FIDIC Contract this statement is made subject to the provisions of clause 65 which deals with 'special risks' (explained below). There is no 'special risks' clause in the ICE Conditions. Incidentally, in the FIDIC Contract clause 20(2), the phrase 'commotion or disorder' is added to riot and these three risks do not apply when only the contractor's employees and subcontractors are involved. Also the design risk must be that of the engineer and a new risk of 'operation of the forces of nature, etc.' appears which will be commented upon later.

This 'special risks' clause in the FIDIC Contract, which relates to the works where war plus kindred risks, nuclear and pressure waves perils are involved, applies to a list of risks identical with the excepted risks in clause 20(2) of the same contract except that the three risks of use and occupation by the employer, design of the engineer, and the 'operation of the forces of nature' which are to be found in clause 20(2) are not included in clause 65. The special risks of war, hostilities, invasion, act of foreign enemies and the nuclear and pressure waves risks need not relate to the country in which the works are being executed. An additional and very wide indemnity is given to the contractor under sub-clause 65(4). It reads as follows:

Increased Costs arising from Special Risks
    (4)   The Employer shall repay to the Contractor any increased cost of or incidental to the execution of the Works, other than such as may be attributable to the cost of reconstructing work condemned under the provisions of Clause 39 hereof, prior to the occurrence of any special risk, which is howsoever attributable to or consequent on or the result of or in any way whatsoever connected with the said special risks, subject however to the provisions in this Clause hereinafter contained in regard to outbreak of war, but the Contractor shall as soon as any such increase of cost shall come to his knowledge forthwith notify the Engineer thereof in writing.

Consequently the words 'any increased cost of or incidental to the execution of the works' and the statement 'howsoever attributable to or consequent on or the result of or in any way whatsoever connected with the said special risks', commit the employer to repay the contractor in extremely wide terms. It has been suggested that these terms are wide enough to enable the contractor to recover any damage or loss by special risks to constructional plant though not mentioned in clause 65 nor covered by clause 20 of the FIDIC Contract.

(iii)   *'a cause solely due to the Engineer's design of the Works'*. By this 'excepted risk' the contractor is relieved both from responsibility for damage, loss or injury due to such a cause and from the obligation to arrange insurance in respect thereof. The problem which faces the contractor is that he has to show that the cause was **solely** due to the engineer's design. Experience has shown that this may often be difficult. Consulting engineers will tend to argue that faulty execution of their design was at least a contributory cause of the damage, and unless such an allegation can be completely disproved, the contractor loses all benefit from this exception and will be held responsible for repairing the damage at his own cost. However, this exception only applies between the employer and the contractor, as parties to the contract, and it does not affect the contractor's or his insurer's rights of recovery from the engineer to the extent of the latter's contributory negligence. See for an example the case of *Pearce v Hereford BC (1968)*.

(iv)   *'operation of the forces of nature as an experienced contractor could not foresee or reasonably provide for or insure against'* is an additional 'excepted risk' in the FIDIC Contract. It does not appear in the ICE Conditions under clause 20. It is clear that the reason for the discrepancy is that the possibility of this risk occurring abroad is greater, therefore the employer is probably forced to accept it as his responsibility. It has been said that the vagueness of the expression just quoted seems to confer a discretion on the arbitrator, in which the borderline between interpretation and sympathy is likely to be confused. It has already been stated in chapter 5 that the excepted risks (with the possible exception of riot) form the basis of the exceptions to the contractors' all risks policy but contrary to popular belief, 'act of God' or 'forces of nature' are not exclusions normally used in insurance policies.

They certainly do not appear in the policies discussed in this book. Admittedly there is the qualification concerning foreseeability or impossibility to 'reasonably make provision for or insure against' the risk. Nevertheless, one must agree that the expression is very vague although the general intention is clearly to saddle the employer with unforeseeable risks and those which a reasonable contractor cannot provide for or insure against. Inevitably there must be borderline cases, and contractors may be encouraged to take chances with a view to increasing their profits always hoping to invoke the exception should damage occur. Incidentally, this 'forces of nature' excepted risk as such is not required to be insured against but it could include risks such as storm and flood which are covered by the contractors' all risks policy. On the other hand, hurricanes, earthquakes, tidal waves and volcanic eruptions and the like are risks to be taken into account when fixing terms for overseas contracts. Thus they are often only partially insurable. See page 169.

Examples of the 'operation of the forces of nature'

In areas where there are frequent natural disasters such as earthquakes or hurricanes insurers take the view that they have to allow for the possibility of a single occurrence destroying large parts of the works up to their substantial completion.

There have been instances of the collapse of steel frames in buildings nearing completion. The cause is often that an excessive strain was thrown on the structure due to incorrect or inadequate bolting of a comparatively small section. Sometimes there can be a combination of forces of nature on the one hand and incorrect or inadequate bolting on the other.

One such case where the steel framework stood up to winds of hurricane force for nine days before the collapse occurred raised a problem for the insurers of the CAR risk as to whether it was the winds or the incorrect bolting in a few small sections which caused the collapse. In that case forces of nature (the employer's risk) were covered but the dispute centred on whether the comparatively small expense of rectifying the defectively executed part (the contractor's expense) was to be excluded in accordance with the policy exception concerning defective workmanship which would be so if the proximate cause of the collapse was defective workmanship and not the forces of nature. It was decided that in view of the time the hurricane operated before the collapse that this was probably the proximate cause, particularly as there were other disasters in the area. Incidentally, provided the insurance covers the peril, the responsibility between the employer and the contractor under the contract does not matter when the insurance is in the joint names as provided by both the ICE and the FIDIC contracts. See the next clause.

There are many coastlines round the world which have a history of battles with the sea thus any contract involving the dumping of earth, sand or rock on such sea beds to build a structure of defence against the sea in some shape or form inevitably involves the loss of a certain amount of material which is swept away by tides and current. While the insurers will wish to exclude inevitable loss or damage, they will be expected to cover the catastrophe. In 1953 disaster was caused by the worst gale on record in the North Sea. A mean wind speed of more than 75 m.p.h. (force 12) was recorded and this is classified as a hurricane according to the Beaufort scale. All down the east coast of England tide water was six to eight feet higher than the tide tables predicted and coastal defences in course of construction suffered catastrophic damage among the other disasters round the coasts of England. While the FIDIC contract would not usually apply in the UK any coastline anywhere in the world, which has a history of suffering from the forces of nature, really needs such cover for its coastal defence works.

Contract works losses due to earthquakes and the like are, compared with other losses from the forces of nature, small in number but the potential for a catastrophe is there and insurers may wish to limit their liability. This can be done in several ways apart from a complete exclusion of the peril. In those countries where there is a high record of minor tremors an increased excess may be applied. Alternatively, if faced with a high aggregation of risk the insurers may apply lower sums insured for earthquake etc claims than apply for other claims. Thus such losses are only partially insurable.

Insurance of the works, etc.: clause 21

21. Without limiting his obligations and responsibilities under Clause 20 hereof, the Contractor shall insure in the joint names of the Employer and the Contractor against all loss or damage from whatever cause arising, other than the excepted risks, for which he is responsible under the terms of the Contract and in such manner that the Employer and Contractor are covered for the period stipulated in Clause 20(1) hereof and are also covered during the Period of Maintenance for loss or damage arising from a cause, occurring prior to the commencement of the Period of Maintenance, and for any loss or damage occasioned by the Contractor in the course of any operations carried out by him for the purpose of complying with his obligations under Clauses 49 and 50 hereof:

> (a)   The Works for the time being executed to the estimated current contract value thereof, or such additional sum as may be specified in Part II in the Clause numbered 21, together with the materials for incorporation in the Works at their replacement value.
>
> (b)   The Constructional Plant and other things brought on to the Site by the Contractor to the replacement value of such Constructional Plant and other things.

Such insurance shall be effected with an insurer and in terms approved by the Employer, which approval shall not be unreasonably withheld, and the Contractor shall, whenever required, produce to the Engineer or the Engineer's Representative the policy or policies of insurance and the receipts for payment of the current premiums.

The differences between the FIDIC Contract in this clause and the same clause in the ICE Conditions are set out below.

(i)   The 'object' of the operative paragraph of this clause (the works, etc.) is detailed at the end of the first paragraph in the case of the FIDIC Contract whereas in the ICE Conditions the works are detailed as early as possible. Otherwise this operative paragraph is almost word for word the same in both contracts, except that the term 'temporary works' has been dropped by FIDIC as the word 'works' is defined as including both temporary and permanent works; see clause 1(1)(e).

(ii)   In the FIDIC Contract paragraph (a) mentions the sum insured for the works as 'the estimated current contract value thereof, or such additional sum as may be specified in Part II in the Clause numbered 21'. In the ICE Conditions there is no Part II and no provision for indicating a sum insured for clause 21, merely a requirement to insure for full value (which really requires some definition, see chapter 3). The FIDIC Contract refers to 'replacement value' which is more explicit, but 'replacement value' can have more than one meaning even to the layman, i.e. as the value of the property was immediately prior to the loss or damage on the one hand or providing 'new for old' cover on the other. In the insurance world a 'replacement' or 'reinstatement' basis of settlement is usually defined in contrast to an indemnity basis of settlement, or in common parlance the former contains a 'new for old' element whereas an indemnity basis is a matter of putting the insured strictly in the same position he was immediately before the loss or damage occurred. Clearly the latter is intended in this case and probably this is usually arranged. Possibly the phrase merely emphasises the necessity for an escalation clause (see chapter 5).

(iii)   The ICE Conditions contain a proviso paragraph which does not appear in the FIDIC Contract. This paragraph provides that it is for the contractor to decide whether to insure or not against the necessity to repair or reconstruct work due to defective materials or workmanship, unless the bill of quantities provides a special item for this insurance. In view of the exception in the contractors' all risks policy this cover, if requested by the bill of quantities, is normally only available to a limited extent, i.e. the exception applies to the cost of repairing, replacing or rectifying the actual property which is defective in material or workmanship which means that only loss or damage to the property protected by the policy which is **caused by** such defective material or workmanship is covered. See chapter 5 for further details concerning this exclusion. Reference should also be made to chapter 3 where this whole proviso is considered in more detail.

(iv)   In the final paragraph of this clause in both contracts the only difference is that when required the policies of insurance and the receipts for payment of the current premiums, in the case of the FIDIC Contract must be produced 'to the engineer or the engineer's representative', whereas in the ICE Conditions the production must be to the employer which in practice no doubt includes the employer's agent, the engineer, although clearly it is open to the engineer in the ICE Conditions to avoid the responsibility if he so wishes.

Damage to persons and property and indemnity by employer: clause 22

Damage to Persons and Property
22.   (1)   Contractor shall, except if and so far as the Contract provides otherwise, indemnify the Employer against all losses and claims in respect of injuries or damage to any person or material or physical damage to any property whatsoever which may arise out of or in consequence of the execution and maintenance of the Works and against all claims, proceedings, damages, costs, charges and expenses whatsoever in respect of or in relation thereto except any compensation or damages for or with respect to:
  (a)   The permanent use or occupation of land by the Works or any part thereof.
  (b)   The right of the Employer to execute the Works or any part thereof on, over, under, in or through any land.
  (c)   Injuries or damage to persons or property which are the unavoidable result of the execution or maintenance of the Works in accordance with the Contract.
  (d)   Injuries or damage to persons or property resulting from any act or neglect of the Employer, his agents, servants or other contractors, not being employed by the Contractor, or for or in respect of any claims, proceedings, damages, costs, charges and expenses in respect thereof or in relation thereto or where the injury or damage was contributed to by the Contractor, his servants or agents such part of the compensation as may be just and equitable having regard to the extent of the responsibility of the Employer, his servants or agents or other contractors for the damage or injury.

Indemnity by Employer
  (2)   The Employer shall indemnify the Contractor against all claims, proceedings, damages, costs, charges and expenses in respect of the matters referred to in the proviso to sub-clause (1) of this Clause.

This clause differs in a number of ways compared with the equivalent clause in the ICE Conditions and the following are the main differences.

(i)   The FIDIC Contract in the introductory paragraph contains no exclusion of the works from the indemnity to the employer for damage to property as in the ICE Conditions. As responsibility for the works is catered for by clause 20 and the insurance thereof by clause 21 this is surprising. Similarly the inclusion in the indemnity of 'surface or other damage to land being the Site suffered by any persons in beneficial occupation of such land' is not specifically mentioned in the FIDIC Contract. While the exact purpose of this quotation from the ICE Conditions (see chapter 6) is not certain it helps to confine the exclusion of the indemnity to the employer to damage which is the inevitable consequence of carrying out the works.

However, the introductory paragraph in the FIDIC Contract qualifies the words 'damage to any property' by the phrase 'material or physical' which apparently excludes consequential or financial losses by third parties from the indemnity.

(ii)   The limitations of the indemnity to the employer in the FIDIC Contract are 'exceptions' whereas in the ICE Conditions they are 'provisos', and the following differences are evident:

(1)   The operation of a proportionate reduction where the employer, his servants or agents are negligent is indicated in a less cumbersome way in the FIDIC Contract. See exception 22(1)(d).

(2)   The other circumstances where the employer remains responsible and no indemnity is required from the contractor are the same in both contracts (although the wording is slightly different and personal injuries as well as damage to property are included in the FIDIC Contract exclusion concerning the unavoidable result of the execution or maintenance of the works in accordance with the contract). The only exception to this statement is that damage to crops on site possession of which has been given to the contractor is not specifically mentioned as the employer's liability in the FIDIC Contract. It is arguable that damage to crops on site in the contractor's possession comes within the exception of damages with respect to 'the permanent use or occupation of land by the Works or any part thereof' which appears in both contracts, with the exception of the word 'permanent' in the ICE Conditions.

(iii)   In sub-clause 22(2) concerning indemnity by the employer the cumbersome method of operating a proportionate reduction in the ICE Conditions, where both parties to the contract have some liability, is avoided in the FIDIC Contract. In fact proportionate reduction for some liability is indicated only once in detail in clause 22 of the latter contract which is obviously better than detailing it twice which is the situation in the ICE Conditions.

A minor criticism of this sub-clause (2) in the FIDIC Contract is that in referring back to the exceptions in sub-clause (1) the wording reads 'the proviso to sub-clause (1) of this clause'. Now there is no 'proviso' (at least that word is not used) in sub-clause (1) and (what is more significant) as the reference is to '**the** proviso' it leaves the reader in doubt as to whether reference is being made to sub-paragraphs (a) to (d) inclusive or only to (d). The introductory words to the statement quoted above from sub-clause (2) do not help to resolve this point but from the history of this clause (particularly bearing in mind the ICE Conditions) it could be argued that it applies only to sub-paragraph (d) but probably it applies the indemnity by the employer to sub-paragraphs (a) to (d) inclusive.

Third party insurance; minimum amount of third party insurance; provision to indemnify employer: clause 23

Third Party Insurance

23. (1) Before commencing the execution of the Works the Contractor, but without limiting his obligations and responsibilities under Clause 22 hereof, shall insure against his liability for any material or physical damage, loss or injury which may occur to any property, including that of the Employer, or to any person, including any employee of the Employer, by or arising out of the execution of the Works or in the carrying out of the Contract, otherwise than due to the matters referred to in the proviso to Clause 22(1) hereof.

Minimum Amount of Third Party Insurance

(2) Such insurance shall be effected with an insurer and in terms approved by the Employer, which approval shall not be unreasonably withheld, and for at least the amount stated in the Appendix to the Tender. The Contractor shall, whenever required, produce to the Engineer or the Engineer's Representative the policy or policies of insurance and the receipts for payment of the current premiums.

Provision to Indemnify Employer

(3) The terms shall include a provision whereby, in the event of any claim in respect of which the Contractor would be entitled to receive indemnity under the policy being brought or made against the Employer, the insurer will indemnify the Employer against such claims and any costs, charges and expenses in respect thereof.

Sub-clause (1) has been tidied up in comparison with sub-clause (1) in the ICE Conditions (with exceptions mentioned in (iii) and (iv) below) in that:

(i) The insurance must be arranged **before** the work commences. Although the effect was the same in the ICE Conditions this wording is better as it brings home to the contractor more obviously that this insurance is not to be left to the last minute.

It might be thought that in liability insurance as the contractors have annual employers' and public liability policies their insurances will always be in evidence but the public liability policy may not contain the indemnity to principals requirement in sub-clause (3) and **both policies will probably have limitations concerning their operation abroad** (see chapters 8 and 10) assuming they are UK policies.

As in clause 22(1) the important words 'material or physical' qualify 'damage, loss or injury which may occur to any property', which appears to exclude claims for consequential or financial loss by third parties from the insurance.

(ii) Reference is to the contractor insuring against his **liability** (as above) which is more correct than the ICE Conditions which merely state that the contractor 'shall insure against any damage loss or injury which may occur to any property or to any person'.

(iii)   The liability in (ii) above specifically includes the property of the employer which is not made clear in the ICE Conditions. One should not necessarily jump to conclusions that it is not intended to include the works in the property of the employer as that property is catered for by clauses 20 and 21. However, this was one of the few criticisms, made earlier of clause 22 of the FIDIC Contract, i.e. that it did not exclude the works, or make clear that it did not apply to the works.

Common sense dictates the necessity of keeping the insurance of the works as a separate matter entirely, especially as responsibilities are indicated between employer and contractor concerning the works in clauses 20 and 21. It is arguable that as clause 22 refers to 'losses . . . which may arise out of . . . the works' that damage to the works is therefore excluded although clearly damage to other property of the employer on the site is to be a matter where the contractor indemnifies the employer. The previous edition of the FIDIC Contract did not exclude the works. In any event a public liability policy will always exclude the contract works in some way. Insurers rightly regard this liability or responsibility as a matter for the material damage policy – the contractors' all risks policy in this case. Thus any request for liability cover for damage to the works is usually met with a firm refusal. Consequently if the correct interpretation is that the liability of the contractor to indemnify the employer includes the works then the construction industry is asking for something which cannot be granted.

What is the position if the contractor destroys the subject-matter of the contract by fire caused by his negligent use of a blowlamp? As the policy is in the joint names of the contractor and the employer (see clause 21) the liability would stay with the material damage insurer, i.e. he could not turn to the contractor for a recovery as the contractor is one of his insureds – at least this is the generally accepted argument. So there should be no difficulty in this direction. One is forced to the conclusion that it is the intention to keep liability for damaging the works as a separate matter to which only clauses 20 and 21 apply and even if this is wrong an attempt by the employer to recover from the contractor will be defeated because the insurance has been arranged in the joint names.

(iv)   Once again there is a reference to 'proviso to clause 22(1) hereof' and the remarks made under paragraph (iii) concerning clause 22 apply.

Sub-clauses (2) and (3) are virtually the same as sub-clause (2) in the ICE Conditions, but once again reference is made in the FIDIC Contract to the engineer or his representative when required, receiving the policies and the receipt for the current premiums, whereas the employer is to receive them in the ICE Conditions.

Accident or injury to workmen; insurance against accident etc. to workmen; clause 24

Accident or Injury to Workmen
24. (1)   The Employer shall not be liable for or in respect of any damages or compensation payable at law in respect or in consequence of any accident or injury to any workmen or other

person in the employment of the Contractor or any sub-contractor, save and except an accident or injury resulting from any act or default of the Employer, his agents, or servants. The Contractor shall indemnify and keep indemnified the Employer against all such damages and compensation, save and except as aforesaid, and against all claims, proceedings, costs, charges and expenses whatsoever in respect thereof or in relation thereto.

Insurance against Accident, etc., to Workmen
(2)   The Contractor shall insure against such liability with an insurer approved by the Employer, which approval shall not be unreasonably withheld, and shall continue such insurance during the whole of the time that any persons are employed by him on the Works and shall, when required, produce to the Engineer or the Engineer's Representative such policy of insurance and the receipt for payment of the current premium. Provided always that in respect of any persons employed by any sub-contractor, the Contractor's obligation to insure as aforesaid under this sub-clause shall be satisfied if the sub-contractor shall have insured against the liability in respect of such persons in such manner that the Employer is indemnified under the policy, but the Contractor shall require such sub-contractor to produce to the Engineer or the Engineer's Representative, when required, such policy of insurance and the receipt for the payment of the current premium.

Sub-clause (1) is virtually the same wording as clause 24 of the ICE Conditions.

Sub-Clause (2) requires the contractor to insure the liability mentioned in the headings above with an insurer approved by the employer. There is the usual sentence concerning the production of policies and receipts for premiums.

Employees of subcontractors are required to be insured by the subcontractor in respect of the subcontractor's liability concerning such persons and this is for the contractor to impose on the subcontractor, and the latter is required to produce the usual evidence to the engineer or his representative. Obviously the employer requires the indemnity given in sub-clause (1) by the contractor to the employer to be insured not only in the contractor's but also in the subcontractor's policy.

There is often no compulsory insurance by law in respect of the employer's (masters') liability to his employees abroad, consequently sub-clause 2 is important.

Remedy on contractor's failure to insure: clause 25

25.   If the Contractor shall fail to effect and keep in force the insurances referred to in Clauses 21, 23 and 24 hereof, or any other insurance which he may be required to effect under the terms of the Contract, then and in any such case the Employer may effect and keep in force any such insurance and pay such premium or premiums as may be necessary for that purpose and from time to time deduct the amount so paid by the Employer as aforesaid from any monies due or which may become due to the Contractor, or recover the same as a debt due from the Contractor.

This clause is practically the same as the one in the ICE Conditions except that clause 24 is included as it contains an insurance requirement in the FIDIC Contract for the reason explained under the previous heading.

Insurance abroad

The following points should be noted:

(1)   While there are very few countries in which construction insurance is legally compulsory (except for liability to employees and to third parties in motor insurance) any contract may stipulate for insurances and in some countries this stipulation may be for a licensed or local insurer. Where governments have formed their own insurance companies these insurers get priority over other insurers (local or foreign).

(2)   In some countries premium collection is delayed, or a proportion of it, by the necessity to establish a premium loss reserve, and this means that usually there is not a realistic rate of interest on such reserve funds.

(3)   A few countries have a discriminating tax on non-admitted insurers and a very few have a form of taxation on losses paid by non-admitted insurers.

The FIDIC form of contract: summary of the main risks remaining uninsured

Once again the same warning is necessary as was given in the summaries at the ends of chapters 5, 8, and 15, namely only the basic policy covers usually issued are being considered.

The contractor's responsibility for the works (clauses 20 and 21)

(1) *Sums insured for works and constructional plant*   The works must be insured 'to the estimated current contract value' of parts executed for the time being and constructional plant and other materials must be insured for their replacement value.

Considering the works, the following difficulties arise:

(a)   While the capacity of the insurance market is normally sufficient to cope with cover up to the full contract value, in the areas where catastrophies such as earthquakes and hurricanes are concerned insurers must allow for such occurrences bearing in mind any other commitments on properties in the same area. In these circumstances a capacity problem may cause difficulty for a contractor particularly if he signs a major contract when a previous large contract has already been awarded.

(b)   Depending on the method of updating the insured value to ensure that the full cost of replacement damage is covered, which is clearly in accordance with the contract requirements (see the methods suggested in chapter 5), problems can still arise if the sum insured is not sufficient to cover the cost of replacement.

(2) *Faulty design*   Clause 20 states that one of the excepted risks is a 'cause solely due to the engineer's design of the works'. To avoid responsibility for damage to the works and to avoid responsibility for arranging the insurance thereof the contractor must show that the cause was **solely** due to the engineer's design. This is usually a very difficult task as there are normally several contributing factors in defective design allegations and not all of them are the result of one party's activi-

ties. Once this happens the contractor will have to meet the full repairing expense unless he can disprove all the allegations except those against the engineer. The contractor's insurers may have a subrogation right of recovery from the engineer, but see under the next heading item 2 entitled 'A cause solely due to the engineer's design of the works'.

As explained in chapter 5 there are two main design exclusions used in the normal contractors' all risks policy and even the widest cover does not pay for the repairing or replacing of the faultily designed part. So the contractor's liability under the FIDIC Contract is not covered completely, assuming he has some design responsibility and even a professional indemnity policy will not cover non-negligent (contractual) liability. Furthermore, the widest cover given under the CAR policy can have a complication in deciding where the division falls between the resultant damage and the faultily designed part which results in delay in settlement.

(3)   *Faulty workmanship and defective materials*   While under the contract the contractor is responsible for paying the cost of making good faulty workmanship and defective materials and for the resultant consequential loss; in contrast to clause 21 of the ICE Conditions he is apparently required to insure these risks as they are not excepted risks. This is so in spite of the fact that (similar to the situation mentioned under the previous heading) the CAR policy does not cover the cost of replacing the faulty part or defective material, it only covers the consequences thereof assuming this latter damage is part of the works.

The employer's responsibility for excepted risks (clauses 20 and 21)

(1)   *The excepted risks*   The contractor has no obligation nor is he required to insure these risks. Thus unless the employer arranges his own insurance for these risks (the majority of which are uninsurable) he will have to pay the cost of repairs out of his own pocket. The excepted risks which are insurable if occurring in certain countries are: riot, commotion or disorder where the acts are 'solely restricted to his (the contractor's) employees or of his subcontractors and arise from the conduct of the works'; pressure waves caused by supersonic aircraft (if required); and 'the operation of the forces of nature' risk. See paragraph (iv) of the commentary on clause 20 in this chapter.

(2)   *A cause solely due to the engineer's design of the works*   It follows from the remarks under the previous heading and the remarks under the heading of 'Faulty design' when considering the contractor's viewpoint that only partial insurance cover is obtainable. Recovery from the engineer has its problems, e.g. the possible application of the 'state of the art' defence, the existence of limiting terms of responsibility, and the fact that without full insurance the engineer might be a 'man of straw'. Very few insurers are prepared to give a limited form of professional negligence cover to the engineer under a type of project CAR insurance covering all the main parties involved in the contract, but as was stated at the beginning of this summary it is not a basic cover but it is a step in the right direction. Hence the reason for mentioning it.

(3) *Contractor's inoperative or inadequate insurance*   Obviously the employer's indemnity is only as strong as the extent to which the contractor's insurance operates, and this is often overlooked completely although too much emphasis must not be placed on the possibilities of it not operating. In any event it is largely out of the employer's hands. See the main possibilities set out in the summary at the end of chapter 5 under the heading 'The employer' in item 3.

(4) *Consequential losses*   As explained in the summary at the end of chapter 5 under the heading 'The employer', item 4, the CAR policy does not cover consequential losses and the employer is likely to suffer considerable loss in this respect particularly as a result of delay in completion. The same remarks apply as under the item 4 just mentioned.

Public liability and the contractor (clauses 22 and 23)

(1) *Limit of indemnity for insurance*   While the contractor's liability is unlimited under the contract there is provision in the appendix to the tender for a minimum limit of indemnity under the public liability policy which is usually for any one accident or series of accidents arising out of one event. It is extremely unlikely that unlimited liability cover will be given by insurers although at common law there is no limit of liability. However, while the contractor will always have some financial interest above the policy limit, by reinsurance and with insurers taking successive layers very large maximum limits can be reached.

Public liability and the employer (clauses 22 and 23)

(1) *The employer's responsibilities because he commissioned the works*   It has been explained in the summary at the end of chapter 8 that the employer is saddled with certain responsibilities arising from the fact that he commissioned the work (see item 1 under the heading 'The employer'). Clause 22 of the International Civil Engineering Contract excepts the contractor from responsibility for injury to any person or damage to third-party property in respect of:

    (a)   the permanent use or occupation of the land by the works or any part thereof;

    (b)   the right of the employer to execute the works or any part thereof on, over, under, in or through any land;

    (c)   injuries or damage to persons or property which are the unavoidable result of the execution or maintenance of the works in accordance with the contract;

    (d)   (to summarise this) injuries or damage from any act or neglect of the employer his agents, servants or other contractors not employed by the main contractor.

While clause 23(3) requires an indemnity to the employer under the contractor's policy it is only in respect of any claim for which the contractor would be entitled to receive indemnity under the policy and items (a) to (d) are not items for which the contractor would be so entitled as the insurance usually follows the contract requirements.

See the steps the employer could take to protect himself from some of these items (although it must always be remembered that unavoidable damage is not covered by liability policies and usually without damage to property, pure economic loss is not covered) in the summary (mentioned earlier under this heading) at the end of chapter 8.

(2) *Consequential losses* The same remarks apply here as were expressed in the summary at the end of chapter 8 under the heading 'The employer', item 2, concerning the wording of the public liability policy.

Chapter 19

General claims procedures

Introduction to claims under contractors' policies

Claims handling is considered under:

(a) *contract works* policies (material damage policies insuring the works under construction);

(b) *public liability* policies (insuring the legal and contractual liability of the contractor for bodily injury to, or loss of or damage to property of third parties);

(c) *employers' liability* policies (insuring the legal and contractual liability of the contractor for accidents to his employees).

The following chapters do not pretend to be a comprehensive claims manual, but they consider some of the problems which face contractors, employers and their insurers when dealing with liability claims and losses under material damage policies.

General claims procedures

It is a basic principle in the handling of any claim that it is first necessary to confirm that the policy operates in respect of the loss or liability concerned.

Insurers will check the following.

(a) *The claimant under the policy is a named insured.* With a contract under the ICE Conditions the claimant will be the employer or the contractor. Subcontractors will not be indemnified unless they are specifically named. Sometimes there can be problems with associated or subsidiary companies of the contractor. Such companies should be named in the policy as joint insureds. Liability policies also normally offer an indemnity to directors, partners or employees of the insured while acting in connection with the business. There may also be an indemnity offered to officers or members of the insured's canteens, clubs, sports, athletic, social or welfare organisations and first aid, fire, security and ambulance services.

(b) *The policy cover is operative.* The insurance must be in force at the time when the loss occurs. It is a question of fact whether an insurance is operative or not. Insurers agree cover, the premium being the consideration for such insurance. Initially cover may be agreed orally, by letter or a formal cover note may be issued. The operation of cover does not depend on the issue of a policy but cover is accepted by insurers subject to their normal policy terms and conditions. Any special terms should be made clear and the amounts of the excess or indemnity should always be specified. The operation of cover does not depend on immediate payment of premium although some insurers may seek an initial deposit. A claim may arise after cover has been issued but before the premium is paid or the policy is prepared. The premium or a substantial deposit should be paid before the claim is paid. Difficulties may arise when an insured is not in possession of a policy document and is not aware of all the terms and conditions. Risks may be excluded which were thought to be covered.

On renewal of a policy insurers normally allow a period of fifteen days in which to pay the premium. If a claim occurs during these days of grace premium payment is still required. It is not a question of extending the original insurance period, a fresh insurance period has commenced and premium is required for that period. If a claim occurs during the days of grace but before payment of premium, cover will not be operative if the insured has already effected a fresh insurance elsewhere or if his actions clearly indicated an intention to effect such insurance.

Where an insurance broker or agent represents the insured and is given credit facilities by the insurer the question of whether cover is operative can sometimes be difficult to ascertain. It is the broker's or agent's responsibility to collect the premium and if the days of grace have expired to establish with the insured whether cover is required or not. The granting of credit by the broker to the insured should not be the concern of the insurer but by this arrangement the broker will make himself responsible for the payment of premium to the insurer. Certainly the position should never arise where the operation of policy cover remains uncertain, but any difficulty can only be resolved by a study of the facts.

(c) *The loss or damage arises during the policy period.* The accident or loss or damage must occur during the policy period. Losses which arise before the inception date of the insurance or after the insurance has lapsed are not covered.

Under a standard policy loss or damage to contract works must occur during the period of insurance which includes the period of maintenance under the contract terms. A short-period policy may be issued for the period of the contract including the maintenance period but the insured should ensure that extensions of the original period are advised to the insurers.

Where an annual contract works insurance is transferred from one insurer to another agreement should be reached about the liability of each insurer. The new insurer may take over liability for all uncompleted contracts or the original insurer may continue to be responsible for such contracts until they are finally completed. Where, however, an annual insurance is allowed to lapse and is not replaced by a new insurance, cover ceases entirely and an insurer would be responsible only for damage occurring when the policy was actually in force.

Liability policies may be arranged on a 'losses occurring' or a 'claims made' basis. The latter basis frequently applies to professional indemnity policies where indemnity is given against claims made against the insured during the insurance period. With general public liability insurance in the United Kingdom, however, the 'losses occurring' basis is the usual method and the indemnity operates in respect of losses which occur during the insurance period.

With public liability insurance, an accident may occur some time after the completion of work; the work may have been done in one policy period and the accident occurs in another. However, provided the policy is in force at the time of the accident, insurers will offer an indemnity. If a policy has been allowed to lapse there will be no protection; insurers may have received premium on the work done but they will not accept responsibility for the claim unless their policy is in force at the time the accident occurs. Where liability insurances are transferred from one insurer to another, it is usual for the new insurer to accept responsibility for accidents occurring during the period his policy is operative, but which may arise out of work done before his policy was effected (known as retrospective liability).

Where a short-period policy has been issued for a specific contract, it is important that provision should be made for liability which may arise following the completion of the contract. Either the insurers should agree to accept liability for future accidents even though their policy may have expired (they may be reluctant to do this because of the problem of allocating reserves in the absence of premium) or provision may be made under an annual policy for the contingent liability arising out of contracts where short-period insurances were effected.

With employers' liability insurances the issues are usually clear-cut; the exact date of an accident can be established and it can be decided whether the policy was in force or not. There may sometimes be difficulties with disease claims (for example, pneumoconiosis, cancer or deafness). A claim may be made on an insurer whose policy lapsed many years previously; indeed, because of a shortage of records it may be difficult to establish the existence of a policy or the period it was in force. However, the insurers will be liable for the period their policy was operative. With diseases of gradual onset, over a period of years there can be many different insurers involved. The fact that some policies have expired does not affect the liability of each insurer, they are still responsible for the injury that occurred during the period their insurance was operative. It is usual for these disease claims to be shared between insurers proportionately to the length of time the claimant was exposed to the disease and the periods the policies were in force. Drivers using noisy vehicles such as earth-moving equipment and operators of road drills are sometimes victims of industrial deafness.

(d) *The subject-matter of the claim is insured.* With the comprehensive cover available under contract works policies there are usually few problems. Where property is lost or damaged the type of property must be covered by the policy. For example, damage to plant is normally insured, but occasionally it may be excluded if an entirely separate plant insurance has been arranged. With liability insurance there must be a potential liability which can devolve upon the insured. If there is no such liability (legal, including contractual) the policy cannot operate, but this statement must not be confused with the liability of the insurer to handle a claim to which there is a legal defence.

(e) *Loss or damage arises out of an insured peril.* Under the ICE Conditions insurance of the contract works is on an all risks basis, but there are excluded risks — both the excepted risks specified in the contract and/or additional risks for which the insurers have not provided cover (for example, wear and tear); see chapter 5. Before paying a claim insurers must be satisfied that the damage does arise from a cause which is insured by the policy.

For example, a policy which insured only against fire risks would not give any cover in respect of damage by storm. Difficulties might arise where the loss or damage arises partly from an insured peril and partly from an excepted peril. The proximate cause of the loss or damage is important (see chapter 2).

Similarly with a liability policy the circumstances of the accident must be considered in relation to the operative clause of the policy and the policy exceptions. Many public liability policies offer indemnity on an 'accidental' basis so it is necessary to establish that an accident has occurred which has resulted in loss or damage which involves the insured in a legal liability.

(f) *The claim falls within the business of the insured.* The insured's business will be shown in the policy. Where a policy is issued for a particular contract, that contract will be described and identified clearly. Where the insured is, for example, described as a civil engineering contractor a wide range of activities will be included. If, however, the insured is described as a building contractor, the construction of civil engineering works such as bridges and dams would not be covered. It is usual for a general description of an occupation to be qualified by the exclusion of certain types of work; the lower the rate paid by an insured, the greater the number of exclusions. In handling a claim, therefore, it is important to check that the policy covers the type of work which produced the claim. For example, a workman employed by a building contractor is injured during the course of demolition work. Is demolition work covered by the policy? Although employers' liability insurance is compulsory and an insurer is not permitted to deny liability because of certain policy conditions, the regulations permit the application of trade endorsements, excluding certain classes of work.

As far as the employee is concerned this is unsatisfactory. The intention of the compulsory insurance laws is that every employer should be insured. If an employer is not insured, either because he has failed to effect a policy or because the policy is not sufficiently comprehensive, he may be unable to meet a large claim for damages. Bankruptcy or liquidation could ensue with the result that the employee would only rank as a creditor and might receive just a small part of his entitlement. It is therefore desirable that trade exclusions should be as few as possible; all activities incidental to the business should be insured.

As yet there is no Employers' Liability Insurers Bureau to meet claims where for one reason or another an employer is not insured and the unfortunate employee has no redress.

(g) *There are no policy exceptions appropriate to the claim.* Policy exceptions may be classified broadly between: (i) types of perils not covered; and (ii) types of property not covered.

There are also three main groups:

(i)   Risks which should be insured under other policies – for example: loss or damage caused to or by aircraft, watercraft or hovercraft; loss of cash, bank notes, securities, money of any type; breakdown of plant. A limited cover is sometimes offered for watercraft, for example, pontoons, rowing boats or small powered craft.

(ii)   Risks which are normally uninsurable – for example: war and radio-active contamination risks; wear and tear, gradual deterioration; sonic boom damage (not liability policies).

(iii) Risks which can be insured on payment of additional premium or subject to certain provisos – for example:
defective design
consequential loss $\Big\}$ under material damage policies.

Policy exceptions are a vital part of the policy wording. They restrict or explain the cover in some way but if the insurer wishes to take advantage of a policy exception then it is his responsibility to prove that the exception operates, not for the insured to prove that it does not.

Wordings vary considerably between different insurers.

(h) *The sum insured or indemnity limit is adequate.* Contract works policies are not normally subject to average. The sum insured may correspond to the value of an individual contract or an annual policy may exclude contracts above or below certain figures. Alternatively an indemnity limit may be applied which is less than the total amount at risk at any one time. Gross under-insurance does not normally arise because the premium for the risk is charged on the insured's turnover. It is necessary to check that the sum insured is adequate to meet the claim.

Under public liability policies there is normally a single accident indemnity limit, that is, a limit which applies to any one occurrence or series of occurrences arising out of any one original cause. Costs and expenses incurred with the insurer's written consent are payable in addition.

There is usually no overall limit during a period of insurance; there can be many different payments arising out of different occurrences.

Occasionally there can be problems in deciding whether a loss arises out of one cause or not. For example, piling damage may occur over a period and it may be difficult to decide whether there have been individual events or whether all the damage arises from one cause. Sometimes, therefore, insurers apply an overall limit to a certain class of work or type of claim.

Occasionally the single accident indemnity limit under a public liability policy may be insufficient to meet the cost of a claim. A policy condition normally gives the insurer the right to pay the insured the maximum sum plus the costs incurred to the date of payment. This enables the insurer to avoid any further costs which might be incurred.

Under the employers' liability compulsory insurance laws a minimum policy indemnity of £2 million is required. It is, however, not the practice of insurers at present, to impose any limit under their employers' liability policies. In other words, policies are issued for an unlimited amount.

(i) *Policy conditions or warranties have been complied with.* The observance of the conditions of the policy is a condition precedent to the insurer's liability. Only a full investigation of the circumstances of the claim will reveal whether a breach of a condition has any bearing on the settlement. (See chapters 5, 8 and 9 for the policy conditions concerned.)

(j) *There is no other policy which is more appropriate to the loss or damage.* Dual insurance which involves an insurance in the same subject-matter, for the same interest, against the same risk, during the same period of time, is comparatively rare. It is one of the functions of an insurance broker to ensure that there is no unnecessary duplication of cover. However, occasionally there are two or more policies which operate for the same risk. Plant, for example, may be insured under contract works policies, engineering policies or motor policies. Which policy operates will depend on the respective terms and conditions.

When a decision is made that policy cover is in order consideration can be given to the liability of the insured and other parties involved, whether at common law, under statute or under contract.

Investigation

Whatever class of policy is involved there is no substitute for an adequate and thorough investigation into both the causes and the amount of the loss, injury or damage. Under the insurances required by the ICE Conditions of Contract (that is, contract works, employers' liability and public liability), a site investigation is usually required unless the claim is for a very small amount.

On receipt of the notification of a claim (a first intimation is often by telephone) the insurers will usually issue a claim form. This gives the insured the opportunity to put down in writing the basic facts of the accident, the names and addresses of witnesses and in the case of liability insurance to indicate whether a formal claim has been made. On the basis of these outline facts the insurers can decide the extent to which they wish to investigate. In a very few cases it may be possible to make a sufficient number of enquiries by letter or telephone but in the vast majority of claims physical action will be necessary.

Investigations will be directed towards establishing the facts and thereby deciding whether the policy operates and to what extent the insured and other parties are responsible. The site will be examined and photographs may be taken. Meteoro-logical, police or fire brigade reports may be obtained and action taken according to these reports. In liability claims information for investigation is often obtained from the claimant's solicitor. It is not always easy to establish the true facts. Sometimes the person or persons responsible may be reluctant to admit the extent of their negligence; whilst no one may lie, the whole truth may not be told. Sometimes there may be no witnesses to an accident; where there are witnesses, there may be a complete conflict of evidence. Investigation must be thorough and complete for only by the establishment of the complete story can the claim be handled efficiently. Knowledge is strength; it may enable an opponent's arguments to be demolished; it will enable the extent of liability to be gauged accurately and for the claim to be settled for the correct amount.

Where witnesses are available they should be interviewed as soon as possible – memories fade very quickly and it is desirable that any discrepancies in witness statements should be resolved at an early date. Where possible the statements from the witnesses should be in writing, signed and dated. The mere recording of an interview by a claims official is insufficient, particularly where the evidence of the witness is vital. The witness should be asked to say only what he saw or knows from his own point of view, not what he may think he saw or knows because questions which are too leading have been put to him by the investigator. The witness should be encouraged by patient enquiry to reconstruct the circumstances of the accident. Often a witness will make a positive statement but when asked to be positive in writing will be far less decisive. In this way one can sort out the good witnesses from the bad ones. With a liability claim the quality of the witness should be assessed; if necessary, would he make a good witness if the case ultimately proceeded to court? There should, however, be no delay in contacting witnesses. In the building and civil engineering trades there is much movement of labour and even after a short time it may be extremely difficult to trace a vital witness.

The essentials of any investigation are, therefore, speed and thoroughness. If a liability case gravitates to solicitors after a period it is unsatisfactory if there are unresolved questions and the investigation has to recommence. It is even more unsatisfactory if a case is heard in court and the insurer's argument is completely demolished by one piece of evidence which had not previously come to light, but which could have been discovered in the initial stages by a more thorough investigation.

Insurers, especially with claims involving property damage (either to the contract works or the property of third parties) may wish to consider the employment of loss adjusters.

Employment of loss adjusters

Small contract works claims are usually settled by insurers as soon as possible with a minimum of investigation. In any event the policy excess will often eliminate the very small claim. Where, however, there is a larger claim involved it is common practice to employ loss adjusters, and this method is also favoured where a risk is scheduled and there are several different insurers involved. The adjuster will investigate the circumstances of the claim and consider the respective liabilities of the parties under the contract conditions; he will assess the damage caused and evaluate the cost of repair if necessary in conjunction with a quantity surveyor or other professional help which may be necessary. Sometimes insureds may make arithmetical mistakes in formulating a claim, sometimes they include items which have not been damaged in the particular incident.

Although the adjuster is engaged by and paid by the insurers, he will act impartially and do his best to settle the claim on a basis which is fair and equitable to all parties. He will take instructions from the insurers on the scope of their policy cover and will negotiate as required to bring about a just settlement.

The insured's brokers may also give assistance in the handling of claims. Adjusters will normally report direct to the insurers – this is essential if there is any dispute between the insured and the insurers, to avoid embarrassment to the brokers who may be under an obligation to disclose to the insured the contents of the adjuster's report.

There can sometimes be a problem regarding the disclosure of reports. In *Waugh v British Railways Board (1979)* the House of Lords held that in order to establish privilege it is not sufficient that a document should come into existence for a series of purposes, one of which is obtaining legal advice. It is necessary that the dominant purpose for which the report came into existence was for the purpose of obtaining legal advice. This question arose in *Victor Melik and Co Ltd v Norwich Union Fire Insurance Society Ltd (1980)* where following a burglary the plaintiffs (the insured) sought disclosure of an adjuster's report. It was held that the dominant purpose of the report was to enable the insurers to ascertain the facts so that they could reach a decision as to whether or not they should rely on a clause of the policy to repudiate liability. Obtaining advice from solicitors was a secondary purpose. In the circumstances of the case the claim for privilege was not made out and the report therefore had to be disclosed.

In this case a question also arose on the interpretation of the term 'without prejudice'. The insurers argued that discussions between the insured and the adjuster had been on a 'without prejudice' basis and that the assessor's report ought not to be admitted as evidence. The judge considered that the discussions were not part of the investigations into liability and the report was not therefore subject to privilege. The use of the term 'without prejudice' by the insurers had meant they wished to avoid it being thought that by investigating the claim they were not reserving their right to repudiate.

The employment of loss adjusters in connection with bodily injury claims is perhaps less satisfactory. Many insurers prefer to carry out investigations by their own staff, perhaps supplemented by expert advice as required (for example, from engineers). In the event of legal proceedings solicitors will be employed to handle matters and to give advice on the legal problems involved.

Investigation of contract works (material damage) claims

Investigation will be directed towards discovering the cause of the loss or damage and the amount it will cost to put matters right. The contract works policy is one of indemnity, so the insured is entitled to be indemnified, neither more nor less. If the insured is not responsible for repairing or reinstating the damage the contract works policy will not operate. Although the contractor under the ICE Conditions is responsible for the works and is required to insure on an 'all risks' basis there are 'excepted risks' (see chapter 2).

For example, although the contract works policy may operate in respect of riot, (there is usually no exclusion for this peril) riot is an excepted risk under the ICE

Conditions. Therefore for a contract under these conditions the employer, not the contractor, is responsible for riot damage. As the contractor is not responsible the contract works policy will not offer any indemnity to the contractor (although it may to the employer). However, the policy usually excludes risks which are the responsibility of the principal under the conditions of the contract. Although the principal may have accepted responsibility under the contract there will thus be no indemnity available to him under the contractor's insurance policy. In this example, therefore, investigation will be directed towards establishing whether there has been a riot or not. The damage may have been caused by malicious persons or vandals in which event the policy would operate and the contractor would be indemnified.

How much it will cost to repair the damage or to replace the lost or damaged property is a question of fact – not always easy to establish. Experts in the particular field may have to be employed, estimates may have to be obtained from the contractor, from subcontractors and others. The experienced loss adjuster will collate and co-ordinate. The aim is to put the insured, as far as possible, in the same position as he was before the loss. However, indemnity is always subject to policy terms and conditions and there are frequently uninsured losses. An accident inevitably involves inconveniences and delays; many of the consequences of an accident are either not insured or are uninsurable.

Third-party property damage claims

Investigations here are directed to the same ends as with the contract works cover – the cause of the loss or damage must be established, the liability of the insured must be decided and the cost of the repair or replacement must be worked out.

Here it is a question of indemnifying the insured against his legal liability and negotiations for settlement are with a third party not with the insured. The insurer owes no special duty to the third party, the duty is to the policyholder, to indemnify him under the terms of the policy and to settle any claim on the best terms that can be arranged. In this respect the interests of the insured and the insurer coincide.

Bodily injury claims

Bodily injury claims fall under two main headings, those made by third parties and those made by employees. Although the legal liability of the employer may arise in different ways, the investigation of the accident and the calculation of the quantum, involves similar procedures.

It is only by establishing the cause of the accident and by the interviewing of witnesses promptly that the extent of liability can be ascertained. The liability of an employer to his employee, for example, can arise out of:

(a) acts of personal negligence by the employer;
(b) failure to exercise reasonable care in the selection of competent servants;
(c) failure to take reasonable care that the place of work is safe;
(d) failure to provide a safe system of working;
(e) failure to take reasonable care to provide proper machinery, plant and

appliances, and to maintain them in a proper condition;
(f)   the negligence of a fellow employee;
(g)   breaches of statutory regulations or duties;
(h)   liability assumed under contract.

Accidents involving third parties may arise out of negligence, nuisance, strict liability, trespass, breach of statutory duty, or from a liability assumed under contract. See chapter 1.

For example, the Health and Safety at Work etc. Act 1974 is a comprehensive piece of legislation and is not confined to accidents to employees. There is a duty on the employee to care for the safety of himself and of other persons, that is, both fellow employees and third parties. Furthermore, manufacturers and suppliers of articles or substances for use at work must ensure that as far as reasonably practicable they are safe when properly used. Employers must conduct their activities so as to ensure that members of the public are not exposed to health and safety risks.

Registration of claims

Whenever a possible claim is intimated to insurers, the problem may arise whether the claim should be registered formally. With a material damage claim under the contract works policy there is usually little difficulty, unless, from the outset, there is clearly no cover under the policy. The claim will be registered and the insurers will create a suitable reserve, either on an individual or an average basis.

With a liability claim, public or employers', the position may not be so straightforward. A formal claim may not have been made against the insured, it may initially be a question of just reporting an accident. The accident may be serious or a minor one, but until insurers have investigated they do not know whether their insured is liable or not. If they register too many accidents as claims and a number of these claims do not materialise, there is a danger that greater claims reserves will be created than are really necessary. Each insurer will have to decide its own policy in this respect and usually there is an attempt to decide when an accident is reported whether a claim is likely to materialise or not. If the accident is obviously a minor one, the injured person has not been absent from work and there is no immediate evidence that the insured has been negligent, then probably there is no need to register the accident as a claim. However, with an accident involving serious injuries, the accident should be registered, for the injured party will usually make every effort to recover some compensation.

Inevitably, in a number of accidents registered, claims will not be made. In other cases claims will materialise for non-registered accidents. Insurers from their experience will try to avoid having too many incidents in either category. Reserves are often calculated on an average basis according to the number of outstanding claims at any one time and it is important that reserves allocated should be on a realistic and factual basis. Too many 'nil' settlements can distort an insurer's statistics.

Incidents not registered are kept in a 'no action' file which can be referred to easily if there are developments at a later date.

Defences to liability

There are several possible defences to a liability claim, for example:

(a) a simple denial of liability by the defendant;

(b) the claimant has been guilty of contributory negligence (see later in this chapter for more detail);

(c) the claim is statute-barred or 'out of time' — normally actions must be commenced within three years for bodily injury claims or within six years for property damage claims;

(d) the claimant was acting outside the scope of his authority or doing an unauthorised act;

(e) the accident was due to the unauthorised act of a stranger;

(f) act of God or inevitability;

(g) *volenti non fit injuria* (to him who is willing there can be no injury);

(h) remoteness of damage;

(i) emergency or necessity.

Negotiation

Bodily injury claims and third-party property damage claims will involve negotiation before settlement can be reached. Negotiation may be with a solicitor or direct with the third party. It is necessary to be well prepared, to be aware of the strength of one's own case and the weaknesses of the plaintiff's case.

If a settlement cannot be reached it usually means that one side has either underestimated or overestimated the value of the claim.

A court action may be necessary to decide either liability or quantum or both. Insurers, however, are very conscious of the costs of legal action and within reason will do all they can to avoid it.

Repudiation of claims

No insurer wishes to repudiate liability to its policyholders; it is much more satisfactory if a full indemnity can be offered within the terms of the policy. Denial of liability to a third party on the insured's behalf is, of course, an entirely different matter — here the insurer is implementing the policy and providing the indemnity to the insured.

Regretably, however, repudiation to the policyholder is sometimes necessary and there are several circumstances under which this might arise:

(a) *Policy not in force.* The insurance may not have been renewed, or although there may have been an intention to renew, the premium remains unpaid, the 'days of grace' period having expired. Although the policy may be current the loss or damage may have occurred before the policy became operative.

(b) *Loss or damage caused by an uninsured peril.* The policy, for example, may exclude defective design risks and loss or damage by this cause would therefore give the insurers the right to repudiate.

(c) *Breach of policy conditions.* There are conditions which are:
  (i) precedent to the policy — these must be observed for the in-surance to be valid from inception;
  (ii) subsequent of the policy — these refer to matters arising after the contract has been completed and affect the validity of the policy from the date of breach;
  (iii) conditions precedent to liability — these only affect the claim which the breach of the conditions concerns.

There is usually a specific condition which states that the due observance of the terms, provisions, conditions and endorsements of the policy in so far as they relate to anything to be done or complied with by the insured and the truth of the statements and answers in the proposal and declaration shall be a condition precedent to liability.

Conditions precedent to the policy

Insurance contracts are based on the principle of good faith and an insured is required to declare all material facts at the inception of a risk, and at subsequent renewal dates. Every circumstance is material which would influence the judgment of a prudent insurer in fixing the premium or determining whether he will take the risk (section 18(2), Marine Insurance Act 1906).

When a proposal form is completed the insured warrants the truth of the statements and answers in the proposal and declaration and therefore a misstatement in the proposal form (whether material or not) gives the insurer a right to repudiate a claim. For example, in *Allen* v *Universal Automobile Insurance Co Ltd (1933)* an insured was asked on a proposal form, 'What was the actual price (of the car) paid by the owner?' He replied £285, whereas he had only paid £271. Because the answer was inaccurate the insurers were held to be entitled to repudiate liability for the claim.

Although it is difficult to imagine such a reason for the repudiation of a claim being accepted today, nevertheless by warranting the truth of the statements and answers in the proposal and declaration, the insured is under a very strict duty and a breach of the warranty gives the insurers absolute powers to repudiate liability. In *Lambert* v *Co-operative Insurance Society (1975)* the insured had signed a proposal form for a policy to cover her own and her husband's jewellery. No questions were asked about previous convictions and the insured did not declare that her husband had been convicted of a crime of dishonesty some years before. Later the husband was convicted for a second offence of dishonesty and this was not disclosed to the insurers when the policy was renewed. It was held that:

(a) the insured was under a duty to disclose the original conviction and the subsequent conviction when renewing the policy;
(b) there was no obvious reason why the rule of disclosure in marine insurance should differ from other insurance;
(c) what was material was that which would influence the mind of a prudent insurer.

It has not been unknown for insurers to take advantage of purely technical reasons for the repudiation of a claim if they really wish to repudiate for some other reason. For example, they may be satisfied that the claim is a fraudulent one, but rather than allege fraud with its attendant risks, they prefer to deny liability on some purely technical mistake or misstatement by the insured. Although misstatements or inaccuracies in proposal forms, or non-disclosure of material information, can give an insurer the right to repudiate a claim, most insurers will not now exercise this right unreasonably. The Statement of Insurance Practice issued by the British Insurance Association on behalf of member companies states, *inter alia*:

Except where fraud, deception or negligence is involved an insurer will not unreasonably repudiate liability to indemnify a policyholder:
  (i)   on the grounds of non-disclosure or misrepresentation of a material fact where knowledge of the fact would not materially have influenced the insurer's judgment in the acceptance or assessment of the insurance;
  (ii) on the grounds of a breach of warranty or condition where the circumstances of the loss are unconnected with the breach.

(This nullifies *Allen* v *Universal* mentioned earlier.)

Insurers have also agreed that payment of claims will be made without avoidable delay once the insured event has been proved and the entitlement of the claimant to receive payment has been established.

This statement only applies to policyholders insured in a personal capacity so it is inapplicable to the bulk of contract works policies. However, most insurers will attempt to apply the 'prejudice' test. Have they been prejudiced by the insured's failure to disclose information, or by the breach of a condition? If they have not been prejudiced then most insurers will handle the claim.

Certain insurers are members of the insurance ombudsman scheme and others are members of arbitration schemes. These offer a means for dissatisfied policyholders to pursue their complaints. However, the service is only available to policyholders insured in a private capacity.

Conditions subsequent of the policy

An insured must advise the insurers of any change of risk during the period of the policy. There is also usually a general condition in policies that the insured will take reasonable precautions to prevent accidents and to safeguard the property insured. This may be classed as a condition subsequent of the policy although it is also a condition precedent to liability.

This condition can sometimes be difficult to interpret and in practice it is only rarely that an insurer will be able to or will wish to repudiate liability on the grounds that an insured has failed to take reasonable precautions. Occasionally following an accident or loss an insurer may have issued instructions or advice about future methods of work or safety precautions. A wilful refusal to accept such advice might justify a repudiation (for example, concerning the fencing of machinery in a joinery shop).

Under liability policies (which offer an indemnity against negligence) it is not practical to say to an insured 'you have been negligent in not taking proper precautions and therefore the policy cannot apply'. The condition has always been interpreted as only operative where an insured behaves blatantly or wilfully after due warning. In *British Food Freezers Ltd* v *Industrial Management for Scotland (1977)* contractors had taken no precautions whatsoever to prevent fire damage during work involving use of oxy-acetylene burning equipment. The court held that the complete lack of fire precautions meant the insured was in breach of a duty imposed on them and this breach excluded the insurer's liability under the policy.

An earlier case, *Woolfall and Rimmer Ltd* v *Moyle (1941)*, had held that delegation to a competent foreman was sufficient to discharge the insured's duty to take reasonable precautions.

The Employers' Liability (Compulsory Insurance) General Regulations 1971 provide for the prohibition of certain conditions in policies of insurance:

Any condition in a policy of insurance issued or renewed in accordance with the requirements of the Act which provides (in whatever terms) that no liability (either generally or in respect of a particular claim) shall arise under the policy, or that any such liability so arising shall cease:

    (a)   in the event of some specified thing being done or omitted to be done after the happening of the event giving rise to a claim under the policy;

    (b)   unless the policyholder takes reasonable care to protect his employees against the risk of bodily injury or disease in the course of their employment;

    (c)   unless the policyholder complies with the requirements of any enactment for the protection of employees against the risk of bodily injury or disease in the course of their employment; and

    (d)   unless the policyholder keeps specified records or provides the insurer with or makes available to him information therefrom,

is prohibited for the purposes of the Act.

It is not therefore possible to repudiate an employer's liability claim because of a breach of certain policy conditions. Whatever the seriousness of the breach the claim must be handled on the employer's behalf. After payment to the employee the wording of the policy often provides for a right of recovery from the employer.

Other conditions may be relevant to the acceptance of a claim. For example, under a contract works policy there may be a condition or warranty which requires non-ferrous metals to be kept in securely locked premises overnight. Theft of such goods from an open yard at night would clearly be outside the scope of policy cover.

Conditions precedent to liability

Policies contain conditions about the steps which an insured must take after the occurrence of an accident or loss or damage. These may be summarised as follows:

    (a)   The insured must give immediate notice of the occurrence.

    (b)   All communications, writs, notices of prosecution, inquest or inquiry must be forwarded immediately to the insurers.

    (c)   No admission of liability must be made or offers put forward in settlement.

    (d)   The insured must co-operate with the insurers and act as reasonably requested.
    (e)   Loss by theft or malicious acts must be reported to the police.

Immediate notice of accident

It is perhaps true to say that one of the major problems insurers face is the delay in reporting accidents.

The insurers require prompt notification in order that they can investigate the loss whilst the evidence is still fresh. Insurers also require prompt notification of claims in order that they can create adequate reserves. 'IBNR' claims (that is, claims incurred but not reported), can be a problem, but this arises more with liability claims than with material damage ones. Where there is material damage to contract works, there is usually little problem for the contractor's pocket is directly affected and reimbursement will be sought as soon as possible. Occasionally damage may occur during the period of the policy, but not be reported until much later.

Provided there is no undue delay, then normally the claim would be handled by the insurers. However, damage may appear some considerable time following completion of work without there being any evidence that the damage occurred during the period of the policy. In such an event insurers would not be liable under the contract works insurance. With claims by third parties and employees, however, there is often serious delay with consequent prejudice to an insurer. After the lapse of many months, or perhaps years, it is not possible to investigate the circumstances of an accident satisfactorily. Witnesses will have disappeared or may be dead. Vital evidence may have been destroyed, inspection of plant or machinery may have little value after the lapse of time. The insurers will be severely handicapped in their handling of the claim and in attempting to refute the allegations of the plaintiff.

Occasionally the first intimation of a claim to the insurer will be the forwarding of a writ served on the employer. Apart from the problem of investigation, unnecessary legal expenses may have been incurred. The action taken by the insurer in dealing with a claim under a public liability policy will depend on the extent to which they have been prejudiced by the insured's failure to report the accident and also the extent to which the insured are at fault. Occasionally the insured may be genuinely unaware of the accident, or unaware that they could be involved. The issue of a third-party notice might be the first intimation.

Insurers generally will not attempt to take advantage of the policy conditions unless they have been seriously prejudiced by the insured's failure to report the accident within a reasonable time.

It is not possible to repudiate an employers' liability claim because of delay in notification, and prejudiced or not, insurers must handle it to the best of their ability and make payment to the employee if the employer has a legal liability.

Immediate forwarding of documents

If correspondence about the claim and writs served on the insured are not passed to the insurers immediately, serious prejudice can arise. Legal expenses might be increased, a court action might not be defended and judgment may be given in default. In such an event most insurers would be reluctant to make any payment under their policy.

In *Farrell* v *Federated Employers' Insurance Association Ltd (1970)* it was held that insurers had been prejudiced because they were unaware of the issue of a writ and the fact that judgment had been obtained against the employer. This was an employers' liability claim, the accident having happened before the operation of the compulsory insurance laws. Despite the serious prejudice an employers' liability claim in these circumstances would now have to be handled as the law intends that an employee's rights should not be prejudiced by the failure of the employer to comply with policy conditions.

With public liability claims, however, breaches of policy conditions which involve prejudice to an insurer afford sound grounds for repudiation of policy liability.

Admission of liability

Insurers insist on complete control of their claims. Sometimes an insured may attempt to settle a small claim without reporting the matter to his insurers. He may fail to settle, finding the sums involved larger than originally thought and then tell his insurers. If negotiations have taken place on the basis that liability has been admitted the insurers may find that the settlement figure is higher than it should be. On the other hand there may be a complete defence to the claim which the insured's admission has destroyed.

Co-operation by the insured

Insurers expect full co-operation from their policyholders. If a claim is to be settled quickly and economically it is essential that the insured co-operates in supplying full details of the accident. The insured must not destroy vital evidence — he must co-operate, where necessary, for example in making machinery available for inspection. He must assist in making witnesses available, he must reply to correspondence promptly, he must not attempt to deal with claims correspondence (for example, letters from solicitors) on his own. He must do nothing which hinders the proper investigation and conduct of the claim.

In the event of property damage the insured must take all reasonable steps to mitigate the loss. Failure to co-operate is rare, most insureds seek settlement of claims as quickly as possible. However, in extreme cases, the failure of an insured to supply necessary information or to comply with insurers' requests could involve a repudiation.

Reports to the police

Thefts and malicious acts should always be reported to the police. This is required partly to ensure that the loss is a genuine one; however, there is always the possibility of the recovery of stolen goods and even the arrest of offenders. If available a police report can be helpful.

Reluctance to report a theft may raise a question about the bona fides of the insured. This is not to suggest that fraudulent theft claims are common under contractors' policies. However, an insured making a grossly exaggerated claim would not welcome a detailed police investigation.

Non-contribution

Many policies contain a clause which states that the insurers are not to be called upon in contribution and shall pay any loss in and so far as this is not otherwise recoverable under any other insurance. This is an important factor to consider before a claim is paid. If there are two or more policies involved where the rules of contribution apply then each insurer will have to share the claim proportionately: *Gale* v *Motor Union Assurance Co Ltd*, *Loyst* v *General Accident Fire and Life Assurance Corpn Ltd (1928)*.

When two or more liability policies are issued insuring the same interest but involving different single accident indemnity limits there may be difficulties in deciding how the settlement is to be apportioned. In *Commercial Union Assurance Co Ltd* v *Hayden (1977)*, Hayden was the representative underwriter of a Lloyd's syndicate which had issued a policy with a limit of £10,000. The Commercial Union had a limit of £100,000 under their policy. Both policies contained a clause that in the event of the existence of another policy covering the same risk the insurer should not be liable for more than a rateable proportion. A third-party claim was settled for £4,425. The Court of Appeal decided that apportionment should be on an 'independent liability' basis and not on a maximum liability basis. This meant that in this case claims up to £10,000 would be shared equally. Over this amount, again applying the independent liability basis, claims would be settled in proportion to the respective liabilities; for example, with a claim of £40,000 the apportionment would be in the ratio of 4:1. The Commercial Union would have been responsible for £32,000 and Lloyd's for £8,000.

Subrogation and recoveries

Having investigated a loss, satisfied themselves that the claim is correctly one for payment and the amount claimed is reasonably correct, the insurers will then consider if anyone else can contribute towards the payment. The principle of subrogation is important and insurers after paying a claim can take over any rights or remedies the insured may have against another party (policy conditions usually permit them to do this before paying). The loss may have been caused by a negligent subcontractor — provided negligence can be proved there should be a good chance of reimbursement. The subcontractor will almost certainly be insured for an indemnity of £250,000 (especially following the Finance (No. 2) Act 1975, although the public liability requirement whereby insurance was a prerequisite for the issue of a tax certificate has now been cancelled by the Finance (No. 2) Act 1980). The indemnity limit under a public liability policy may still be inadequate to meet the claim. Fire damage especially may be particularly expensive. The liability insurers may, for example, deny liability because of a breach of their fire warranty or because their premium has not been paid. Public liability insurance is not compulsory and the contract works insurers have many hurdles to overcome before they are reimbursed.

Subrogation often involves the public liability insurers of a subcontractor. *Balfour* v *Barty King and Others (1957)* is an example of this type of subrogation action although the property insurers were not covering a building in course of erection in that case. Nevertheless, the public liability insurers of the third party had to reimburse the property insurers for their outlay. Furthermore, Balfour's case involved an employer's property insurers recovering from a contractor's public liability insurers and it also illustrates the absolute liability of the employer for the contractor's activities, as well as for negligence for spreading fire to property other than that on which the contractor is engaged — in accordance with the *Rylands* v *Fletcher* rule. However, the wording of any subcontract may alter the common law situation just described.

Occasionally there may be another insurance policy which can also deal with the loss, for example, an engineering policy for plant. If plant is hired in under the Model Conditions for the Hiring of Plant and damage has been caused by the negligence of the owner's driver, the argument may be put forward that because of the Unfair Contract Terms Act 1977, the indemnity given by the hirer is unreasonable and should not apply. This is probably a forlorn hope between two parties of equal bargaining power.

Contributory negligence

Contributory negligence arises where both the defendant and the plaintiff are negligent. Before 1945 at common law if an injured person was guilty of contributory negligence, the defendant escaped altogether. Under the Law Reform (Contributory Negligence) Act 1945 blame is now apportioned and the plaintiff's contributory negligence only has the effect of reducing the amount of damages recoverable. In *Jones* v *Livox Quarries Ltd (1952)* Lord Denning said:

A person is guilty of contributory negligence if he ought reasonably to have foreseen that if he did not act as a reasonable prudent man, he might hurt himself and in his reckonings he must take into account the possibility of others being careless.

The principle of contributory negligence applies to all torts. The court has to find and record the total damages and then deduct a percentage according to the degree of contributory negligence appropriate to the circumstances. If a claim does not proceed to court it is open to the parties to agree between themselves an appropriate deduction for contributory negligence.

Civil Liability (Contribution) Act 1978

If there is a possibility of another party being responsible for a payment it is preferable if the extent of their liability is agreed before the payment to the plaintiff is made. It is then not so much a question of recovery as a joint payment on the basis of joint liability. This, however, may not always be possible, in which event, after payment, recovery may be attempted for some or all of the amount involved.

Where there is a joint liability it is usually possible to join the other party in the action. However, one person may be liable under contract and the other in tort. For example, a house may fall down and two persons are at fault, the architect who has breached his contract and the local authority whose building inspector negligently approved the work of the house. Before the passing of the Civil Liability (Contribution) Act 1978 no contribution could be sought by the architect from the local authority as legislation only provided for contribution between joint tortfeasors. However, the position has now been remedied. The Act makes provision for contribution between persons who are jointly or severally or both jointly and severally liable for the same damage and in certain other similar cases where two or more persons have paid or may be required to pay compensation for the same damage. Any person liable in respect of any damage suffered by another person may recover contribution from any other person liable in respect of the same damage (whether jointly with him or otherwise). The amount of contribution recoverable is such as shall be found by the court to be just and equitable having regard to the extent of that person's responsibility for the damage in question. The right to recover compensation under the Act applies whatever the legal basis of the liability, whether tort, breach of contract, breach of trust or otherwise.

Where any person becomes entitled to a right to recover contribution in respect of any damage from any other person, any action must be brought within two years from the date on which the right accrued.

Another example showing a gap in the common law which this Act remedied occurred in *McConnell* v *Lynch-Robinson (1957)*. The plaintiff engaged an architect to draw plans and supervise the building work of a new house for P. By a separate contract the plaintiff engaged a builder to undertake the work. The builder in breach of contract, failed to provide an adequate damp course and the architect, in breach of contract failed to notice the error. The plaintiff expended a substantial sum to rectify the error and sued the architect, who was unable to join the builder as a third party as the plaintiff's claim lay in contract and not in tort and contribution could not be ordered.

Policy excesses

An insured may be required to be responsible for a certain amount of each claim – this is known as an excess. A contract works (material damage) policy will invariably carry an excess, which will be deducted from the amount of the final claims settlement. If payment to a third party is involved the insured should pay the amount of excess to the insurer before the insurer settles.

The level of excess will vary according to the individual contractor and the type of work he is doing. A small contractor's contract works policy may be subject to an excess, or deductible as it is sometimes termed, of say £50 or £100. It is common to have a higher excess for risks such as storm, tempest, flood, subsidence and collapse. More hazardous work again will involve a higher excess and a figure of several thousands of pounds could apply to a sea defence contract or harbour work, for example. Insurers like to have excesses to encourage contractors to take care, to reduce the claims costs and cut out the administrative work in handling too many small claims. Insureds accept excesses because it enables their premiums to be reduced or makes it possible for risks to be accepted at premiums they consider are reasonable. The excess applies to each and every incident, that is, a claim or series of claims arising out of one event, and it is therefore important to establish the number of events which relate to the claim being made.

Whether one excess or many excesses apply can make a considerable difference to the final claims cost but it is not always easy to decide on the number of events. For example, suppose a wall is damaged by frost and the bricks begin to flake or spall. If frost had occurred on several nights during the preceding month, how should the claim be settled? Should one excess be deducted or should the excess apply to each night of frost or to damage over a lesser period? There is no simple answer and each claim can only be decided on the individual facts involved.

Plant may be hired out and returned months later for extensive repair with the allegation by the hirer that damage was due to the action of vandals. Basically responsibility would remain with the hirers under the CPA Conditions but could damage have occurred in a single incident? If an insured is unable to establish any accidental cause, then in this example it must be accepted that the machine was badly used, damage was caused by wear and tear and there is no cover under a contract works policy.

A public liability policy also frequently bears an excess for property damage claims, either on a general basis or for a particular type of claim.

An excess is particularly suitable where there tends to be a number of smallish claims of a similar type, or where it may be felt that some claims are almost inevitable bearing in mind the type of contract being undertaken. An example is damage to underground property caused by a road and sewer contractor.

Again, the wording usually refers to claims arising out of one event or it may refer to occurrences arising out of one original cause. Occasionally an 'aggregate' excess might apply, that is, the excess will relate to all claims which arise during the period of insurance or during one contract. For example, piling work often causes vibration damage to buildings, damage may be cumulative over a period of weeks or months. It is not usually possible to say exactly when damage occurred, the driving of each individual pile may have added to the damage, the driving of one particular pile may have been the principal cause of the trouble. For pile driving operations, therefore, it is common to apply the excess overall to the particular contract concerned. This avoids the problem of having to decide on the number of occurrences or events.

It is not usual to apply an excess to bodily injury claims – partly because insurers wish to control such claims from the outset. However, occasionally a large contractor may wish to insure what is virtually the catastrophe risk and is prepared to accept responsibility for claims of all types up to a substantial figure.

Occasionally a policy is subject to co-insurance clause where a fixed percentage of each claim settlement becomes the responsibility of the insured. Another variation is a franchise where the insurer does not become involved until the claim exceeds a certain figure and then is responsible for the total amount.

Whichever method is adopted to give the insured a financial interest in claim settlements, it is still essential for the insured to report accidents, loss or damage as soon as possible and for the insurers to control the settlement. In practice difficulties can arise where there is a relatively substantial excess under a policy and the insured attempts to settle an apparently straightforward property damage claim without reference to his insurers. If the claim really is straightforward insurers will not know about it, but many simple accidents can produce complications. For example, damage to an underground cable may involve only a modest repair bill. On the other hand the bill might run into many thousands of pounds if, for example, a Central Electricity Generating Board main cable is cut. Apart from the physical damage to the cable there may be legitimate economic loss claims.

Ex gratia payments

Sometimes there is no cover under the policy for a claim but a payment is made by insurers on a 'without prejudice' basis. Such payments are made entirely at the discretion of individual insurers. They may be influenced by economic factors. A payment may be made to preserve goodwill if the circumstances of the claim are such that it is felt that otherwise hardship would accrue. Sometimes the extent of cover may not have been explained adequately to the insured or there might be some ambiguity in the policy wording.

If the policy does not cover a certain risk and cover would not normally be available for that risk, it would be unusual for an *ex gratia* payment to be available.

Ex gratia payments are for the benefit of the insured not for any third party. Payments may be made to a third party on the insured's behalf, but no special duty is owed to the third party. The insured may be given the benefit of the doubt over a policy wording, thus enabling the policy indemnity to operate and for the claim to be paid. Sometimes the *ex gratia* payment will take the form of a contribution towards the settlement, leaving the insured to find the balance.

Chapter 20

Contract works claims

Policy cover

Before considering the nature of the claims involved under a contract works policy it is necessary to recall the extent of the insurance cover available.

A contract works policy is a material damage insurance, which, to meet the requirements of the ICE Conditions, offers protection on an all risks basis. Although cover is inevitably subject to certain exceptions and limitations, policies generally are wide in their scope with the consequence that claims can arise and do arise under every conceivable and inconceivable heading. Generally, however, the loss or damage must involve an element of fortuity. A contract works policy is one of indemnity, that is, in accordance with recognised insurance principles, the insured, as far as possible, is to be placed in the same position after the loss as he was before; he is to be indemnified, neither more nor less. The insured under the policy may just be the contractor but more usually the contractor and the employer are joint insureds, both are entitled to the benefits of the policy in so far as they can apply. The ICE Conditions require insurance in the joint names. Large projects (particularly for overseas risks) may involve a form of package insurance where the employer, the contractor, the subcontractors and perhaps even the architects and consulting engineer are all jointly indemnified.

Types of claim

All or any of the items insured can be lost or damaged in a wide variety of circumstances, act of man, act of God, or sometimes apparently act of the devil. Dams burst, bridges fall, buildings collapse, goods mysteriously disappear, the causes seem endless. The type of claim may vary according to the type of work or the nature of the construction involved.

Fire damage
For buildings in course of erection or for work involving alterations to existing structures, fire (including explosion) is probably the most common cause of damage and it is also likely to be one of the most costly. Fire can damage the buildings being constructed, the contractor's temporary works, the temporary buildings on

the site, the unfixed materials, plant and employees' tools. Temporary buildings by their very nature, that is, of wooden or light construction, present a considerable hazard; fires frequently destroy canteens and site offices with all their contents. There are little fires and there are big fires but they are all damaging; they delay the completion of the works with possible serious inconvenience to both the employer and the contractor, not forgetting the insurers who will ultimately have to pay for the loss.

The reduction of fire damage is a national problem. Total fire damage for the twelve months ended September 1982 was approximately £405 million compared with £339 million for the preceding twelve months. Regretfully, on contract sites the hazards are not always appreciated; lack of supervision, ignorance, inexperience, all play their part but as in other areas, good housekeeping and good management produce results. The larger contractor will employ safety officers and should have the organisation to impose satisfactory site security control.

The smaller subcontractor may not be so well placed and many fires on contract sites are caused by subcontractors using their blowlamps in confined spaces with disastrous results. The availability of fire extinguishers and water at the site of operations, together with the presence of fire watchers, is important; a fire can often be extinguished before it has taken hold if prompt action is taken. If, however, the fire extinguishers are at the other end of the site then by the time they are brought to the scene it may be too late.

Additional hazards can arise from the occupation of existing structures or from the proximity of other buildings containing hazardous materials. Fire can spread from adjoining buildings just as it can spread to adjoining property. The adjacent plastics factory or timber yard is a peril for both the public liability and the contract works insurers.

Fire damage involves a tremendous waste of effort, of resources and of money. It can mean a loss of jobs, and production, loss of export orders; at best there is inconvenience and delay whilst the damage is being reinstated. At worst there is a loss of life involving pain, grief and suffering of both the victims, relatives and friends (not that such claims would come within the scope of a CAR policy).

Flood damage

Flood (which may be defined as a large movement or irruption of water), if on an extensive scale, can be a disaster for a large number of people. It is very difficult for a contractor to introduce satisfactory protective measures against flood, especially if the work is in the nature of a flood protection scheme. Temporary works may be damaged by flood waters, coffer dams may collapse, partly constructed works may be swept away.

On sewer contracts, flood can be a major hazard, trenches can collapse making extensive re-excavation necessary. Even if the trench does not collapse then the flooding may scour the trench bed, wash away concrete or distort pipes which have already been laid; the debris and flood water will require to be pumped out. It is not unusual for the pumps themselves to fall into open trenches and be damaged.

There had been great concern about the possibility of the flooding of the River Thames and contingency plans were in being for the expeditious handling of insurance claims under all classes of policies. The flood protection barrier has been completed recently.

Wet risks

Risks involving potential sea or water damage (referred to as 'wet' risks), can frequently result in very expensive claims and for this reason underwriters impose substantial excesses for work of this type. The power of the sea should never be underestimated. On civil engineering projects, for example, with the construction of sea defence works or jetties, there is a danger that the whole of the works will be damaged by the force of the sea. Sometimes the work can only be undertaken at low tide, and the incoming tide will damage partly completed structures. Special cements or concretes are available for this type of work but claims are by no means unusual.

Harbours may be comparatively sheltered, but some are more exposed than others. A combination of high tides and strong winds can be disastrous to work in progress. Site cabins may be swept out to sea; plant may be ruined by submersion in salt water.

Apart from the hazards of the elements, there is the risk from ships — temporary works may be damaged by passing vessels; rather like the 'hit and run' motorist, it may be difficult to identify the culprit.

Other perils

Buildings in course of erection are frequently damaged by storm and high winds, gable walls are blown down with resultant collapse of roofs. Bursting and overflowing of water pipes can be troublesome especially if the leak remains undetected for many months. Severe earthquake is unusual in the United Kingdom but there are regular minor earth tremors which perhaps do little more than shake the windows in a building. The possibility of a more serious tremor cannot be entirely ignored. (The most serious earthquake in recent times was in Colchester in 1884.)

'Aerial devices' (a named peril in the Standard Form of Building Contract) might, for example, involve a piece of a space project falling onto a contract site; not very likely, but the insurers must be prepared for every possibility.

Tunnelling work involves many hazards — for example, claims may arise from rockfalls and rockbursts, collapse of tunnel sides, inrush of water, inrush of soil or rock, fire and explosion (especially when compressed air is used), heave or subsidence of ground surface.

Theft on contract sites is a major cause of trouble; there are big thefts in one incident or smaller thefts spread over a period of time. Vandalism or malicious damage is another national problem to which no one has found the answer but which can be serious for the contractor. Fires caused by vandalism account for 30% of all fires and more than 50% of all large loss fires. Site control can help but in certain parts of the country vandalism slows down completion of works and it makes it almost impossible for the contract to be handed over in a satisfactory state.

Subsidence damage is another source of claims which can be very troublesome to handle and which may involve complicated legal and technical issues.

Perils affecting plant

Claims for lost or damaged plant are extremely common and with the amount of plant being used on sites this is not surprising. Theft claims are frequent even for plant with an apparently specialised use. Excavators which sink into the mud can be very costly to extricate and repair. Toppling cranes are even more expensive and pontoons sink never to appear again.

Plant may be insured on a specified sum insured basis and some insurers may import an average condition in order to ensure they are receiving adequate premium. On the other hand, there may be no definite sum insured for plant, all plant used incidental to the contract being deemed to be insured up to its full value. The contract works policy will not normally cover breakdown risks but accidental extraneous damage is insured.

The Model Conditions for the Hiring of Plant, operative from September 1979, are the result of several years' negotiations between the plant hire industry, civil engineering employers and the Office of Fair Trading. They take into consideration the provisions of the Restrictive Trade Practices Acts and the Unfair Contract Terms Act 1977. Nevertheless, the conditions impose onerous liabilities on the hirers and a substantial responsibility is passed to them in respect of loss or damage to plant under their control. It is preferable if hirers insure their liability on an all risks basis.*

Cost of claims

Very large sums of money can be involved if something goes badly wrong. For example, some years ago on a dry dock project the sheer strength of a clay stratum was measured too high by a specialised firm of engineers. Later measurements made during construction by a contractor's engineer indicated values approximately 25% lower but no action was taken. An enormous quay wall 400m long almost completed and back-filled, slipped and became a total loss of about $10 million.

Even with so-called straightforward building work the cost of putting right a defective building can be very large. Fire damage to an almost completed structure can be enormously expensive.

However, there is little doubt that many losses of all types could be avoided or reduced in cost, given a more professional approach to risk management on the construction site.

Investigation and handling of claims

Subject to the general procedures commented upon in chapter 19, insurers or their adjusters will investigate the circumstances of the loss and consider the legal and contractual responsibilities of the employer, the contractor, the subcontractors, the engineer, the architect, and possibly other third parties who may be involved.

* For further details see *Model Conditions for the Hiring of Plant (1979) Implications for Insurers* by F.N. Eaglestone, published by P.H. Press Ltd, Waterloo Road, Stockport 7.

When a claim is made then immediate reference to available contract conditions is necessary to ascertain what liabilities are imposed upon the respective parties. If there is no written contract in force then liability will require to be decided on the basis of the common law or statute law.

In deciding upon the extent of a contractor's liability the actual contract wording must be considered and investigated; even a so-called standard wording sometimes involves major amendments.

A failure to detail or specify insurance requirements does not mean that a contractor or subcontractor will not be liable for a loss. The property on a contract site may be the contractor's own and under his control; it may belong to a sub-contractor and the contractor may or may not have control. The property may belong to the employer and interim payments may have been made for work done. The contract conditions should specify responsibility in the event of a loss and detail who is to effect insurance and the nature of the insurance to be taken out. A contractor, irrespective of contract conditions, is generally responsible for completing the contract and handing the work over to the employer in a satisfactory condition as this is what he undertakes to perform. This may be regarded as an implied term in any contract but it is also frequently an express condition.

In *Charon (Finchley) Ltd* v *Singer Sewing Machine Co Ltd (1968)* there was damage to contract works by vandals. The courts held that if the obligation on the plaintiffs to complete the contract involved doing over again work which had been wilfully damaged, this harsh liability was one they had to shoulder.

Indemnity clauses

There is a difference, however, between the indemnity clauses (which term includes both liability and responsibility clauses as defined in chapter 1) in a contract and the insurance clauses. Indemnity clauses are frequently very wide in their scope and may extend considerably beyond the cover given under a policy or what is available on the insurance market. A contractor may therefore be responsible for matters for which he has no insurance protection, for example, the replacement of defective workmanship, or for delays caused by strikes or non-availability of materials.

Damage to subcontract works

Who pays for damage to subcontract works and for loss or damage to subcontract materials?

The ICE Conditions of Contract say:

The Contractor shall take full responsibility for the care of the works from the date of commencement thereof . . . and shall insure in the joint names of the Employer and the Contractor against all loss or damage from whatever cause.

See clauses 20(1) and 21.

Policies may be issued in the joint names of the employer, the contractor and all subcontractors, for their respective interests, but more usually in the joint names of the employer and contractor only. Can the subcontractors claim the benefits of the main contractor's policy? Generally they cannot unless they are specifically named as joint insureds.

It is necessary to consider the ICE form of subcontract. This should set out the respective responsibilities of the parties. Part I of the fifth schedule will specify the insurances the subcontractor is to effect.

Part II of the schedule provides details of the main contractor's insurances. The main contractor can decide whether to make the benefits of his contract works policy available to his subcontractor. If he does not, then because under clauses 13(2) and 14(1) the subcontract works are at the risk of the subcontractor until the main works have been completed under the main contract — 'and the subcontractor shall make good all loss of or damage occurring to the subcontract works prior thereto at his own expense' — it is clearly necessary for the subcontractor to insure the subcontract works against all risks.

The conditions do, however, suggest there should be no unnecessary duplication of insurance. Thus if the subcontractor is given the benefit of the contractor's policy he should rely on this.

Negligent subcontractors

A subcontractor may damage his own contract works, but he may also damage other parts of the works which do not form part of his subcontract. Where damage has been caused by a negligent subcontractor to other parts of the work a main contractor is entitled to proceed against the subcontractor for the cost of repair of the damage.

Subrogation

If all subcontractors are included as insureds in the main contractor's policy the right of subrogation by the insurers of the policy is lost so far as recovery against any of the subcontracting insureds is concerned (and insurers may require additional premium because of this).

In England this question does not appear to have been litigated*, but it was considered in the USA in *General Insurance Co of America* v *Stoddard Wendle Ford Motors (1966)* when the following statement was made:

The courts have consistently held in the builders' risks cases that the insurance company having paid a loss to one insured cannot as subrogee recover from another of the parties for whose benefit the insurance was written even though his negligence may have occasioned the loss there being no design or fraud on his part.

A similar view was taken in the case of *Commonwealth Construction Co Ltd* v *Imperial Oil Ltd and Wellman Lord (Alberta) Ltd (1977)*. The court felt the word 'insured' in the policy referred to the group called 'the insured' and not merely to the owner and general main contractor. The policy makes it clear that it is only to the rights of the group against outside parties that the insurers are entitled to proceed by way of subrogation.

However, one party to a contract may in principle pursue a claim in tort against

* Now see *Petrofina UK) Ltd* v *Magnaload Ltd (1983)* pages 70 and 90.

the other contracting party. This is not to say a claim will necessarily succeed. The case of *Raymond Batty* v *Metropolitan Property Realisations Ltd (1978)* is relevant. Land on which a house was built subsided and the house itself, although not damaged, would suffer damage in later years. Damages for economic loss were recovered and the claim by the house owner had to be shared equally between the builders and the developers.

Claims problems with subcontractors

Difficulties may arise with contract works claims in respect of subcontractors' work. Insurers usually receive premium for a risk on the total contract price and it may be argued that without question they should deal automatically with the claim under their policy. This is a matter for each individual insurer to decide. If, however, there is a formal contract which regulates the position and a policy is effected to meet the terms of the contract, many insurers will wish to deal with the matter on a strictly legal basis. They will not offer indemnity to a subcontractor, only to their own insured. A subcontractor may therefore find himself liable for repair or reinstatement of work and be unable to claim under the main contractor s policy or any other insurance.

Period of maintenance

Once the contract works have been handed over to the employer and the period of maintenance has commenced, as defined in clause 49(1), the contractor's and the insurer's responsibilities are much reduced. Under the ICE Conditions of Contract, there is a period of fourteen days after the issue of a certificate of completion when it is the responsibility of the contractor to insure. Normally, the insurer will be liable during the period of maintenance only for damage arising from a cause occurring prior to the commencement of the period of maintenance or for loss happening during the period of maintenance caused in the course of operations for the purpose of rectifying defects.

The contract conditions should make clear the levels of responsibility, but in the event of a loss, there are sometimes difficulties. Works may be occupied before a certificate of completion is issued. The employer may introduce materials or start processes without the consent of the contractor or his insurers.

The case of *English Industrial Estates Corpn* v *George Wimpey and Co Ltd (1973)* (referred to later), illustrates the difficulties, with the contractor having to foot the bill, possibly without insurance, because of the policy exception concerning use or occupancy by the employer.

Repair of damage and contractor's right to profit

Where damage has occurred it is usual for the contractor to repair the works himself, although in theory these repair works could be put out to tender and handled by an independent contractor. This would usually be a much more expensive way of handling the claim and far less satisfactory to all concerned. If the work was done by an independent contractor that contractor would include a profit element in his tender price. The question arises therefore, can the insured who is doing the repair work and being indemnified under the contract works policy make a profit out of the operation? The damage originally may have been caused by his negligence. Is it right morally or otherwise that he should be entitled to make a profit in these circumstances and if he does make a profit should it be at a lesser rate than his usual percentage? There is no simple answer to this question. Some insureds may agree to do the job at cost in order to minimise their claims loss ratio especially if they are partly to blame for the loss or damage. Others may seek full payment on the basis that they are entitled to a normal profit for the extra work. A contractor cannot be expected to work at a loss. He has heavy overheads which themselves may have increased because of the claim, he has wages to pay, he must at least cover the cost of labour, materials and overheads in putting right the damage. The amount of profit element if any, to be included, is a subject for discussion by the adjuster but it cannot be ignored.

Where repair work is done by a subcontractor, the main contractor will usually add an 'on cost' charge of say, 5% to cover the costs involved in the supervision of the work and to pay for the administrative charges incurred.

When goods are replaced the adjuster will check the original purchase invoices to ensure that the amounts being claimed represent a reasonable indemnity. Where necessary depreciation will also have to be taken into account.

Rises in costs due to inflation may also have to be considered, especially where there is a considerable time-lag between the date of original purchase and the date of replacement.

The amount of work which has been previously certified as having been completed will be a relevant factor especially if there is a total or substantial loss. There may also be materials purchased and received but not incorporated in the destroyed works and therefore undamaged. The onus of proof must always rest with the insured.

Additional working costs

Another problem may arise over additional working costs in repairing the damage. Extensive overtime may be necessary and/or additional work at holiday times. It will be necessary for the adjuster to check labour rates, but the cost of reasonable overtime would have to be paid. There may be other costs related to bringing the contract back to the position it was immediately before the loss or damage; there may be additional plant to be hired or even special plant to be brought onto the site to enable a particular operation to be performed. The adjuster may have to establish which items are reasonable or not and extensive negotiations may be necessary.

In remote or isolated parts of the country it may be difficult to employ extra labour and there may be physical problems about transporting materials quickly. The costs of repair might be higher than on a normal contract site.

Remedial measures

After damage has occurred it may be necessary for the insured to take emergency steps to minimise the loss and prevent further damage. For example, additional shoring may be necessary if subsidence has occurred or is likely to occur. The policy may not cover all expenditures of this type. However, it is in everyone's interests that steps should be taken to minimise further losses and reasonable additional costs will therefore be regarded by most insurers as forming a legitimate part of the claim.

Different construction methods

Sometimes the reconstruction work may involve different building methods. For example, a site area may have collapsed because the contractor failed to shore the sides adequately or perhaps he should have driven sheet piling and failed to do so. He may have adopted a cheaper method of work through ignorance or with the deliberate intention of saving costs.

The cost of re-doing the work correctly will be greater than in the original estimates; however, insurers should not be saddled with the extra expense of doing the work properly and the cost of the claim should be adjusted to take these matters into account.

Value added tax

The adjuster must also consider the question of VAT and whether it is correctly applied to the cost of the claim. Most large contractors will be registered and therefore any VAT due can be offset by them and will not be the responsibility of the insurers. The supply by the builder of a major interest in all or part of a building or its site is zero-rated. The supply by a contractor to his client of any services in the construction, alteration or demolition of any building or of any civil engineering work is also zero-rated. The term 'construction alteration or demolition' includes extensions, but does not include services of repair or maintenance if it leaves the building or civil engineering work essentially as it was. However, the supply by a subcontractor to a main contractor or by one contractor to another of services in the course of construction, alteration, repair, maintenance or demolition of any building or of any civil engineering work is taxable at the standard rate.

VAT rules can be complicated and where necessary the assistance of the Customs and Excise Commissioners should be sought. The problem of 'repair or maintenance' was considered in *A.C.T. Construction Ltd* v *Customs and Excise Commissioners (1981)*. The appellants had carried out work on the foundations of a number of buildings which had been damaged by subsidence. The original foundations did not comply with modern building regulations and correcting work could not be carried out merely by repairing or replacing the original foundations. Additional foundations were constructed to underpin the building, leaving the original foundations unaltered, and VAT was assessed on the basis that it was work of maintenance. On appeal to the House of Lords in 1981 (upholding the view of the Court of Appeal in 1980), it was held that where work of alteration to a building resulted in the building being substantially and significantly different in character from that which it was before the work was done, the work was not work of repair or maintenance and was therefore zero-rated for VAT. It was a question of degree depending on the facts of each case whether the work done had altered the character or nature of the building, but relevant considerations were whether the work had substantially altered the life of the building, significantly affected its saleability or market value or made good a building which prior to the work was defective because it did not comply with modern building regulations.

Following the case of *Parochial Church Council of St Luke's Great Crosby* v *Commissioners of Customs and Excise (1982)*, the Customs and Excise have reviewed the VAT position for reinstatement work in buildings. They have said that if a building is reduced by fire, or other perils, to ground level, or if no more than a shell remains, the work of reconstruction may be zero-rated. Also, if more than the shell of the building is left undamaged, the work of reconstruction may be broken down into separate elements for tax purposes. This means that any work which would have been zero-rated as an alteration, if it had been undertaken before the fire, may be zero-rated if it is carried out afterwards — so long as it can be separately identified. Until now, all reconstruction work (short of rebuilding from ground level) was regarded as work of repair and maintenance, and thus was standard-rated for VAT purposes.

VAT rules were changed in the 1984 Budget. See appendix 6.

Plant

In claims settlements it may sometimes be necessary to distinguish between temporary works and constructional plant. A problem might arise in connection with the classification of non-mechanical plant such as scaffolding or shuttering. A decision will largely depend on whether the equipment is expected to be reusable elsewhere. Non-mechanical equipment which is specifically designed and constructed for a particular project or for which the total cost (rather than just the cost of hire) is included in the contract price, will generally be regarded as temporary works.

Site shortages

In the building and civil engineering trades there are always shortages on sites. Items go missing, they disappear without trace, barrow-loads of bricks, cement, timber, glass, nails, where does it all go? Some people may build garages with the material they 'win' from a contract site. Others may just 'borrow' a piece of wood to put up a shelf.

The contract works insurer does not wish to pay for the minor thefts which regularly occur, hence the excess, which may be increased according to the vulnerability of the site.

The policy may have a condition to the effect that losses must relate to a specific occurrence which has been notified to insurers, or stock shortages or losses discovered only on inventory-taking may be excluded. For example, a contractor may undertake intermittent checks on unfixed materials during the course of a contract and during the check he may discover that some hundreds of tiles are missing. He does not know why they are missing, all he knows is that he is short of this figure. They may have been stolen in one incident, on the other hand they may have been pilfered on thirty or forty different occasions. The excess should apply to each individual theft so it is up to the insured to prove or attempt to prove the loss and to identify the number of separate incidents involved. This can be a difficult area and often an insured does not appreciate the problems involved. Usually the matter is one for compromise.

Scaffolding often present similar problems; at the close of the contract a contractor dismantling his scaffolding may find he is 25% short. He may never have had the scaffolding on the site or he may have moved it to another site and forgotten about it, but he is short and makes a claim. Occasionally someone may have seen an illicit lorry disappearing down the road the day before but usually there is no proof that the scaffolding was ever stolen, it is just a question of stock shortages. A contract works policy does not pay for unexplained shortages and an insured must produce some reasonable proof of the loss.

Apart from pilfering there may be other reasons why stock or equipment are missing when an inventory is taken, for example:

    (a)   goods never delivered to the contract site, perhaps misappropriated before delivery;

    (b)   incorrect or inaccurate invoices, that is, accounting errors;

    (c)   incorrect calculations of work which has been done;

    (d)   materials damaged by defective workmanship and loss not reported to employer;

    (e)   fraud on the part of the employer (not usual — inefficiency is more common).

Removal of debris

Following loss or damage there may be a good deal of clearing up to be done on a site, debris to be removed, perhaps some demolition and existing structures to be made safe. Most policies give cover for these contingencies and the costs involved will form a legitimate part of the claim.

The reinstatement of the damaged work may also involve new plans and specifications, the employment of architects and surveyors. These costs will also form part of the claim, excluding, however, the fees incurred in preparing the claim under the policy.

Off-site storage

Some insurers offer cover for all goods off-site wherever they may be. Others are more restrictive and confine the main cover to goods on contract sites with a limited extension for goods elsewhere.

There may be a liability under contract conditions for goods stored off sites after certification for payment, for example, under clause 54(3) of the ICE Conditions. Under this clause ownership becomes vested in the employer but the contractor is responsible for insuring the goods. Claims settlements should not present any problems provided that the policy cover is operative and that the ownership of the goods is clearly identified.

Sometimes there may be a duplication of cover, for example, the insured's fire or burglary policy on premises may also insure certain goods which are destined for contract sites. It is necessary to check carefully which is the more appropriate policy. If two different insurers are involved it may be necessary to share the claim.

Transit risks

Most contract works policies cover the contract works, goods and materials in transit to and from the contract site by road, rail or inland waterway. Theft from motor vehicles is common and there can be the problem of shortages on delivery. Vehicle accidents are also common and damage may also be caused by fire.

Goods may be carried in the contractor's own vehicles or by independent haulage contractors. Where loss or damage involves an independent carrier, there may be a possibility of recovery but much will depend on the wording of the contract conditions between the contractor and the carrier.

Sum insured or indemnity limit

The contract works policy will pay a claim up to the specified amount of indemnity. Where a policy is arranged on a single contract basis, the indemnity will be the value of the contract plus any amount which may be decided upon for the inflation element. Where an annual policy is involved, the insured will select a figure which represents the maximum value of any one contract. The policy frequently contains an 'escalation' clause whereby provision is made for a percentage increase in the indemnity limit.

A contract works policy is not normally subject to average and most insurers will expect the premium to be paid on total contract values. They may, however, be prepared to accept, or may prefer, a lower sum insured per contract, enabling insurance to be arranged on a MPL basis (maximum probable loss). This does not affect the handling of a claim except in so far as the indemnity provided is insufficient to meet the cost of a claim.

There may be separate and lower sums insured for certain items, for example, there could be a limit applied to non-ferrous metals. These have proved very attractive to thieves in the past and depending on the circumstances of the risk, insurers may wish to apply restrictive limits.

A separate limit usually applies to tools, plant and temporary buildings. There may also be a separate limit for personal effects and tools of the insured's employees, not exceeding say £100 in respect of any one employee.

Inflation

Inflation has been with us for a number of years and it is unlikely to disappear, although rates have varied at different periods in history.

Inflation was one of the causes of the fall of the Roman Empire; for example, between AD 258 and AD 275 prices rose by nearly 1000%. The collapse of the German mark at the end of the First World War meant that a wheelbarrow of money was needed to buy a loaf of bread. Insurers do not yet have problems of this magnitude, nevertheless the problem of inflation can complicate the settlement of claims.

The cost of rebuilding an extensively damaged structure can be substantially more than the original contract price. Further inflation may take place during the rebuilding period so that the final cost of the work may bear little relation to the original estimates. Contract works policies are not subject to average and the insurers will pay up to the amount of the specified indemnity limit. Many policies incorporate an escalation clause which provides for an automatic increase in the indemnity limit. This offers at least some protection and at the end of the day, the insurers will receive premium on the total amount of work done; nevertheless, at a time of rapid inflation, there can be a gap between the indemnity offered by a policy and the cost of repair or reinstatement.

The contractor must consider the possible impact of inflation on rebuilding costs in the event of serious damage to almost completed works. Contracts for periods in excess of twelve months are not now usually arranged on a fixed price basis so a contractor in arranging insurance must take into account fluctuations in contract price, the possible impact of inflation up to the time of damage, and if there is delay, the impact of inflation from the time of recommencement of the works to completion.

The example on page 214 shows the possibilities of catering for the problem by extension of the contract works policy with a separate sum insured in respect of the inflation factor. This is geared to a rate of inflation suggested by the insured and agreed to by the insurers. During the last decade the monthly rate of inflation in building costs has fluctuated between 1 and 2%.

The example assumes a constant rate of inflation which in practice will not apply. It also assumes a similar period of time for rebuilding work, which may not be so. Whilst the figures are a little arbitrary they do illustrate how serious the effects of inflation can be.

Damage to existing structures

It is necessary to distinguish between the damage to the contract works and damage to the existing structures. The contract works policy normally only covers the contract works and materials that the insured is putting in or erecting. Damage to the existing structure (which is third-party property) would normally be the concern of the public liability insurers who would offer an indemnity to the insured for legal and contractual liability in respect of accidental loss or damage to the property temporarily occupied for the purpose of the reconstruction or repair.

Example (in round figures)

Estimated contract price	£1 million
Period of contract	2 years
Final contract price	£1.2 million
Contractor's policy sum insured	£1 million +20%

A total loss takes place in month 17 when
£900,000 of work has been completed. Five
months' delay for replanning and reordering.

The £900,000 original work will now cost
£1,391,400 to replace, i.e. £900,000 × 2%
to the power of 22.

(1) Contract works policy

Loss limited to sum insured of		£1,200,000

*(2) Reinstatement of damaged works catering for
inflation excess of (1)*

Cost of reconstruction	£1,391,400*	
Less final contract price had no damage occurred	£1,200,000	
		£191,400*
		£1,391,400

(3) Cost of uncompleted work due to inflation

Uncompleted work £300,000 will now cost £463,800		
Policy pays 'additional' construction costs (if arranged)	£463,800	
Less original cost	£300,000	
Additional cost		£163,800

Total cost of contract works

Original works	£900,000	
Reworking	£1,391,400	
Undamaged and delayed balance of work	£463,800	
		£2,755,200

Who pays

(1) Contract works policy		£1,200,000	
(2) Reinstatement of damaged works excess of (1)	Insurers if suitable inflation covers arranged	£191,400	
(3) Uncompleted work		£163,800	
			£1,555,200
Sum principal originally contracted to pay			£1,200,000
			£2,755,200

* At other stages in the contract this additional cost would be appreciably more.

As a matter of administrative convenience it is common for the contract works and public liability insurance to be arranged under the same policy. This does not usually involve any more comprehensive cover than if individual policies were issued.

In the event of damage to the existing structure an employer may be protected to only a limited extent by the contractor's public liability insurance. He may have to prove negligence, or the occurrence of an accident or that the damage arose out of the execution of the work. To remove this problem, on occasions the contract works policy may be extended to cover also the existing structure on a full all risks basis. In practice this can work quite well, but of course additional premium is involved and all parties should be very clear about their intentions.

Other exceptions

Defective design

The exception usually reads: '. . . loss or damage due to fault, defect, error or omission in design plan or specification'.

The ICE Conditions of Contract (which gave birth to the contract works policy) exclude damage due to fault, defect, error or omission in the design of the works and most insurers have incorporated an exception on these lines in their policies. Some insurers apply a complete exception, others only apply the exception to the part which is defective and cover the consequences of the failure of the defective part (see chapter 5).

For example, a concrete beam may have been designed defectively and it may not be strong enough to support the building. The fracture of the beam may involve the collapse of all or a part of the building. Whether the complete collapse is excluded or merely the defective beam which broke, depends on the policy wording.

Other insurers may offer an indemnity to the insured provided the insured is not engaged in designing. Many contractors now offer a complete 'design and build' service, but almost any contract, however simple, will involve some design element. The contractor will have to decide how to put materials together; methods and procedures of erection all need consideration. Most insurers intend to exclude only the 'professional risk', but in considering the application of the exclusion to a claim it is necessary to examine the exact policy wording carefully to establish the intention of the underwriters.

It is argued that if there is a defective design then this is a matter for the professional indemnity insurers of the architect or engineer concerned. Rates of premium for contract works policies do not contemplate a professional design element. At least that is the theory.

There are many examples of claims involving a defective design element. For example, a long gable wall in a factory was designed without a sufficient number of intermediate supports and the completed wall was not strong enough to withstand gale force winds. Some years after completion a storm blew the wall down, falling bricks did extensive damage to the machinery and contents of the building and the force of the wind tore through the building and blew down the gable wall at the opposite end. The actual steel framed structure of the building remained intact. The contractors accepted that the design was defective and did not press a claim under their contract works policy. There was, however, a substantial claim for damaged machinery, contents and consequential loss for interrupted production at the factory. This claim was handled by the public liability insurers.

In practice difficulties can arise over the interpretation of the exclusion and its application to claims. It will be a question of fact whether the design was defective or not but it is important to note that the exclusion does not refer to negligent designing, it is defective design which is excluded.

There is an Austrilian case on this subject which is of considerable interest: *Queensland Government Railways and Electrical Power Transmission Property Ltd* v *Manufacturers Mutual Insurance Ltd (1969)*. A railway bridge in Australia was being constructed to replace a bridge which had been swept away by flood waters. Prismatic piers similar to the original piers but stronger were being erected when they were overturned by flood waters after exceptionally heavy rain. The insurers denied liability contending that loss was due to the faulty design of the piers. The court held that loss was due to faulty design, and that it was erroneous to confine faulty design to personal failure or non-compliance with standards which would be expected of designing engineers. A faulty design can be the product of fault on the part of the designer, but a man may use skill and care, he may do all that in the circumstances could reasonably be expected of him and yet produce something which is faulty because it will not answer the purpose for which it was intended. His product may be faulty although he be free of blame. The policy exclusion was not against loss from negligent designing, it was against loss from faulty design. For further details see chapter 5 and the case of *Pentagon Construction (1969) Co Ltd* v *United States Fidelity and Guarantee Co (1978)*.

If, however, the design has not been negligent then it will probably be difficult for the contractor or the employer to have any recourse against the engineer involved. In *Greaves Contractors* v *Baynham Meikle (1975)* the engineer had guaranteed the result desired by his client which was sufficient to make him liable apart from his professional negligence but this was an exceptional case (see the end of this chapter for further details under the heading 'Recoveries').

Design in relation to subsidence   Subsidence of buildings has been a particular problem in recent years, often due to the shrinkage of clay soil and/or the action of tree roots. These causes do not necessarily affect the contract works insurer but if damage does occur during the period of active work or during the period of maintenance, then there is a potential claim. The cause of the subsidence must be identified. Subsidence may be due to inadequate foundations – perhaps there was made-up ground and piles should have been used. Perhaps the foundations were not deep enough, perhaps they were wrongly designed. A contractor may undertake ground tests with bore-holes and be quite satisfied with the results. However, an unknown pocket of peat may be lurking below the site, the ground may sink with the result that a building may be severely damaged. Can the contractor claim under the contract works policy for the repair of the damage? The contractor may not have been negligent but there is little doubt that the foundation design was defective otherwise the building would not have subsided. On the basis of the aforementioned Australian case it would seem that the insurers are not obliged to pay a claim of this type although, of course, much will depend on the wording of the exclusion concerned.

Defective workmanship

The policy exclusion usually reads: 'the cost of repairing replacing or rectifying property which is defective in material or workmanship'.

This may form part of the defective design exclusion although it is regarded by most insurers as a non-insurable risk. A contractor is expected to pay for the cost of his own defective work; this is a trade risk. However, the **consequences** of defective work are not excluded. For example, the bills of quantities may specify a certain mix of concrete but the contractor's employees through carelessness may use the wrong mix. A defective floor might have to be relaid and the contract works policy would not pay for the cost of that. However, it would pay for other damage to the works caused by the defect; if, for example, the floor collapsed and damaged other parts of the work under construction, then that would be covered.

In another claim the design of a storage warehouse involved galvanised mild steel ties to be shot-fired to the face of the gable steel portal and the brickwork to be built to incorporate these ties in appropriate courses. The engineer's drawings called for the metal ties to be provided at 450mm centres along all the steelwork. However, there was lack of supervision on the site and the necessary ties were not used. At a time of strong winds double access doors at one end of the new warehouse were opened to remove equipment and the opposite gable wall of the warehouse started to collapse. The issue was complicated by the fact that the insurers did not come on risk until the period of maintenance had commenced. The adjusters considered that the defective workmanship exclusion applied, but it was arguable that the whole work was not defective and that payment should be made for the consequences of the absence of ties. In the event, the insurers paid the claim but shared it with the insurers of the building who provided storm cover. No payment was made for the additional costs of extra wall ties which did not form part of the original structure.

In connection with defective design and faulty workmanship the Canadian case of *Pentagon Construction (1969) Co Ltd* v *United States Fidelity and Guarantee Co (1978)* is of interest. The interpretation of the following wording was considered:

This insurance does not cover: (a) Loss or damage caused by: (i) faulty or improper material or (ii) faulty or improper workmanship or (iii) faulty or improper design.

(For details see chapter 5.)

Claims after completion of work

The existence of, or the results of, defective design or defective workmanship may not be apparent for some time after the completion of the work and the period of maintenance (usually twelve months) may have expired. What is the position if damage is reported some years after the completion of the contract? Who is liable and can the contractor recover under his contract works policy?

A contractor can remain liable for defective work or deficiencies in buildings and these can be the subject-matter of proceedings provided breach of contract, negligence, or breach of statutory provisions can be established. However, a civil engineering contractor (unlike a building contractor) does not often have to consider the Defective Premises Act 1972 or the National House-Building Council's scheme. Therefore it is right here to consider negligence as even the ICE Conditions do not operate after completion of all work.

Limitation

The original Statute of Limitations in 1623 prescribed six years as the limit for actions founded on simple contract or tort, although these terms were unknown in the seventeenth century. In connection with property damage the period remains unchanged. Section 2 of the Limitation Act 1980 states: 'An action founded on tort shall not be brought after the expiration of six years from the date on which the cause of action accrued'. This Limitation Act is a consolidating measure only.

When there is an accident which causes damage to property, clearly the action accrues from the date of the accident. However, in other circumstances the date on which the cause of action accrues may not be so obvious.

Time for taking proceedings (at least as far as defective buildings are concerned) formerly ran from the date the loss was discovered or could reasonably have been discovered by the plaintiff, not from the date the work was completed: *Eames London Estates Ltd v North Hertfordshire District Council (1980)* following the formula proposed in *Sparham Souter v Town and Country Developments (Essex) Ltd (1976)*. In the latter case building work was completed in September 1965 but proceedings were not commenced until October 1971. Two or three years after completion several cracks appeared in the brickwork of houses and they became uninhabitable. The alleged reason was that the foundations were inadequate to support the load, due to the negligence of the developers/builders and also the negligence of the local authority surveyor in passing the work as satisfactory.

The Court of Appeal held that a cause of action for negligence accrued not at the date of the negligent act or omission but at the date when the damage was sustained by the plaintiff. Thus a person who bought a house which developed defects and caused him loss or damage could bring an action at common law against a local authority alleging negligence in inspection and certification of the original work, many years after the alleged negligent acts. The owner's cause of action did not accrue until he discovered the damage or ought with reasonable diligence to have discovered it.

However, there was a doubt as to the extent to which the House of Lords confirmed this view in *Anns v Merton Borough Council (1977)*. Blocks of flats were completed in 1962 and in 1970 structural movements began to occur. Writs were issued in 1972. The fact that defects only became apparent more than six years after the negligent inspection by the Council was not a bar to the action. However,

Lord Salmon expressed the view that where a cause of action depended upon proof of the occurrence of damage, the date when the cause of action accrues for limitation purposes should not necessarily be the date when the damage was first reasonably discoverable by the plaintiff. When the damage occurred would be a question of fact to be determined by evidence.

The question was again considered by the House of Lords in *Pirelli General Cable Works* v *Oscar Faber and Partners (1983)*. It was held that a cause of action in tort in respect of damage to a building caused by negligent design or workmanship accrued for the purpose of the Limitation Acts when the damage came into existence. The date of accrual was not to be postponed to when the damage was discovered or ought with reasonable diligence to have been discovered. The House said their decision was unreasonable and unjustifiable in principle but section 26 of the Limitation Act 1939 made it impossible to come to any other conclusion.

The Limitation Act 1963 had introduced amendments for personal injury claims but it appeared that Parliament had deliberately left the law unchanged as far as actions or damages of other sorts were concerned. Legislation can be expected to remedy the position but in the meantime if plaintiffs do not discover the damage within a period of six years from the date the damage came into existence then they are unable to make a claim.

Protection for future liability

The contract works insurer would not normally be involved in liability for damage which occurs after the policy has lapsed, that is, outside the period of active work or the period of maintenance. If, for example, there is subsidence damage to a building which arises many years after completion, the contract works policy cannot usually be involved, for it will no longer be current for the contract in question. If, however, the subsidence is, for example, due to the contractor's negligence in putting in inadequate or faulty foundations, the contractor may be liable to the employer for the repair of damage to the remainder of the building. Although a liability resulting from negligence, this risk will not usually be covered under a public liability policy because the latter document normally excludes the repair of contract works erected by the insured where damage is an inherent part of the contract and not due to some later independent extraneous cause. Unless special arrangements are made there can thus be a gap in cover between a contract works policy and a liability policy.

It is reasonable that a contractor should have protection for future liability. Some insurers may be prepared to amend their public liability policy so that after completion of a contract, the contract works cease to be contract works for policy purposes and the insured will be indemnified in respect of his legal liability for damage to buildings erected. Such an extension would still not cover the cost of replacing the insured's defective workmanship but the cost of repairing the consequences of the defective work would be insured. For example, a policy would not pay for the replacement of defective foundations due to bad workmanship but it could pay for other damage to a building arising consequent upon the defective foundations, where the damage arises after the completion of a contract and falls outside the period of maintenance.

In practice it may be very difficult to distinguish between the causes of the damage, especially when a number of years have elapsed. However, evidence was available in *Sharpe* v *Sweeting and Son Ltd (1963)* where the defendants constructed a house for a local authority. The plaintiff's husband was the first tenant of the house. A negligently reinforced concrete canopy fell on the plaintiff and caused her injury some eight years after the premises were built. The builder was held liable but the public liability policy of the builder (operating at the time of the injury) would cover only the bodily injury not the damage to the property, that is, replacement of the canopy for which the builder was liable.

Another possible method of giving protection is to arrange for the period of maintenance cover under the contract works policy to be extended.

Occupation by employer

The exception usually reads:

. . . loss of or damage to the contract works or any part thereof after such works or such part have been or has been completed and delivered up to the Principal or taken into use or occupation by the Principal except to the extent that the Insured may remain liable
  (i)   under the maintenance conditions of the Contract
  (ii) during a period not exceeding fourteen days after the issue of a Certificate of Completion and which is the responsibility of the Insured to insure.

The ICE Conditions exclude loss due to use or occupation by the employer, his agents, servants or other contractors (not being employed by the contractor) of any part of the permanent works. Temporary occupation may occur before the issue of a certificate of completion but it seems reasonable that the contractor should not be responsible if loss is due to occupation by the employer who may introduce hazardous material or work processes; risks which were not contemplated by either the contractor or his insurers, hence the exception relating to occupation of the work by the employer.

In the case of *English Industrial Estates Corporation* v *George Wimpey and Co Ltd (1973)* (involving a contract under the standard Form of Building Contract), a principal did in fact occupy and use part of the works but no certificate of partial completion had been issued by the architect. Fire severely damaged the premises and the plaintiffs contended that at the time of the fire the building and other contents were at the risk of the defendants by nature of provisions in the bills of quantities. This was disputed by the defendants who also maintained that for the purposes of clause 16 (now clause 18) possession of the relevant parts of the works had not been taken with the consent of the contractors. It was held by the court that the plaintiffs had not taken possession of the relevant part of the works for the purpose of clause 16. The effect of this decision is that the wording of the bills of quantities would not override the printed contract conditions and this case again emphasises the need for thorough acquaintance with the intentions and effects of the various conditions in the contract.

Consequential loss

Contract works policies usually exclude consequential loss of any nature whatsoever and this is also an area which can produce problems in connection with claim settlements.

There are many headings under which loss may occur and insurers attempt to exclude economic loss which flows from the original damage to the property insured. The contract works policy offers a material damage cover; loss of profits cover and similar items are best handled by means of separate insurances — when cover is available.

Contracts may be delayed because of the operation of an insured peril or other factors, such as strikes or bad weather, but financial loss incurred by the contractor or employer is outside the scope of the standard contract works policy. For example, the insured contractor may have incurred legal fees in resisting or pursuing claims arising out of the loss or damage against a third party. Plant may have been damaged and additional hiring fees incurred under a plant-hiring agreement. The employer may claim against the contractor for delay, loss of production and additional costs incurred in renting other premises. He may have suffered severe loss of profits because the new premises are not ready for occupation. There may be extra charges for storing materials which cannot be delivered to the site because of loss or damage; additional interest charges may be payable because of extended bank overdrafts.

The contract works policy will not provide cover for these items. Some contract works policies may offer a limited cover in some areas (see chapter 5 concerning the policy exclusion), but generally the consequential loss items will require to be deleted from the amount of the claim. However, this is not to say that difficulties will not continue to arise.

Under this heading also may be considered fines, penalties and liquidated damages and some policies may have a specific exclusion for these items. For example, the contract conditions may contain penalty clauses if the work is not completed on time. Delay may be due to the contractor's inefficiency or delay on the part of subcontractors. On the other hand it may be due to the operation of a peril insured under the contract works policy. It is normally possible to arrange insurance for the latter contingency, perhaps as an extension of the contract works policy but more usually by means of a separate insurance.

See the example given (in round figures) earlier in this chapter.

Interpretation of consequential loss

The term consequential loss may be difficult to interpret; insured and insurers may have different views. Consequential loss, legally, has been held to mean something which is not a direct and natural result. In *Millar's Machinery Co Ltd* v *David Way and Son (1934)* there was a contract which said: 'We do not give any other guarantee and we do not accept responsibility for consequential damages.' The Court of Appeal rejected the argument that this provision excluded the supplier's liability to pay the buyer's expenses incurred in obtaining other machinery to replace a defective machine.

In *Croudace Construction Ltd* v *Cawoods Concrete Products Ltd (1978)* the plaintiffs commenced an action to recover damages for breach of contract arising out of alleged late delivery and defects. They also sought to recover the cost of loss of productivity and additional costs of delay in executing the main contract works. The terms of the contract included a condition, 'We are not under any circumstances to be liable for any consequential loss or damage caused or arising by reason of late supply or any fault, failure or defect in any materials or goods supplied by us.' It was held that the exclusion of consequential loss or damage in the defendant's conditions did not cover any loss which directly and naturally resulted in the ordinary course of events from late delivery. Thus the cost of men and materials being kept on site without work was recoverable as damages and not excluded as 'consequential'.

In the context of the contract works policy which clearly specifies the material damage risks which are insured, an exclusion of 'consequential loss of any nature whatsoever' seems adequate and clear-cut. Possibly the words 'direct or indirect' in the exclusion may be even better. Nevertheless, in the aforementioned case the court commented that: 'the word "consequential" provides a fertile ground for philosophical debate as do other words involving the concept of causation'. Costs which have to be incurred as a part of the reinstatement will be paid, that is, those which are a direct and natural result of the damage. For example, reasonable over-time payments could be allowed but probably not the employment of a different and more costly method of work.

Recoveries

As regards contract works policies there may be another party who is totally or partially responsible for the damage and after indemnifying their insured within the terms of the policy insurers will naturally seek recovery if at all possible, although in practice opportunities tend to be limited. Subcontractors are an obvious target; although they may be responsible for their own subcontract work under contract, they will also be responsible for damage to other parts of the contract caused by their negligence.

Who else can be negligent — what about the engineer or architect? It is possible to proceed against an engineer or architect on the grounds of their professional negligence and professional indemnity insurers often complain about their poor claims experience in this area.

Although specialist engineering contractors owe no higher duty of care than other professional experts, they may breach that duty where they fail to design a building fit for the purpose for which they knew it would be required. In *Greaves and Co Contractors* v *Baynham, Meikle and Partners (1975)* the structural engineers had to design a warehouse in the knowledge that on the first floor fork-lift trucks would be involved in the storage and moving of oil drums. After being put into use the first floor began to crack because it did not have sufficient strength to withstand vibrations caused by the fork-lift trucks. The main contractors were absolutely liable to the employers for the high cost of the remedial work; they claimed from the engineers who were held to be in breach of the implied warranty that the floor would be fit for the purpose for which it was required. They were also held liable for professional negligence.

Another example of a claim involving engineers is where the roof of a building collapsed when some laminated wood beams supporting the roof failed. The beams made of substandard wood were not properly glued in the manufacturing process, but the defects would only have been discovered by elaborate laboratory tests. There was a claim for physical damage to the building and contents plus loss of profits. The normal contract works policy would pay for the damage to the remainder of the building under construction but it would not pay for the replacement of the defective beams, nor would it pay for damage to occupied buildings or the contents thereof. In fact, the beam manufacturer was held to be partly liable for the loss and the engineer was also partly responsible because he failed to take action when it was obvious the beams were under stress.

Other recovery possibilities

In connection with the property of employees, that is, loss of or damage to tools or clothing, there may be a possibility of recovery under householders' comprehensive policies, but in practice this is not a fruitful source of enquiry. Workmen are rarely insured adequately and in any event, they tend to be unco-operative and to ignore correspondence.

In the event of damage following a riot the possibility of making a claim under the Riot (Damages) Act 1886 should not be ignored.

Occasionally there may be impact damage by a motor vehicle or by an item of plant used by another contractor with the possibility of recovery from motor insurers.

Chapter 21

Public liability property damage claims

The general principles of claims settlements have been discussed in chapter 19. Property claims can involve payments of a few pounds only but if valuable property is extensively damaged the cost might run into many millions of pounds. Hence the need for an adequate single accident indemnity limit under public liability policies. In general construction work, damage caused by fire or water is very common. With road and sewer contractors, damage to underground pipes and cables is perhaps the most serious problem. Excavations which are left unguarded and/or unlit also involve many claims. Removal or weakening of support to adjoining roads or structures can result in heavy payments; pile driving or blasting operations frequently produce vibration damage. Plant often damages other plant or buildings. The list of possible claims and causes is virtually endless.

Liability

It is necessary to consider both policy liability and legal liability. Some of the general principles about policy liability have already been discussed but in relation to third-party insurance it is essential to consider the wording of the operative clause in the policy. Policy wordings vary between insurers but a typical policy might offer cover as follows:

In the event of accidental

  (a) bodily injury to any person and/or
  (b) loss of or damage to property and/or
  (c) obstruction or trespass

occurring within the territorial limits, the company will indemnify the insured in respect of all sums which the insured shall be legally liable to pay as damages arising in respect of such event.

See appendix 1, section 2 of the policy.

Accidental

Was the damage caused accidentally or not? What is meant by the term accidental? An accident has been described as 'an unlooked for mishap or an untoward event which is not expected or designed', or 'an untoward occurrence which has adverse physical results'. However, Viscount Haldane L.C. pointed out in *Trim Joint*

District School Board v *Kelly (1914)* that the word must always take its meaning from its context. Thus when the pupils murdered the master it was accidental from the insured master's viewpoint and the policy (a personal accident one) operated and the insurers had to pay. In *Gray* v *Barr (1971)* which involved a claim under a personal liability policy, the insured visited his wife's lover carrying a loaded shot-gun. The gun went off during a struggle and the lover was killed. The Court of Appeal held that the word accident does not include injury which is caused deliberately or intentionally, the death was not an accident since it was foreseeable, although unintended. The insured was unable to recover under his policy which offered indemnity against legal liability for injury caused by accident.

Some insurers omit the word 'accidental' and instead incorporate an inevitability exclusion; for example, damage which results from a deliberate act or omission of the insured, or damage which could reasonably be foreseen to be inevitable, having regard to the nature of the work to be executed or the manner of its execution. However, one problem is then exchanged for another, for it is still necessary to decide what caused the damage and whether such damage could reasonably have been avoided.

Clause 22(1)(b)(iv) of the ICE Conditions of Contract excludes damage which is the unavoidable result of the construction of the works in accordance with the contract. Whatever wording may be used it is the insurers' intention that the policy should only offer cover against unexpected events. If damage is caused deliberately or it is inevitable then the policy will not operate. It is always necessary to decide whether the event which caused the damage comes within the scope of policy cover and within the broad definition of 'accidental'. For example, if a contractor to reach a contract site is obliged to cross a corn field, it is inevitable that the corn will be damaged and therefore there is no indemnity under a public liability policy. Generally, a gradually operating cause will not be regarded as accidental. For example, demolition work over a period of weeks may result in dust damage and interference with business in adjoining premises. Most insurers would not regard this as accidental, either because there would be no specific event causing the damage or because some dust damage is probably unavoidable in demolition work. If, however, there was a collapse during the demolition which resulted in dust drifting into adjoining premises this would be different. There would be an unexpected event which resulted in damage.

Property

What is meant by the term property? Property is a very comprehensive term inasmuch as it is indicative and descriptive of every possible interest which a party can have (see *Jones* v *Skinner (1835)*).

The Trustee Act 1925 defines property as thus:

Property includes real and personal property and any estate and interest in any property real or personal and any debt and anything in action and any other right or interest whether in possession or not.

What about financial loss if there is no physical damage to property? A financial loss is property if there is a right of action to recover it. Property includes a sum of money under the Debtors Act 1869. (See *R* v *Humphries (1904).*) Financial loss is an 'intangible' within the definition of property.

Most insurers intend to cover only physical damage to tangible or material property. Intangibles such as copyrights, patents, trademarks and the like are not usually covered. To make their intentions clear, therefore, some insurers may offer cover only for loss of or damage to material property with occasionally an extension on a limited basis for pure financial loss. However, the ICE Conditions of Contract under clause 22 refers to:

all losses and claims **for injuries** or **damage** to **any** person or **property whatsover** (other than the Works for which insurance is required under Clause 21 but including surface or other damage to land being the Site suffered by a person in beneficial occupation of such land).

There can therefore be a gap in the cover required by contract conditions and the cover offered by an insurance policy. Whilst most damage claims do involve material property, the nature of the property damaged must be considered in relation to the wording of the policy. Problems are most likely to arise out of financial loss where there is no physical damage. Obstruction is a common cause but as illustrated this is a risk which may be provided for specifically. Clause 29(2) of the ICE Conditions requires that all work shall be carried out without unreasonable noise and disturbance and requires that the contractor shall indemnify the employer. Noise and disturbance may or may not be accidental, but if they involve financial loss to a third party, there will be a potential liability on the contractor. Whether there is a liability on the insurer will depend on the policy wording. A strict wording which referred only to 'material property' would not offer any protection.

However, with public liability insurance much depends upon the interpretation placed on its policy by the insurers. Even with similar wordings insurers may produce different interpretations for claims purposes.

Legal liability

The policy may refer to legal liability or liability at law — it does not cover any moral liability. Cover is not limited to the negligence of the insured and all forms of liability should be covered: negligence, nuisance, trespass, strict liability, liability under statute and under contract, and vicarious liability. Liability can only be decided after an investigation of the facts.

The extent of liability which may be assumed under contract depends on the exact wording of the contract concerned. It may also partly depend on statute law, for example, indemnity clauses and exemption clauses now have to be considered in the light of the Unfair Contract Terms Act 1977. No exclusion or restriction of loss or damage by negligence is permitted unless reasonable, although the 1977 Act does not affect the insurance contract itself. It is worth emphasising that the contractual liability extension under a policy does not extend the basic policy wording; all the terms, conditions, exceptions of the policy remain unchanged. For example, if a policy excludes damage to property in the charge or under the control of the insured an extension to include contractual liability does not remove the original exclusion; property in the insured's charge or control is still not covered.

Limitation

It is important that any claim should be brought within the statutory time limit. This is normally six years from the date of accident for property damage claims. (See comments in chapter 20. Also for personal injury see chapter 22.)

Property damage payments*

Following a decision on policy and legal liability, the amount of the damage will be assessed. It is usual for payment to be supported by accounts. The third party will be indemnified; as far as possible he will be placed in the same position after the loss as he was before. The insurers will have decided whether to make a cash payment or whether repair or replacement is possible. Whatever method is adopted the third party should not make a profit out of his loss. It is therefore necessary to take into account wear and tear, depreciation and salvage. If repairs are effected it is probably better if the claimant gives his own instructions, then payment may be made direct to the claimant on completion of the claim.

VAT should not be overlooked and where appropriate should be deducted from the amount of settlement.

Where there is injury to animals veterinary reports are essential both as regards value, the extent of injury and where necessary cause of death. Owners of horses, for example, almost invariably set an inflated value upon their animals. Purchase price, they argue, is not relevant for they always believe they have bought a bargain.

Assessment of damages

The general object underlying the rules for the assessment of damages is, so far as is possible by means of a monetary award, to place the plaintiff in the position which he would have occupied if he had not suffered the wrong complained of, be that wrong a tort or a breach of contract.

In *Livingstone* v *Rawyards Coal Co (1880)* Lord Blackburn said:

> The measure of damages is the sum of money which will put the party who has been injured or who has suffered in the same position as he would have been in if he had not sustained the wrong for which he is now getting his compensation or reparation.

In the case of a tort causing damage to real property the measure of damage may be the diminution in market value at the time of loss. It is then necessary to consider the capital value of the property in an undamaged state. Alternatively the cost of repair or reinstatement may be considered. As a general rule in English law damages for tort or breach of contract are assessed as at the date of the breach (*Miliangos* v *George Frank Textiles Ltd (1976)*). The date when damage occurs is usually the same day as the cause of action arises. Although this may be a general principle it cannot be applied rigidly. For example, the damage may be concealed by a wrong-doer and not reasonably discoverable until some time afterwards.

---

* Reference should be made to *Insurance Law Reports*, Volume 2, *Building damage – Indemnity settlements*, published by George Godwin. The introduction is particularly informative on this point.

Furthermore, in the case of building damage repairs cannot usually be put in hand at once, specifications may have to be prepared and estimates obtained. When foundations of a structure have been disturbed by vibration, for example, it may be prudent to wait until it is reasonably certain that further damage will not occur. Nevertheless, the owner of the damaged property should not delay commencement of the work unnecessarily; he must take reasonable steps to mitigate his loss for no person is entitled to recover the amount of an avoidable loss. In recent times inflation has been a problem and if repairs are delayed for a number of years costs will increase enormously.

This is illustrated in the case of *Dodd Properties (Kent) Ltd and Another* v *The City of Canterbury (1980)*. The plaintiffs were owners of a garage and the defendants had erected a large multi-storey car park close by. Damage was caused by piling work and the contractors admitted liability but only shortly before the trial. The issues were the extent of the damage and the basis of assessment of the amount of damages. Damage had occurred in 1968 but the case was not heard until 1978. 1970 was the earliest date at which repairs could have been effected, bearing in mind the need to ensure there was going to be no more damage and the fact that a contractor had to be found to do the work. The question was should damages be paid at 1970 prices or 1978 prices? There was a considerable difference, the estimated cost of repair having risen from £11,375 to £30,327. There was also a claim out of prospective interruption of business whilst repairs were carried out; the 1970 figure was £4,108, but in 1978 £11,957. If the owners had effected the repairs in 1970 it would have involved them in a measure of financial stringency. Furthermore, they were reluctant to spend money on a building before being sure of recovering the cost from the defendants. It would not have made commercial sense to spend money on a property which would not produce corresponding additional income. So long as there was a dispute on liability or the amount of compensation, the owners would only keep the building weatherproof and in working order. The Court of Appeal found for the plaintiffs and the 1978 figures applied. Although a plaintiff is under a duty to mitigate his loss, he is not obliged in order to reduce the damages to do that which he cannot afford to do. In *Clippers Oil Company* v *Edinburgh and District Water Trustees (1907)* it had been held that in relation to the duty to minimise damage the tortfeasor must take his victim as he found him, including any lack of means. In all the circumstances therefore, the court found in the Dodd properties case that:

(1)   When a building is damaged by a tortious act, the cost of repair was to be assessed at the earliest date when having regard to all the circumstances those repairs could reasonably be undertaken.

(2)   It made commercial good sense to postpone carrying out repairs until the action was heard and the liability of the defendants established. The cost of repairs should be assessed at the date of the trial.

In any property damage claim it is desirable that repairs should be effected as soon as possible, for delay will almost inevitably involve a higher expenditure. However, as the aforementioned case illustrates, in the event of dispute the courts will consider all the circumstances and will not necessarily apply rigid rules regarding the appropriate date at which damages should be costed.

Indemnity

The principle of indemnity has already been mentioned. Neither the insured nor a third party may make a profit out of the loss, as far as possible they have to be placed in the same position after the loss as they were before. In practical terms this may be difficult and there is no rigid formula which can be applied. To a large extent each case must be considered on its own merits.

Leppard v *Excess Insurance Co Ltd (1979)* concerned the application of the principle of indemnity under a fire policy and whilst not directly relevant to a public liability claim it does illustrate the problem of what constitutes an indemnity. The plaintiff purchased a cottage from his father-in-law with the object of selling it and dividing the proceeds between himself and his father-in-law. The original asking price was £12,500 later reduced to £4,250, although £4,000 would have been accepted. Before any sale took place the property was totally destroyed by fire. The cottage was insured for £14,000. The plaintiff contended that the actual loss suffered was the cost of reinstatement. The insurers contended that the plaintiff was entitled only to the market value of the cottage at the time of the fire (agreed at £3,000 — selling price less site value). The court held that the plaintiff could recover only the market value, that is £3,000. The value of loss at the date of loss was a question of fact depending on all relevant factors including factors foreseeable at the date of loss as being likely to affect the value of the thing insured if the loss had not occurred. This case does not alter the principle of indemnity but it illustrates the difficulties which can arise over the values of property. Putting a person back into the same position that he was in before a loss is not always easy.

A similar problem arose in *Reynolds and Anderson* v *Phoenix Assurance Co Ltd (1978)*. An old malting was insured, its market value being only £5,000. It could have been replaced by a modern structure which would have been adequate for the purpose, for about £50,000. However, the cost of replacement less depreciation was £243,000, but in view of the rise in costs at the date of hearing plus architects' and surveyors' fees and VAT, £343,320 was awarded. The court decided this was the true measure of indemnity as the insured maintained it was their intention to reinstate the building.

It appears, therefore, that where there is an intention to reinstate a claimant is entitled to have his loss settled on that basis taking into account depreciation. Otherwise indemnity can be satisfied by payment of the market value. In fact an out-of-court settlement of about two thirds of the award was agreed under threat of appeal which was based on the cost of a similar building.

For stock in trade and similar items, indemnity is normally represented by market value. For a retailer that will be the cost of replacement to him, not the selling price to the public. Deductions may have to be made for obsolete or otherwise unsaleable stock. For a manufacturer indemnity will be the cost of raw materials, plus production costs, plus appropriate overheads. There should be no allowance for profit made on sale of stock.

Interest on damages

Section 3 of the Law Reform (Miscellaneous Provisions) Act 1934 provides that in any proceedings tried in any court of record for the recovery of any debt or damages the court may, if it thinks fit, order that there shall be included in the sum for which judgment is given, interest at such rate as it thinks fit on the whole or any part of the debt or damages for the whole or any part of the period between the date when the cause of action arose and the date of judgment. In connection with property damage claims the court has a discretion and interest only applies to those cases where a judgment is given. In other words where a case is settled out of court, a plaintiff is not **entitled** to receive interest. An agreed settlement can, of course, reflect the interest which a plaintiff might have obtained if the case had proceeded to trial. See now the Administration of Justice Act 1982, section 15.

The basis for awarding interest is that the defendant has kept the plaintiff out of his money, the defendant has had use of it himself, so he ought to compensate the plaintiff accordingly.

Where repairs to property are paid for by the plaintiff, and action is taken against a defendant who is not prepared to admit liability, a considerable time might elapse between the date of the cause of action and the date of judgment. It seems equitable in such circumstances that the settlement should include an allowance for interest. Where, however, repairs are not effected until after the judgment, the question of interest will not arise for there can be no question of the plaintiff having been deprived of money.

In the case of *Dodd Properties* v *Canterbury CC* referred to earlier, delay increased the cost of repair, but as the money had not been spent interest was not payable. In that case the plaintiff's alternative ground of appeal related to the appropriate calculation of interest, but in the event the issue did not arise.

Many claimants will include interest payments in their claim, often as a ploy in the hope of securing a higher settlement. Insurers are not obliged to pay these amounts; to what extent they do so depends on the whole circumstances of the claim. If their insured is liable for the damage insurers will be reluctant to incur the costs of a court action and a settlement may well take possible interest into account.

(Note: personal injury claims involve different considerations — courts are **obliged** to award interest on damages.)

Property of employees

The policy will cover the legal liability of the contractor for accidental loss or damage to the property belonging to employees. Such property is not regarded as being in the insured's charge or control. Often an employee's clothing is damaged in an accident involving personal injury and as a matter of convenience such a damage claim, for a relatively small amount, is handled under the employers' liability policy.

Generally an employee will be responsible for the safety of his own property, but the contractor will be responsible for his negligence or the negligence of his employees; for example, an employee's cycle may be crushed by the negligence of a bulldozer driver — a fellow employee.

Goods contained in site huts are frequently lost or damaged by fire or theft. The question of negligence may have to be considered but it is more usual for such items to be insured under contract works material damage policies where 'all risks' cover is provided up to a stated limit (often £100) per employee.

Damage to property in the ground

A very large number of claims arise from this cause and the overall cost involved is a matter of great concern to both insurers and the different authorities. It is the general practice of insurers to apply an excess to this type of damage. An insured (depending on the level of the excess) is given a direct incentive to avoid damage. With the amount of mechanical plant in use and the large quantity of services buried in the ground it is perhaps almost inevitable that some damage will occur. Regrettably, however, some contractors do not take the precautions they should; a few ruthless ones may even say it is quicker just to excavate and repair the damage afterwards. Their insurers will not share this view and contractors with a consistently poor claims record will find themselves heavily penalised.

The Public Utilities Street Works Act 1950

This Act attempts to control excavation work and lays down obligations on those referred to as 'operating undertakers'. The Act refers to works (other than works for the purposes of a railway undertaking) executed in a street or in controlled land in exercise of a statutory power.

Section 26(2) of the Act reads:

Operating undertakers shall not begin any works to which this section applies which are likely to affect apparatus to which this section applies of owning undertakers (other than works relating only to a service pipe or service line or an overhead telegraphic line, in this section referred to as 'excepted works') until they have given to the owning undertakers notice of their intention to execute undertakers' works, indicating the nature of the works and the place where they intend to execute them, and three days have expired from the date on which the notice was given.

There is a qualification that emergency works may be begun provided notice is given as soon as reasonably practical.

Section 26(3) requires the operating undertakers to give to the owning undertakers reasonable facilities for supervising the execution of the works.

Section 26(4) lays down further obligations on the operating undertakers:

(a) compliance with the reasonable requirements of the owning undertakers for the protection of apparatus;

(b) provision of proper temporary support for apparatus where the works include tunnelling or boring under apparatus;

(c) when electric lines are being laid, ensuring that such lines are effectively insulated and do not touch other apparatus.

Section 26(6) provides for the payment of compensation to the owning undertakers equal to the expense reasonably incurred in making good the damage caused by the execution of works by the operating undertakers or by their failure to comply with section 26(4).

There is a qualification that operating undertakers will not be liable for damage which would not have been sustained but for misconduct or negligence on the part of the owning undertakers or their contractors or any person in the employ of the owning undertakers or their contractors.

Section 26(8) states that failure to comply with the requirements of sections 26(2), (3) or (4) may result on summary conviction in a fine not exceeding fifty pounds unless the undertakers prove that the failure was attributable to their not knowing of the existence or not knowing of the position of the apparatus of the owning undertakers in question and that their ignorance thereof was not due to any negligence on their part or to any failure to make some inquiry which they ought reasonably to have made.

Reasonable inquiries   It is debateable to what extent these statutory requirements reduce the incidence of damage but the last phrase of section 26(8), 'failure to make some inquiry which they ought reasonably to have made', emphasises where the real problem lies.

Before any contractor commences excavation it seems reasonable that he should be aware of what lies beneath the ground. If there are cables and pipes he should know their respective depths, exactly where they run and their relationship to one another. Unfortunately this detailed knowledge is only secured on occasions. The contractor may say: 'This appears to be a virgin field — there is nothing here, it is safe to dig.' So often he is proved to be wrong. Alternatively the contractor may secure a plan which is old or not sufficiently detailed and he takes a chance without undertaking preliminary test holes. The moral hazard is important, for all too often a contractor may decide he cannot afford the time for proper investigation and he tells his workmen to carry on and hope for the best.

A contractor must therefore make reasonable enquiries before he starts to excavate. In practice this will mean consulting available plans of cables and pipes; the various authorities, gas, electricity, water, sewage, telephone will all need to be approached. The local authority may also be able to supply information. Sometimes the local authority supplies a combined services plan. Sometimes plans are old or they may be inaccurate. Often there is a denial of liability as to the plan's accuracy. Where appropriate the contractor should carry out test bores to establish the line of a pipe; it cannot be assumed that all pipes run in a straight line. Where it is known that underground services are very close to the line of excavation, it may be imprudent to use mechanical diggers or excavators, it may be necessary to hand dig certain stretches.

Cost of repair

If despite all the precautions adopted and care taken by a contractor damage still occurs then the contractor's public liability policy will be available to meet the cost of repair, subject to a deduction for the amount of excess. A few claims may be inexpensive, for example, a damaged water pipe may result in only a modest bill. At the other end of the scale there are certain types of oil filled main electricity cables which are enormously expensive to repair. Water may get into the cable, a repair is usually impracticable and it may be a question of replacing a long length of the cable. The same problem can arise with telephone cables where it is usually necessary to replace the damaged part with a completely new length.

Even although an existing cable may be fairly old, it is not usually practical to secure a deduction for depreciation. Authorities will not usually accept this argument. It is also very difficult to check items on the account, for example, in connection with labour charges. With a nationalised industry undertaking the repair themselves, there is little room for discussion.

However, the Electricity Board have an agreement with their employees' union not only to pay time and a half or double time for night work but also to pay those men for the next day when they are off work. This can amount to three times the normal rate which is exorbitant as far as the negligent contractor is concerned. This is particularly so as the hourly rate not only includes the employees' pay but overheads as well. In these circumstances it is often possible to obtain a considerable reduction.

The Telegraph Act 1878

Section 8 of this Act reads as follows:

Where any undertakers, body, or persons by themselves or by their agents destroy or injure any telegraphic line of the Postmaster General, such undertakers, body or person shall not only be liable to pay to the Postmaster General such expenses (if any) as he may incur in making good the said destruction or injury but also if the telegraphic communication is carelessly or wilfully interrupted shall be liable to a fine not exceeding twenty pounds per day for every day during which such interruption continues.

This Act is a rare example of where Parliament has provided for an absolute liability to be imposed in respect of damage to the apparatus of a statutory undertaker irrespective of negligence by the defendant. The Post Office are relieved of the duty of establishing negligence and the proof of absence of negligence is not a defence to an action. In *Postmaster General* v *Beck and Pollitzer (1924)*, the defendant's servant while driving a motor lorry accidentally and without negligence on his part ran into and damaged a fire alarm post belonging to the Postmaster General. The defendants were held liable under the Telegraph Act.

Liability under this statute was considered in an earlier case, *Postmaster General* v *Liverpool Corpn (1923)*. The local authority had done nothing except to lay their electric cable perfectly properly under the ground. Through nobody's fault there was a leak of electricity from the local authority's cable which caused an explosion and damage. The Post Office had stupidly laid their cable in close proximity to and in fact touching the main which had been laid by the local authority. The House of Lords felt strongly that the council should not be saddled with the cost of the claim and that it would be wrong to apply the Telegraph Act to these circumstances.

However, in a later case, *Post Office* v *Hampshire CC (1979)*, the Court of Appeal decided that the real *ratio* of the Liverpool case was not that the negligence by the Post Office affords a defence but where the act occasioning the damage is not that of a third party but of the Post Office itself the Act has no application. It is necessary to decide who it was that caused the damage.

The problem was also considered in *Post Office v Mears Construction Ltd (1979)*. A local authority was making highway improvements. Following a request the Post Office marked a copy of the local authority's plan of the proposed works with details of existing Post Office underground cables. A disclaimer was added: 'The information given is compiled from records and is believed to be correct. There may, however, be departures from the course(s) and depth(s) shown.' The plan was inaccurate in that the scaled distance of one cable showed it in one place to be six feet or more from its actual position and the cable was shown to run beneath the footpath and not, as was the fact, beneath the adjacent grass verge. The local authority passed the marked plan to the contractors. Relying on the plan, the contractors instructed their driver to avoid the cable beneath the footpath. While digging in the verge the excavator struck the cable.

The Post Office brought an action under section 8 of the Telegraph Act 1878, claiming the defendants were absolutely liable for the damage. The contractors admitted responsibility under section 8 but alleged that the Post Office owed a duty to give accurate information. They counterclaimed by way of damages for negligence for such expenses as they might be ordered to pay and to set off the amount claimed by the Post Office. It was held that there was no duty on the Post Office to supply a plan which would do more than indicate to the local authority the rough layout of the cables and the obligation was not extended by reason of the knowledge that the plan would be passed to the contractors and relied upon. It was clear the plan could not be relied upon in detail. In any event the disclaimer of accuracy on the plan absolved the Post Office. Accordingly the Post Office's claim was allowed and the contractor's counterclaim was dismissed.

Post Office v Hampshire CC (1979) has already been referred to briefly. Following the flooding of a road, council employees attempted to drain the water away by prodding a crowbar and fork into the grass verge beside the carriageway. They brought up a piece of porcelain duct which suggested the presence of a Post Office cable below the verge. A Post Office engineer was called to the site who indicated that the cable ran under the carriageway. The council employees resumed their work three feet away from where it had been indicated the cables lay and struck a Post Office cable under the verge and not under the carriageway. The Post Office brought an action under section 8 of the Telegraph Act alleging the council were absolutely liable for the cost of repair. The Court of Appeal held that the requirements of section 8 were fulfilled once it was shown that the council employees had by their voluntary act damaged the Post Office cable. However, the Post Office's engineer had been negligent in informing the council's employees that the cable ran under the carriageway and not the verge. The Post Office had suffered damage and incurred expense through its own fault. There was ground present for the application of the doctrine of circuity of action. The council were entitled to rely by way of defence on facts which indicated that if they had brought a counterclaim they would have been entitled in law to recover from the Post Office as damages for negligence the sum claimed against them by the Post Office.

Pearson J. in *Ginty* v *Belmont Building Supplies Ltd (1959)* commented on circuity of action:

Circuity of action would arise in this way. Suppose that the plaintiff said that his employer committed a breach of statutory obligation whereby damage was caused to him and he was entitled to recover damages from his employer. The employer would reply that by the contract of employment the employee owed a duty to his employer who therefore was entitled to recover damages against the employee and that the amount of the damages which the employer was entitled to recover was equal to the amount of the damages which the employee was supposedly entitled to recover against the employer. If that were the position the litigation would go round in a circle and for that reason there is in my view a valid plea of circuity of action. The plea of circuity of action is not usually found in these days because that situation is usually sufficiently provided for by the modern provision for set-off and counterclaim. But it is a valid plea . . .

In *Post Office* v *Hampshire CC (1979)* the court decided that the law of negligence has developed since 1923 to the point where the potential injustice which may be created by section 8 of the 1878 Act can be remedied in most if not all cases by the application of the law of negligence. Where the damage is caused solely by the negligence of the public utility in question they should be unable effectively to rely on their statutory right of compensation.

Negligent misstatements

In *Hedley Byrne and Co Ltd* v *Heller and Partners Ltd (1963)* it was held that there can be a liability for negligent misstatements which cause financial or physical loss. The Hedley Byrne principle was considered in both the Mears Construction and the Hampshire CC claims.

In the Mears Construction claim it was argued by counsel that although there was no contractual relationship between the parties and the defendants did not seek information from the Post Office, nevertheless the Hedley Byrne principle operated. It was held that there was no duty to supply a plan which would do more than indicate to the local authority roughly where the cables lay. It was clear the plan could not be relied upon in detail and in any event the disclaimer whether read alone or together with the plan, absolved the Post Office from any special duty on the basis of Hedley Byrne.

With the Hampshire CC claim, although the matter was not considered in detail (the judge said he was absolved from the necessity of getting into the morass of Hedley Byrne), counsel for the Post Office accepted that a Hedley Byrne situation was involved. This meant that the Council could have made a successful counterclaim based on the argument that the Post Office employees had by their negligence in not pointing out where the cables really ran caused the Council's employee to put his crowbar through the cable. This aspect was not pursued and the Council succeeded on the circuity of action point.

Undoubtedly the Telegraph Act 1878 remains a very valuable statute for the Post Office. Where damage is caused by a third party, negligence does not have to be proved. Where, however, the damage is a result of the negligence of the Post Office the automatic right to compensation is not available.

Application of policy conditions

If a contractor fails to investigate the layout of underground services before he commences work, can it be said that he is in breach of the policy condition requiring the insured to take reasonable precautions to prevent accidents? With a liability policy which offers indemnity in respect of negligence, insurers cannot deny liability because the insured has been negligent. What are reasonable precautions will be a question of fact in relation to the circumstances of the loss. Available plans should be inspected, the requirements of the Public Utilities Street Works Act 1950 should be adhered to, owners of underground services warned of the intention to excavate and where there is doubt, mechanical plant should be used circumspectly.

The question of reasonableness was considered by Lord Diplock in *Fraser* v *B.N. Furman (Productions) Ltd (1967)*:

The condition cannot mean that the insured must take measures to avert dangers which he does not himself foresee, although the hypothetical reasonably careful insured would have foreseen them. That would be repugnant to the commercial purpose of the contract for failure to foresee dangers is one of the commonest grounds of liability in negligence . . . What in my judgment is reasonable is that the insured where he does recognise a danger should not deliberately court it by taking measures which he himself knows are inadequate to avert it. In other words it is not enough that the insured's omission to take any particular precautions to avoid accidents should be negligent, it must be at least reckless, i.e. made with actual recognition by the insured himself that a danger exists not caring whether or not it is averted. The purpose of the condition is to ensure that the insured will not refrain from taking precautions which he knows ought to be taken because he is covered against loss by the policy.

Where there is a blatant and wilful disregard of all precautions, insurers might be able to deny liability, but few claims come within this area. Frequently an insured inspects some plans but misses others; often the plans are inaccurate or misleading. When in doubt the insured may sink trial holes but these may be in the wrong place. More often than not damage to property in the ground arises out of inefficiency, lack of liaison and reluctance to have men and plant standing idle.

A few insurers may attempt to specify in their policies what they mean by reasonable precautions in relation to damage to property in the ground. They may require pre-excavation inspection of plans, warning to owners of services, compliance with statutes, hand digging where appropriate. However, they still have the problem of deciding what to do in the event of a minor breach of the conditions and in practice they are no better off than those insurers who just rely on a general reasonable precautions clause. The only advantage of a special clause is that it brings to the notice of the insured the need to take certain precautions before work commences.

Most insurers have found that the only effective way of ensuring that reasonable precautions are taken is the imposition of an excess. The size of the excess will be related to the type of work an insured carries out and his previous claims record. The excess will have to be sufficiently large to give the insured a real financial stake in claims. A figure of less than £250 per claim is probably inadequate. Although this may solve the problem of the smaller claim and reduce the number of incidents reported, it will still not avoid the claim which costs many thousands of pounds.

Mechanical plant

A high percentage of damage to property in the ground claims arise out of the use of mechanical plant. Such plant may be:

(a)   owned by the insured and driven by his driver;
(b)   hired in by the insured and driven by his driver;
(c)   hired in by the insured and driven by the owner's driver.

Accidents may arise out of the negligence of:

(a)   the insured or his employees;
(b)   the driver supplied by the owner;
(c)   the owner.

Investigation of the accident will reveal which circumstances apply.

Where plant is hired in it is necessary to establish the conditions which apply to the hire. Most of the plant-hire contractors are members of the Contractors' Plant Association who use standard conditions, known as 'Model Conditions for the Hiring of Plant'.* These conditions make the hirer responsible for loss of or damage to the plant from whatever cause and provide for an indemnity to the owner for injury to persons or property caused by or in connection with or arising out of the use of the plant.

Where operators are supplied with the plant such drivers are to be regarded as servants or agents of the hirer, who alone is to be responsible for claims arising in connection with the operation of the plant. There are certain exceptions and qualifications, for example, when an owner accepts responsibility for the erection of plant or for accidents happening during transit from owner to site.

It is necessary to study the actual wording of the contract in relation to the circumstances of the loss and to decide the liabilities of the respective parties.*

Damage to the plant (being property in the insured's charge and control) will not be covered under a public liability policy but provision should be made under the contract works policy.

However, where plant being operated under the insured's control causes injury or damage, it is reasonable that the insured should accept responsibility for the loss. In connection with claims (and damage to property in the ground claims in particular) the problem may arise as to whether the driver was operating under the insured's directions or was he engaged on his own activities? A driver may, for example, be given instructions to dig in a certain place and at a certain level but may fail to obey those instructions and indeed may deliberately disobey his orders. The owners of the plant undertake to supply a person competent in operating the plant; what is the position if an incompetent person is supplied? Presumably the owners are in breach of their contract. The CPA Conditions clearly make the owners responsible. A competent driver may still be negligent; an underground pipe may be damaged

* For further details see *Model Conditions for the Hiring of Plant (1979) Implications for Insurers*, by F.N. Eaglestone, published by P.H. Press Ltd, Waterloo Road, Stockport 7.

because of the driver's negligence or because the insured failed to warn the driver of the presence of the pipes or issued incorrect instructions.

However, the intention of the Conditions is that the hirer (the contractor) should accept responsibility for most of the accidents arising out of the use of the plant; this will include accidents to the driver and accidents caused by the driver, irrespective of the driver's negligence.

In *Arthur White (Contractors) Ltd* v *Tarmac Civil Engineering Ltd (1967)* the hirer was held responsible for the negligence of the driver, (not only in respect of his operation but also the maintenance of the hired-in plant), it being decided that it was the intention of the indemnity clauses to apply in the circumstances of the accident*. The effect of the indemnity clause is that the hirer takes over a part of the employer's liability and public liability risks of the owner. Although employees of the owner are deemed to be servants of the hirer for the purpose of the contract this does not affect the legal status of these persons and in law a third-party relationship exists.

The Conditions clearly put the owner's drivers under the direction and control of the hirer and in the event of an accident this certainly helps to establish the areas of responsibility. Occasionally plant hire may be undertaken without a formal contract and this can raise problems. *Mersey Docks and Harbour Board* v *Coggins and Griffiths (Liverpool) Ltd (1947)* laid down that the true test of whether the general employer or the hirer is responsible for the acts of servants is 'whether or no the hirer had authority to control the manner of the execution of the relevant acts of the servant'.

Even although no formal contract may be entered into an owner's contractual conditions may be incorporated into a contract of hire on the basis of a common understanding between the parties that such conditions applied.

In *British Crane Corporation* v *Ipswich Plant Hire (1974)* both parties were engaged in the business of hiring out earth-moving equipment. The defendant, who was involved in drainage work on marshy ground, hired a crane from the plaintiff. Soon after the crane had been delivered the plaintiff sent the defendant a copy of their conditions of hire, which provided, *inter alia*, that the hirer would be reponsible for all expenses arising out of the crane's use. Before the defendant signed the form the crane sank into the marsh, although through no fault of the defendant. The plaintiff's claim was allowed. It was held that the plaintiff's conditions of hire applied since both parties were in the trade and of equal bargaining power; on the evidence both the defendant and the plaintiff understood that the plaintiff's conditions of hire would apply.

Contract conditions may have to be considered in the light of the Unfair Contract Terms Act 1977 and the decision in *Smith* v *South Wales Switchgear Ltd (1978)* (see chapter 1). For example, condition 13 of The Model Conditions for the Hiring of Plant dealing with the hirer's responsibility for loss and damage does not include any reference to negligence. In relation to property damage, therefore, the question arises as to whether it is reasonable for the hirer to have to accept respon-

---

* According to *Phillips Products* v *Hampstead Plant Hire (1983)* in the particular circumstances of that case such a clause (number 8 of the CPA Conditions) was unreasonable under the Unfair Contract Terms Act 1977.

sibility for the damage from whatever cause, and also whether in the absence of any specific provision for negligence the hirer must accept responsibility for the owner's negligence. The owner's negligence might arise out of the defective repair of plant or the negligence of a fellow employee to the operator, for example.

It is a basic principle of the common law of contract that parties to a contract are free to determine for themselves what primary obligations they will accept. It seems that commercial contracts between two equal parties can contain effective clauses either excluding or restricting liability for negligence in tort or in breach of a contractual duty of care. The test of reasonableness may have to be applied but businessmen entering into contracts are presumed to be capable of apportioning the risks and deciding their intentions. Sometimes plant may be hired in on a sub-contract basis, that is, to do a certain operation. Plant hired in this way would not normally be under the insured's direction and control and the respective liabilities of the parties would be regulated by a subcontract form. It would probably be the subcontractor's responsibility to check on plans and the layout of services, although in practice he might depend on information supplied by the main contractor.

Economic loss

Another problem which often follows damage to property in the ground claims concerns economic loss or consequential loss. An electric cable may be damaged by contractors and clearly the cost of the repair of the cable is their responsibility. However, following the damage the electricity supplies to an adjoining factory may be cut off, production may be interrupted and goods being processed, damaged. Disruption may be on a much wider scale, whole towns may be put out of action for many hours; at best there may be just inconvenience, at worst there may be accidents resulting in extensive damage to property and injury to persons.

The public liability policy of the contractor will cover his legal liability for consequential or economic loss which follows injury or damage which is insured by the policy. Liability for economic loss has been the subject of much legal discussion. The courts have admitted that it is not always possible for the law to be logical. It has to evolve practical and workable rules in the light of experience. The common law has always developed by experience rather than logic and by dealing with situations as they arise in what seems a reasonable way.

In *Electrochrome* v *Welsh Plastics Ltd (1968)* the defendants' servant negligently drove their lorry so as to collide with and damage a fire hydrant. Water escaped and the supply had to be stopped for several hours. As a result the plaintiff's factory had its water supply cut off and this caused the loss of a day's work. The main and hydrant were owned not by the plaintiff but by the owners of the industrial estate. It was held there was no actionable wrong as the duty of care not to damage the hydrant was owed, not to the plaintiff but to the owners of the hydrant. If the damaged hydrant had been owned by the plaintiff all the loss flowing therefrom would have been the responsibility of the contractors; it is the indirect economic loss which is excluded.

The problem of economic loss following damage to an electricity cable was considered in *S.C.M. (United Kingdom) Ltd* v *W.J. Whittall and Son Ltd (1970)*. Power was cut off for several hours and damage was caused to molten materials in machines which solidified owing to lack of electric heat. Lord Denning said that:

> The law is the embodiment of common sense or at any rate it should be. In actions of negligence when the plaintiff had suffered no damage to his person or property, but had only sustained economic loss, the law did not usually permit him to recover that loss. The reason lay in public policy. When an electric cable was damaged many factories might be stopped from working. Could each claim for their loss of profit? No. It was not sensible to saddle losses on that scale on to one sole contractor. The risk should be borne by the whole community rather than on one contractor who might or might not be insured against the risk. There was not much logic in that but it was the law.

The contractors were held liable for the material damage done to the factory owners and the loss of profit truly consequent upon it but not for any other economic loss.

The problem was again considered in *Spartan Steel and Alloys Ltd* v *Martin and Co (Contractors) Ltd (1972)*. Contractors negligently damaged an electric power cable while digging up a road. They were held liable for physical damage to property and any economic loss flowing from that damage. They were not liable for loss of production in the factory as a direct consequence of the power supply being cut off. When the power failed metal had to be poured out of the furnace to avoid it solidifying; this physical damage assessed at £368 was recoverable. If that particular melt had been completed the plaintiffs would have made a profit of £400 and that amount was also payable. During the fourteen and a half hours when the power was cut off the plaintiffs would have been able to put four more melts through the furnace; by not being able to do so they lost a profit of £1,767. The amount was not recoverable.

In the past the courts have considered the questions of foreseeability and remoteness and the spheres of duty. The issues are not always straightforward and judges do not always agree. In the Spartan Steel case Lord Davies dissented. He concluded that an action lay in negligence for damages in respect of purely economic loss provided that it was reasonably foreseeable and a direct consequence of failure in a duty of care.

There is a feeling that it would be much easier to decide future cases on the basis that there is a liability for all economic loss flowing from loss or damage to the plaintiff's property, thus paying for all loss of profit of £1,767 in Spartan's case. In practice there has been considerable difficulty in deciding where to draw the line once loss or damage to the plaintiff's property is caused and economic loss follows.

In the Spartan Steel case the court drew a somewhat arbitrary line and doubt has now been raised about this decision. The recent case of *Junior Books Ltd* v *Veitchi and Co Ltd (1982)* has been referred to in chapter 13, although there was no damage to property or likelihood of injury to persons or property in the future, the respondents were able to recover the cost of re-laying the floor which included the consequential losses, for example, the cost of moving machinery, the fixed overheads, etc.

Prior to this case it had already been well established that a reasonable anticipation of physical injury to persons or property was not a *sine qua non* for the existence of a duty of care. It had also been established that where a duty of care exists through the presence of such reasonable anticipation and it is breached, then even though no such injury has actually been caused because the person to whom the duty is owed has incurred expenditure in averting the danger, that person is entitled to damages measured by the amount of that expenditure (see *Hedley Byrne and Co Ltd* v *Heller and Partners Ltd (1964)* and *Anns* v *Merton London Borough Council (1978)*). Lord Keith did not consider the Junior Books case to be one for seeking to advance the frontiers of the law of negligence, principally because he felt that the deterioration of the flooring was not damage to the respondent's property such as to give rise to a liability following directly within the principle of *Donoghue* v *Stevenson (1932)*.

However, the issues raised are not always clear-cut and Lord Fraser commented that if and when such other cases arise they will have to be decided by applying sound principles to their particular facts. The judges are not always of one mind and in the Junior Books case Lord Brandon gave a dissenting opinion.

Consequential loss and policy cover

The public liability policy does not usually refer in any way to consequential loss, but loss which flows from damage to property is covered if the insured has a legal liability. As illustrated under the previous heading in the damage to property in the ground claims an insured may have no legal responsibility for pure economic or financial loss which results from his negligence.

Liability for consequential loss will only be covered if the policy wording provides for cover for damage to the property concerned. If there is no property cover there is no consequential loss cover. For example, contract works the insured has erected may be damaged by subsidence because of the insured's faulty workmanship. The building may have to be closed whilst repairs are effected, with a resultant loss of production claim by the owner. The public liability policy may exclude damage to contract works erected by an insured and therefore there would be no cover for the loss of production claim. Much, however, will depend on the actual wording employed. Claims arising out of defects in structures completed by an insured perhaps many years previously, may involve a situation where there is no insurance protection in force either under public liability or contract works insurance. The policy may exclude the cost of making good faulty workmanship, not the liability which flows from it. For example, an insured may install defective radiators or install sound radiators defectively. The public liability policy will not cover the cost of repairing the radiators or putting right the faulty installation. It will, however, take care of the damage caused by the radiators, for example, leakage of water damaging carpets or stock, plus any consequential loss which may arise from that damage. In the Junior Books case referred to earlier, the contract works policy would not offer any indemnity because of the usual policy exclusion regarding the replacement of defective work. As there was no damage to other property the consequential loss involved would also be excluded.

The average public liability policy only covers material damage to property so in the absence of specific cover for financial loss a contractor would have to shoulder this risk himself. A similar case involved defective flooring where because of the soft cement, tiles laid on top had been damaged by the weight of filing cabinets. The floor had to be re-laid and the damaged tiles replaced and there were consequential losses flowing from the interruption of business. An average public liability policy would not cover these risks except possibly for the damage to the tiles.

Dust damage

Claims involving dust damage are usually difficult to handle. The policy will indemnify the insured against his legal liability for accidental damage; whether the damage is accidental or inevitable will be a question of fact. Such claims may be connected with other claims for general disturbance and interference with a person's business. Dust may be entering shop premises from the contract site, with the result, for example, that customers do not wish to enter the shop. The stock may have been reasonably protected by the owner or by the contractor, with the result there is no property damage claim, only a claim for financial loss. If the stock has been damaged by dust, insurers will have to decide on the accidental aspect; if there is no evidence of accident, there will be no policy indemnity. If the damage is accidental, then adjusters will be employed to assess the damage. Whilst clothing for example, may be cleaned or sold as shop soiled there will be a considerable reduction in value. Similarly with upholstered furniture there will be considerable depreciation. With wood furniture on the other hand, there may be little or no damage.

Where there is only financial loss with no damage to material property, there may be no policy indemnity. If insurers restrict their cover to damage to material property, pure financial loss is excluded. If they just refer to property, there may be cover, for financial loss may be regarded as property if there is a right of action to recover it. Certainly an insured can be legally liable for financial loss where there is no specific material property damage. In *Benjamin* v *Storrs (1874)* the constant presence of vans and the intermittent urination of horses made the plaintiff's coffee house incommodious by obstructing the light and fouling the air. There was no injury to material property but the plaintiff was able to maintain an action.

Pollution

Pollution may be considered in relation to the air, ground and water. Noise may also be regarded as a form of pollution. In relation to construction work in the building and civil engineering trades, claims are relatively few. A contractor is unlikely to pollute the atmosphere to any material extent and the opportunities for ground and water pollution are also limited. However, work is frequently undertaken in or near canals, rivers and streams and obviously care must be taken. Chemicals and waste materials should not be poured away or just dumped. Many waterways contain valuable fish which are easily killed by relatively small amounts of noxious substances. Even sand if allowed to be washed away in quantity can choke the fish.

The Control of Pollution Act 1974 makes provisions with respect to waste disposal, water pollution, noise, atmosphere pollution and public health. For example, it is an offence for any person to cause or knowingly permit:

(a)   any poisonous noxious or polluting matter to enter any relevant waters;
(b)   any matter to enter a stream which either directly or indirectly or in combination with other material tends to impede the proper flow of the water in a manner leading or likely to lead to a substantial aggravation of pollution due to other causes or of the consequences of such pollution;
(c)   any solid waste to enter a stream or restricted waters.

Public liability policies issued to contractors do not normally exclude damage caused by pollution. There is, however, the intention that damage should involve an accidental element, the damage should be fortuitous and unexpected; the policy will not cover deliberate or inevitable damage, that is, non-accidental damage.

Noise

The Control of Pollution Act 1974 again deals with this subject. For the purposes of the Act noise also includes vibration. Where a local authority is satisfied that noise amounts to a nuisance it must serve a notice which, *inter alia*, will:

(a)   require the abatement of the nuisance or prohibit or restrict its occurrence or recurrence;
(b)   specify other steps which will be necessary;
(c)   specify time limits for complying with the notice.

Local authorities have extensive powers to control noise on construction sites. They may, for example:

(a)   specify the plant or machinery which are or are not to be used;
(b)   specify the hours during which the works may be carried out;
(c)   specify the level of noise which may be emitted from the premises in question.

There is a **BSI** Code of Practice for noise control on construction and demolition sites.

Persons intending to carry out work on construction sites may apply to the local authority for prior consent to a noise level and the authority in reaching a decision will take into account any code of practice, the best practicable means, suitable alternatives, and the protection of persons in the locality.

There is no doubt that the use of heavy construction plant, for example, excavators and pile driving equipment, can cause great inconvenience to surrounding property owners either by noise or by vibration. Property may be damaged and a person's business disrupted. In dealing with a claim the difference between policy liability and legal liability must be appreciated. Financial loss or inconvenience without damage to property may not be covered under a public liability policy. The policy would certainly not cover any fine which might be imposed by the local authority for the continuance of the nuisance. Whether there will be any legal liability on the contractor for nuisance is a question of fact. There must be substantial interference with the use or enjoyment of the property of another. Mere annoyance is insufficient.

Removal and weakening of support

During construction work, excavating can cause damage to surrounding property. Foundations may be undermined if adequate support is not given to nearby structures. The driving of piles can cause vibration which again can weaken foundations and cause cracks in buildings or other structures. Piling work may also interfere with ground conditions; on sandy ground there may be disturbances affecting nearby property. De-watering plant may have to be used in certain conditions and the extraction of water and sand can sometimes have disastrous results.

There is no liability for water percolating from neighbouring land even if the process is increased (*Langbrook Properties Ltd* v *Surrey CC (1969)*), but if soil is removed as well there would probably be a liability for subsidence so caused.

Occasionally rock may have to be removed involving the use of explosives for blasting purposes. This can result in vibration damage to buildings some distance away — shock waves can travel through the rock.

The extent to which piling operations cause vibration varies tremendously according to the type of piles and the type of ground involved. When work is involved close to or even inside existing structures, a type of bored pile is usually employed, and there are many systems available which are designed to reduce vibration to a minimum. In certain circumstances, however, it seems almost inevitable that some damage will be caused to existing structures by piling work. Both the legal and policy liability will need consideration. The contractor may not have been negligent and the contractor under the ICE Conditions, for example, will not be responsible for damage which is the unavoidable result of the construction of the works in accordance with the contract (clause 22(1)(b)(iv)).

The public liability policy issued to the contractor will normally involve an 'accidental' element and these considerations can raise problems in connection with claims settlements. However, most insurers anticipate the problem by accepting a degree of inevitability and imposing a substantial excess for claims arising out of pile driving operations. In any event damage is rarely completely unavoidable, for a slightly changed method of work will usually reduce a risk to acceptable levels.

There may be a problem about the application of the excess, has there been more than one accident or event? Should the excess apply on an aggregate basis over a contract or over a period? It is up to the individual insurer to lay down the terms he considers appropriate to the risk involved. However, an aggregate excess applicable to each contract usually provides a satisfactory solution.

Existing damage

Owners of property adjacent to piling operations frequently make nuisance claims or claims which are not completely genuine. The excess will take care of nuisance claims such as minor cracks in plaster. Other claims may involve damage which was already in existence before the commencement of the contract. Claims require careful investigation to decide the cause of the damage and the extent to which a contractor is responsible. Existing damage cannot be paid for, but such damage may have been worsened by the contractor's operations. It may be that repairs are now essential, but that previously there was no urgency. Such claims are essentially ones for negotiation between the parties, to decide the extent of the respective liabilities. Prudent contractors will arrange for a schedule of dilapidations to be prepared before work commences. This will record existing damage, cracks in walls and ceilings of adjoining property or the property being altered. This information will enable insurers to refute claims which are not genuine or are exaggerated. Pre-accident photographs are helpful.

Indemnity limits

Some insurers apply separate indemnity limits to removal or weakening of support claims, although this practice is now less common. There may be a lower limit for 'any one accident', than for other risks or the period of indemnity limit may be for the same amount as the single accident indemnity limit. These features are not usually important in claim settlements unless the limits which have been selected are totally inadequate to meet the cost. The same problems then arise as with any other claim where the indemnity limit is inadequate to meet the liability. It is undesirable that there should be separate and lower limits of indemnity for certain types of risk.

Fire damage

Fire caused by contractors is one of the most costly causes of claims. Although the owners of the damaged property may claim from their fire insurers, those insurers will in turn seek indemnity from the negligent contractors.

In *H. and N. Emanuel Ltd* v *Greater London CC (1971)* the Council employed a firm of contractors to demolish some prefabricated buidings on land that it owned. The contractor disposed of some of the debris by burning it and sparks from one of the bonfires set fire to surrounding buildings. The Council were found liable for the damage, notwithstanding that the contractors had been expressly forbidden to burn rubbish on the site. It was held that an occupier was liable for the escape of fire caused by the negligence not only of his servant but also of his independent contractor and anyone else who was on his land with his leave and licence.

Presumably if similar circumstances arise under the terms of an ICE Contract, the contractor would be liable to indemnify the employer and in turn insurers would indemnify the contractor under the contractual liability extension to the public liability policy.

If the claim was made against the contractor not under the contract but because of his negligence, the basic policy cover would operate.

It is not always possible to establish the exact causes of a fire and insurers frequently employ independent technical experts to investigate and report.

Gas pipes may be damaged during excavation work and escaping gas may result in fire and explosion damage. It is necessary to prove that the contractor is responsible, and if several contractors are working in the vicinity this may not always be possible. On the other hand if the contractor is using welding equipment and damage occurs there is usually little defence to a claim.

Recoveries

After paying a property damage claim insurers will seek recovery from any other negligent party or they will take advantage of the terms of any contract which may be available.

A possible line of action is recovery from subcontractors. Much will depend on the wording and intention of the subcontract form – if there is one. The subcontractor will usually be responsible for his own acts of negligence and he may or may not be adequately insured. The question of liability is independent of the existence of an insurance policy. However, if the subcontractor is an uninsured 'man of straw' chances of recovery will be remote.

Another possibility is recovery against a manufacturer if the damage was caused by an item of defective equipment.

Sometimes insurers have agreements with other insurers to share claims and avoid disputes on liability and where appropriate these agreements should be invoked. Some insurers also have agreements with various authorities regarding damage to property in the ground claims, whereby liability is accepted in fixed proportions, say 75% for the contractor, 25% for the authority. Again this may avoid disputes on liability. The possibility of dual insurance should also be considered. Sometimes there may be public liability insurance under a plant policy, a motor policy, and a public liability policy. If there is dual insurance the claim would have to be shared equally.

The excess should be recovered from the insured before the claim is paid, but in practice this may not always be done. Before the claim is filed, therefore, the amount of the excess must be paid by the insured.

Chapter 22

Bodily injury claims

Claimants

Bodily injury claims made by employees will be handled under employers' liability policies, those made by third parties (that is, claimants who are not employees) will be dealt with under public liability policies. It is a question of fact whether a person is an employee or a third party, although occasionally there may be a dispute about the status of an individual. For example, with the employment of labour-only subcontractors in the construction industry there are often areas of doubt. Most insurers will say in their policies that labour masters and persons supplied by them or persons employed by labour-only subcontractors or self-employed persons shall be for the purposes of the policy deemed to be under a contract of service or apprenticeship with the insured. This only means that the claims by such parties will be handled under the employers' liability policy.

This, however, will not alter the legal status of an injured person. It may be important to decide the exact status, for generally a greater duty of care will be owed to an employee than is owed to a third party. *Ferguson* v *John Dawson and Partners (Contractors) Ltd (1976)* dealt with this problem of status. The defendants were building contractors and the plaintiff when taken on was told 'there were no cards — we were purely working as a lump labour force'. It was held that the nature of the relationship between the plaintiff and the defendants should be determined by looking at the contractual terms governing the reality of that relationship and the parties could not by a declaration that the plaintiff was to be deemed a self-employed labour-only subcontractor alter what was the true nature of their contract. In an earlier case *Market Investigations Ltd* v *Minister of Social Security (1969)* it was said that the distinction between 'employees' and 'independent contractors' was fundamentally a matter of economic reality. The question for the court to answer was: 'Is the person who has engaged himself to perform those services performing them as a person in business on his own account?'

However, in *Massey* v *Crown Life Insurance Company (1978)* it was held that the terms of an agreement could alter both the legal situation in which the parties stood and the legal consequences and it was open to the parties to agree what the legal relationship between them should be. This is distinguished from the Ferguson case, for the agreement there was so unspecific and wanting in accuracy that there was nothing to go by except the way in which the parties had acted and that bore all the hallmarks of a master and servant relationship. In *Ready Mixed Concrete* v

Minister of Pensions (1968) it was held that a contract of service (that is, employment) exists if the following three conditions are fulfilled:

(a)   The servant agrees in consideration of remuneration to provide his own work and skill in performance of service for his master.

(b)   The servant agrees that in performance of that service he will be subject to the other's control in a sufficient degree to make that other the master.

(c)   The other provisions of the contract are consistent with its being a contract of service.

How claims arise

There are endless causes of accidents and in relation to accidents to employees the construction industry has a relatively poor record. For example, out of a total of 435 fatal accidents at work in 1980, 128 related to the construction industry. The extensive use of machinery, the use of scaffolding, working at heights, lifting of heavy weights, all these contribute to the fact that being an operative in the construction industry is a relatively dangerous occupation. Nevertheless, there has been an improving trend over the last ten years. Construction sites are not usually open to the general public, but other contractors and their employees working on the site will be regarded as third parties. Furthermore, there is always the risk of injury to legitimate visitors, for example, lorry drivers delivering goods, architects and employees of the local council. Passers-by on adjacent public roads may be injured by falling debris. Road and sewer contractors are vulnerable to claims from the public; excavations and heaps of sand are potential dangers to both vehicles and pedestrians.

In the event of accident the extent of liability is a question of fact. A breach of a statutory regulation which results in an accident to an employee will usually mean that an employer has no defence to a claim. However, merely the fact that an accident has occurred does not mean that an employer is automatically liable. Several defences may be available. It may, for example, be decided that an employer has not been negligent. In *Bolton v Stone (1951)* Lord Porter said:

Nor is the remote possibility of injury occurring enough; there must be sufficient probability to need a reasonable man to anticipate it. The existence of some risk is an ordinary incident of life, even when all due care had been, as it must be, taken.

In *Black v Carrick (Caterers) Ltd (1980)* an employee was injured whilst lifting trays of cakes and bread. She claimed damages for breach of statutory duty; under section 23(1) of the Offices Shops and Railway Premises Act 1963 no one in the course of work shall be required to lift, carry or move so heavy a load as to be likely to cause injury. The Court of Appeal dismissed the claim holding that the plaintiff had not been 'required' within the meaning of the Act to lift the trays. The employers had not been negligent and the accident was due to ill fortune. There is still such a thing as a pure accident even in the employer's liability field.

Smith v *George Wimpey and Co Ltd (1972)* concerned an employee who was going down a path which was steep with loose bits of rubble. He put his foot on a piece of rubble, slipped and cut his hand on a bit of concrete slab. He claimed damages against his own employers and also against the main contractors. His employers owed him the ordinary common law duty to take reasonable care for his safety and they also had a statutory duty under the Construction (Working Places) Regulations 1966 to provide 'so far as is reasonably practical, suitable and sufficient safe access to and egress from every place at which any person at any time works'. The main contractors and occupiers of the site owed him the 'common duty of care' in the Occupiers' Liability Act 1957 to see that the visitor will be reasonably safe in using the premises. They were also under a duty to lay out the subcontractors' work and co-ordinate their activities with reasonable care so that there would not be unnecessary danger. However, the court held that the main contractors did not owe the subcontractors' employee any statutory duty under the Construction Regulations. This is an example of an accident where both the employers and main contractors owed a duty to the employee, but where the employers had a greater duty because of their statutory obligations.

The fact that both the employer and the main contractor had certain obligations to the workman did not mean that compensation would be paid automatically. The judge found that the path was reasonably safe for a workman using it with reasonable care for his own safety. Slipping on loose rubble was just one of those accidents which could occur at any time without negligence. For that type of eventuality compensation was provided for by the social security benefit; the workman's claim was therefore dismissed.

Limitation

The law regarding limitation of actions has been consolidated in the Limitation Act 1980. The first statute regarding limitation was passed in 1623 and prescribed six years as the limit for actions founded on simple contract or tort (although these terms were unknown at the time). The comprehensive statute, the Limitation Act 1939, retained this six-year limit. However, this period was reduced to three years for personal injury claims under the Law Reform (Limitation of Actions, etc.) Act 1954. The time ran from the date when the loss or damage was suffered by the plaintiff irrespective of his knowledge of such loss or damage. However, this involved a certain amount of injustice particularly in relation to insidious diseases such as pneumoconiosis. In *Cartledge* v *E. Jopling and Sons Ltd (1963)* a man was held to be statute-barred before he even knew he had contracted the disease. The position was amended by the Limitation Acts 1963 and 1975. The basic rule is still that an action for damages arising out of negligence, nuisance or breach of duty where the damages claimed consist of or include damages in respect of personal injury, must be brought within three years.

If the person injured was at the time of the injury under a disability (for example, he was a minor or suffering from insanity) the period only begins to run from the date he ceases to be under a disability or dies, whichever is the earlier. Thus, if a child of two is injured, nineteen years elapse before any action is statute-barred, that is, three years after the child's eighteenth birthday.

The period of three years is applicable from:

(a)   the date on which the cause of action accrued; or
(b)   the date of knowledge (if later) of the person injured.

Where the injured person dies and the action is brought for the benefit of his estate under the Law Reform (Miscellaneous Provisions) Act 1934 the period of three years is from the date of death or the date of the personal representative's knowledge.

Where the action is brought under the Fatal Accidents Act 1976 the period is three years from the date of death or the date of knowledge of the person for whose benefit the action is brought.

Section 14 of the Limitation Act 1980 states that references to a person's date of knowledge are references to the date on which he first had knowledge of the following facts:

(a)   that the injury in question was significant; and
(b)   that the injury was attributable in whole or in part to the act or omission which is alleged to constitute negligence, nuisance or breach of duty; and
(c)   the identity of the defendant; and
(d)   if it is alleged that the act or omission was that of a person other than the defendant, the identity of that person and the additional facts supporting the bringing of an action against the defendant;

and knowledge that any acts or omissions did or did not, as a matter of law, involve negligence, nuisance or breach of duty is irrelevant.

With diseases of gradual onset it is often some time before an employee realises he has contracted a serious complaint or that he has a right of action against his employer. The current legislation removes the injustice which might have existed previously.

Furthermore, under section 33 of the Limitation Act 1980 (formerly section 2D of the Limitation Act 1939 as amended by the 1975 Act), the court has considerable discretion to override time limits where it is equitable to do so. The court will have regard to all the circumstances of the case: the length of and reasons for the delay on the part of the plaintiff; the effects of the delay; the conduct of the defendant; the duration of the disability of the plaintiff arising after the date of the accrual of the cause of action; the extent to which the plaintiff acted promptly and reasonably once he knew that he might have a right of action and the steps taken by the plaintiff to obtain medical, legal or other expert advice; and the nature of any such advice he may have received.

The onus of showing that in particular circumstances it would be equitable to make an exception lies with the plaintiff, but subject to that, the courts' discretion to make or refuse an order is virtually unfettered. This was the decision in *Firman* v *Ellis (1978)*. In that case a writ had been issued before the expiry of the primary limitation period but it had not been served or renewed within a year. A new writ was then issued and an application made for an order under section 2D of the Limitation Act 1939 (as amended by the 1975 Act).

However, in *Walkley* v *Precision Forgings Ltd (1979)* the House of Lords decided that where a plaintiff had started an action within the three-year limitation period he could not unless there were the most exceptional circumstances bring a second action based on section 2D. This appeared to overrule the *Firman* v *Ellis* decision.

Nevertheless, in a later case *Thompson* v *Brown Construction (Ebbw Vale) Ltd (1981)*, the House of Lords upheld the principle of unfettered discretion to allow an action for damages for personal injuries to proceed after the expiry of the ordinary three-year limitation period where it would be equitable to do so. In that case a writ was not issued until thirty-seven days after the expiration of the three-year time limit. Lord Diplock pointed out that where a court exercised its discretion to allow a trial to continue under section 2D the defendant must always be prejudiced as his defence that the action was statute-barred was automatically defeated. Although he might have another good defence he would have to spend time and money in establishing it. On the other hand, if it were held that the case were statute-barred, the plaintiff assuming he had a good case would be at a disadvantage. He might be able to sue his solicitors for professional negligence but he would have to instruct new solicitors, there would be delay and he would be liable for costs up to the date of the court's refusal to give a direction under section 2D.

In practice, therefore, the courts can weigh the respective merits of each case; they have a discretion but it is one which must be exercised soberly and within set rules.

Medical reports

The medical report is an essential feature in the assessment of quantum in connection with a bodily injury claim. Reports are either obtained by both the plaintiff's and the defendant's solicitors or one side agrees to the other's report. Sometimes there may be a difference of opinion between the medical experts and the courts will have to reach a decision on the evidence available. For example, in *Joyce* v *Yeomans (1981)* the plaintiff's medical evidence alleged that the epileptic attacks were due entirely to the head injury received in the accident. The defendant's consultant considered that the attacks arose from an epileptic focus in the posterior temporal region of the brain present since birth. The defendant's view was favoured.

It is simpler if the medical evidence can be agreed between the two parties before the case is heard in court. This means that medical reports should be exchanged. This may avoid the attendance of doctors who are often reluctant to spend their time giving evidence. In *Worrall* v *Reich (1955)* it was said, 'in cases of personal injuries it is proper and desirable that medical reports should be exchanged to the greatest possible extent'. In *Lane* v *Willis (1972)* Sachs L.J. made the comment:

There is far too much reluctance in this matter of exchanging – far too much manoeuvring behind the scenes – far too much (especially on the part of defendants) trying to hold back a report until the moment of trial.

In *Clarke* v *Martlew (1972)* it was held that the seeking of a medical examination by the defendant was a privilege, the privilege should not be accorded unless he was prepared to act fairly by it and fairness required that he should show the medical reports to the plaintiff. If the defendant sought to have the action stayed for the purpose of having the plaintiff medically examined, it was reasonable and just that as a condition of the stay he should give an undertaking to make the medical reports available to the plaintiff.

The problem was again considered in *McGinley* v *Burke (1973)*, where it was held that although a plaintiff could make it a condition of submitting to a medical examination that the defendant should provide him with a copy of the report, he could only do so where he was willing to offer in exchange on a basis of reciprocity his own equivalent report.

In the past plaintiffs' solicitors commonly requested the attendance of their doctor at the medical examination arranged by the defendant. However, in *Hall* v *Avon Area Health Authority (Teaching) (1980)* it was said that:

To require a medical consultant to be present to act as chaperone, although he had no personal function to perform in the course of the examination by the defendant's surgeon seemed a serious addition to the costs of litigation and also an inexcusable requirement.

Provision was made in the Civil Evidence Act 1972 for rules of court relating to the disclosure of medical reports. Rules of the Supreme Court came into existence in June 1974 which say that in an action for personal injuries, unless the court considers there is 'sufficient reason' for not doing so, it shall direct that the substance of the evidence be disclosed in the form of a written report or reports to such other parties.

However, the 'sufficient reason' for not disclosing may include a situation where the pleadings contain an allegation of a negligent act or omission in regard to medical treatment or where the expert evidence may include an expression of opinion either as to the manner in which the injuries were sustained or as to the genuineness of the symptoms of which complaint is made.

These rules appear to give the courts some discretion and may partly overrule the decision in *Clarke* v *Martlew*. Thus examination of the plaintiff by the defendant's doctor may not always be conditional on the disclosure of a medical report.

Usually when insurers do not want to exchange reports it is because of some gloomy or adverse feature in their own report. Nevertheless, whatever the reasons, if a claim is to be settled out of court (and this is usually desirable) an exchange of medical evidence is always helpful and is usually preferable.

The proper assessment of damages requires that the plaintiff's medical condition should be stable. In practice in a serious injury case several medical reports will be obtained by each side.

Damages

Damages may be divided into two main headings: special damages and general damages.

Special damages

These are damages which do not necessarily follow the tort and which must be pleaded specially and proved, for example, hospital expenses, loss of earnings, travelling expenses, cost of employing a deputy. Loss of earnings to the date of accident fall under the heading of special damages, whilst future loss of earnings are regarded as general damages.

Special damages are thus capable of exact calculation and insurers should always obtain full details and necessary proofs where appropriate. Wage details (pre-accident average weekly earnings, net of tax and DHSS contributions) should be certified by employers and receipted accounts obtained. It is not unusual for inflated claims to be put forward in this area, so a careful check should be made on all the items involved.

As deductions have to be made in respect of certain social security benefits the plaintiff receives (see later), the details of such benefits should be obtained from the claimant's solicitor so that the appropriate deductions can be made from the loss of earnings and profits items of damages.

General damages

These are damages awarded by a court for an infringement of a legal right which cannot be assessed precisely but only generally in terms of cash, for example, pain and suffering, future disability (which involves loss of earnings and loss of amenity), costs of future care, loss of pension rights, loss of expectation of life and other expenses still to be incurred.

In *H. West and Son Ltd* v *Shephard (1964)* it was said:

All that judges and courts can do is to award sums which must be regarded as giving reasonable compensation. In the process there must be the endeavour to secure some uniformity in the general method of approach. When all this is said it still must be that amounts which are awarded are to a considerable extent conventional.

The judges accept that it is not possible to offer perfect compensation or to put the plaintiff back in his original position; all they can do is to take a reasonable view and give what they consider under all the circumstances to be fair compensation.

To some extent there is a judicial tariff, a price range for each type of injury, but the individual judge has a great deal of discretion to take into account the circumstances of each case and to vary the award accordingly. The loss of an eye, for example, will justify an award of £X. The value of X will depend on a variety of factors of which perhaps the most important is the value of money at the time of loss. In the early 1970s, for example, the conventional award for the loss of an eye was about £5000. In the early 1980s an award for a similar loss would be about £10,000. In the early 1990s, assuming inflation continues at its present rate, we can expect awards of £20,000 plus.

The amount of damages will be influenced by factors such as the extent of future loss of earnings, but most individuals can live successfully with one eye and this type of injury lends itself to a tariff pricing.

How damages are assessed

On several occasions the courts have indicated that the law as to damages for personal injuries is in urgent need of reform. It has been commented that the assessment of damage is an exercise in guesswork (see *Joyce* v *Yeomans (1981)*). The current method involved is to itemise the various heads of damage costs of future care, loss of future earnings, pain suffering and loss of amenities, as if they were separate causes of action, and then add the items together and award the total sum as damages, together with interest where appropriate. The breakdown of damages in this way inevitably means that higher damages are awarded than would be if several items were grouped together.

The necessity to apportion the damages under different headings was imposed by the Administration of Justice Act 1969. Under section 22 the award of interest became compulsory in personal injury cases. However, different rates of interest applied to various items of the award, hence the need for apportionment. Damages for future financial loss do not attract interest at all.

Greater liabilities imposed by legislation or by judicial precedent, coupled with the ravages of inflation, have caused the amounts paid in damages to rise substantially. Many judges feel that matters are getting out of hand and that in many areas awards are excessive. Following the decision in *Gammell* v *Wilson (1981)* for example, a deceased's estate could recover damages for loss of earnings or other income relating to the shortened period of the deceased's life as a result of the defendant's negligence. An estate swollen with the damages award might pass to persons in no way dependent on the deceased. This is difficult to justify and the Administration of Justice Act 1982 remedies the situation. Section 4 of the Act provides that damages for loss of income in respect of any period after the deceased's death cannot be claimed by the legal personal representatives.

The Act also makes several other changes.

Section 1 abolishes loss of expectation of life claims. The rights of a surviving victim are safeguarded as the court must take into account, in assessing damages for pain and suffering, any suffering caused or likely to be caused by consciousness that the expectation of life has been reduced.

Section 2 prohibits certain claims for loss of services, for example, loss of the services or society of a wife.

Section 3 amends the Fatal Accidents Acts. The class of dependants now includes former spouses and persons who were living with the deceased in the same household for at least two years immediately before the date of death and dwelt there during the whole period as husband or wife of the deceased.

To offset the vanished loss of expectation of life claim a Fatal Accidents Acts action may now include a claim for damages for bereavement, for the benefit of a remaining spouse, parents or children of the deceased. If a dead child is illegitimate only a claim by the mother is allowed. The sum awarded under this heading is £3500 but the Lord Chancellor has power to vary the amount by order.

Section 5 enables a defendant to offset any saving to a victim suffering personal injuries, attributable to maintenance wholly or partly at the public expense, in a hospital or similar institution.

Section 6 deals with personal injury cases where there is proved or admitted to be a chance that at some definite or indefinite time in the future the injured person will as a result of the act or omission which gives rise to the cause of action, develop some serious disease or suffer some serious deterioration in his physical or mental condition. The court will have the power to issue a declaratory judgment in his favour, which in the event of deterioration of his condition, will enable the plaintiff to apply for a review of the original award. This section could have far-reaching effects for insurers. Extra reserves will have to be created for such claims and the number of 'outstanding' claims will be increased substantially.

With the exception of section 6 the Act came into force on 1 January 1983. Section 6 comes into operation on such day as the Lord Chancellor may by order appoint.

The advances in modern medicine have enabled injured persons to live for many years after an accident whereas formerly they would have died. Mere existence is prolonged but the mind may have gone. Lord Denning in *Lim* v *Camden Health Authority (1979)* considered that:

Fair compensation requires there should be ample provision in terms of money for comfort and care during the lifetime of the sufferer such as to safeguard her in all foreseeable contingencies including the effect of inflation, that if he or she has any dependants they should be compensated for any pecuniary loss which they suffer by reason of his or her incapacity and inability to earn, just as if he or she had died and compensation was being awarded under the Fatal Accidents Act. Beyond that there can be conventional sums for pain suffering, loss of amenities but these should not be too large seeing they would do the plaintiff no good and only accumulate during her lifetime to benefit others who survive after her death.

He could see no justification for awarding large sums for future loss of earnings. His view was not entirely endorsed by the House of Lords who considered that reforms were a matter for the legislature.

On the question of inflation Lord Scarman in *Cookson* v *Knowles (1978)* thought that an attempt to build into lump sum compensation a protection against future inflation was seeking after perfection. In the great majority of cases inflation should be disregarded because:

(a)   It was pure speculation whether inflation would continue at present or higher rates.
(b)   Inflation was best left to be dealt with by investment policy.
(c)   It was inherent in a system of compensation by way of a lump sum immediately payable that the sum be calculated at current money values leaving the recipient in the same position as others who had to rely on capital for their support to face the future.

Whether tort is the best method of settling compensation cases is a debatable issue and one which only Parliament can solve. The real difficulty lies in trying to offer compensation in money for the loss of life, limb or health. Money cannot really compensate, so inevitably the exercise is something of an artificial one. With property damage the principle of indemnity can be applied, money can usually repurchase what has been lost or destroyed. When it is a question of trying to put a monetary value on such intangible items as pain and suffering or loss of amenity a court is attempting to measure the immeasurable. All the courts can do is to award such a sum of money as will put the injured party in the financial position he would have occupied if he had not suffered the accident.

Some awards, following very serious injuries, are now approaching a figure of £500,000 in the United Kingdom and it is felt that a complete reappraisal of the methods of awarding damages is long overdue.

Once liability towards a plaintiff has been established the methods employed in the assessment of damages are the same irrespective of whether an employee or a third party is involved.

Deductions from special damages

In calculating special damages for loss of earnings a plaintiff is not necessarily entitled to a sum which represents the loss exactly. Whilst compensation is provided it is not intended that a plaintiff should profit because of the accident and many collateral benefits may be deducted in whole or in part.

There are two general areas in which post-accident receipts are disregarded: the proceeds of insurance and the proceeds of benevolence. In *Parry* v *Cleaver (1970)* a policeman injured in a traffic accident had a police disablement pension payable during life and the question arose whether he had to give credit for such pension. It was decided that the employer and the employees were both contributors to the fund and that it would be wrong that the policeman could gain nothing from the benevolence of his employers and associates while the only person to gain would be the wrongdoer. The pension was not an equivalent to the wages which had been lost.

There is, however, a good deal of overlapping between the social security and tort systems and there are some areas where doubts and difficulties may arise. Social security benefits are in part paid for by the claimant's contributions and it has been argued that because of the insurance element a claimant should not be denied the benefits which might accrue. However, the present tendency of the courts seems to be to allow a full set-off for many of the collateral benefits. A defendant is entitled to offset any refund of income tax that a plaintiff might be entitled to in respect of tax paid during that particular fiscal year.

Unemployment benefit   Unemployment benefits may be regarded as a combination of insurance and national benevolence. The question as to what extent such benefits should be deducted from damages awards was considered in *Parsons* v *B.N.M. Laboratories (1964)*. The Court of Appeal held that such benefits should be deducted.

The problem was again considered in *Nabi* v *British Leyland (UK) Ltd (1980)* and it was suggested that the House of Lords decision in *Parry* v *Cleaver (1970)* overruled the decision in the Parsons case. It was, however, decided that there was no inconsistency between the two decisions. Sums received by an injured workman as unemployment benefit should be taken into account in assessing damages to be awarded against negligent employers. Damages awarded for consequent loss of earnings should be reduced accordingly.

Supplementary benefit In addition to unemployment benefit, supplementary benefit must also be deducted from special damages for loss of earnings. In *Cackett* v *Earl (1976)* it had been held that unemployment benefit, industrial rehabilitation allowance and supplementary benefit must all be deducted. The problem was again considered in *Plummer* v *P.W. Wilkins and Son Ltd (1981)* where the plaintiff received unemployment benefit for the statutory period, then a supplementary allowance. Supplementary allowances were not discretionary, a qualifying recipient was entitled to them as of right. The court held that it would be unfair for the plaintiff to receive double compensation. From one source or another he had received that to which he was entitled to put him in the position he would have been in if the accident had not occurred. The amount of supplementary benefit was therefore deducted from the agreed amount of special damages, that is, the items for loss of profits or earnings.

The Court of Appeal considered the matter in *Lincoln* v *Hayman and Another (1982)*. The question regarding deduction of an amount for supplementary benefit raised a question of principle to which there was no decision binding on the Court of Appeal. However, *Parsons* v *B.N.M. Laboratories Ltd (1964)* was binding and as this held that unemployment benefit was deductible from the plaintiff's damages, it followed that supplementary benefit should also be deductible. If it was not deducted the plaintiff would achieve double recovery and this was contrary to the basic principle of damages as compensation for loss actually suffered.

Income tax In making payments a plaintiff's liability to income tax and surtax must be taken into account. In *British Transport Commission* v *Gourley (1955)* the House of Lords pointed out that failure to pay regard to a claimant's income tax position would be to over-compensate. Damages are not punitive nor are they a reward – they are simply compensation.

Industrial injury and sickness benefit The Law Reform (Personal Injuries) Act 1948 requires that an award must be reduced by one half of the value of any rights which have accrued or probably will accrue in respect of industrial injury or disablement benefit under the National Insurance (Industrial Injuries) Act 1946 or in respect of sickness benefit under the National Insurance Act 1946 for five years beginning when the cause of action accrued. However, an increase in an industrial disablement pension in respect of the need of constant attendance may be disregarded for this purpose.

Where damages are subject to the above deductions, the same Act caters for contributory negligence; the total damages will be assessed, one half of the industrial injury or sickness benefit then deducted and the scaling down for contributory negligence or other cause applied, in that order. For example, if total damages are

£5000, national insurance benefit £800 and contributory negligence one quarter, the calculation may be shown as follows:

$$\frac{£5000 - £400 \ (\frac{1}{2} \text{ of } £800)}{3/4} = £3,450$$

If the benefit takes the form of a gratuity or lump sum this is treated as benefit for the period covered by the assessment. The court can only take into account benefit which will accrue within five years from the time when the cause of action accrued. In *Hultquist* v *Universal Pattern and Precision Engineering Co Ltd (1960)* where a gratuity was assessed for life (thirty-five years in that case) only 5/35ths of the gratuity was taken into account when damages were assessed.

Pain and suffering

Loss of amenity is commonly included with pain and suffering under one head of damages. However, Lord Roche said in *Rose* v *Ford (1937)*: 'I regard impaired health and vitality not merely as a cause of pain and suffering but as a loss of a good thing in itself.'

Damages may therefore be awarded for the temporary or permanent loss of the pleasures or amenities of life. In assessing the amount of damages, the individual circumstances of the plaintiff are important; the extent, for example, to which he is prevented from engaging in recreational activities. Age is also a factor to take into account, a lower assessment may be applied to an older person or someone who is already a permanent invalid.

The amount awarded for pain and suffering will depend upon the nature of the injury, how quickly recovery takes place and the extent to which there is any continuing disability. Future pain and suffering must be taken into account as well as the suffering occurring immediately after the accident. Where a claimant fully recovers from his injuries within a few weeks and there is no remaining disability a relatively small award will be made under this heading.

Walker v *John McLean and Sons (1979)* considered the appropriate level of award for pain, suffering and loss of amenity. A motor cyclist aged sixteen and a half had been involved in an accident due to the fault of contractors and became a paraplegic. The rate of inflation had increased dramatically between 1973 and 1978 and the courts considered that many awards had not maintained their true value. Although awards are not made on the basis of a retail price index, changing money value could not be ignored. The court awarded £35,000 for pain suffering and loss of amenity.

In *Joyce* v *Yeomans (1981)* a boy of ten received a head injury after being knocked down by a car and afterwards began to have epileptic fits. The Court of Appeal awarded a figure of £6000 for the pain suffering and loss of amenities.

However, the examination and analysis of comparable awards can be misleading because the facts in each case differ. Other items in the total award may also influence the assessment.

The mathematical basis   In calculating damages for future loss and future expenses (that is, in respect of the cost of future care or the loss of future earnings), the time-honoured method is to use a multiplicand and a multiplier. The amount of the annual loss under the appropriate heading is calculated and thus provides the multiplicand. To this figure the judge applies a number of years' purchase and this is the multiplier. The multiplier will reflect the period during which the loss is likely to continue but in practice the figure seldom exceeds sixteen or seventeen. The lump sum when awarded does, of course, earn interest and this is why the multiplier will be for a lesser period than the estimated period of loss. Also taken into account is the accelerated receipt of the money as part of a global award and the vicissitudes of life, for example, the risk of an early death from other causes.

The multiplier decided upon will be based on the age of the plaintiff. Theoretically this method enables the plaintiff to spend a part of his capital each year, leading to complete extinction by the time retirement age is reached. There has been some attempt by plaintiffs to substitute this method by the employment of actuarial systems but the courts have resisted this approach. It seems to be considered that the traditional method is the fairer one. Actuarial calculations are based upon averages, whereas each individual injury case varies and requires individual treatment.

In *Auty Mills Rogers and Popow* v *The National Coal Board (1984)*, the judge resisted the introduction of actuarial evidence in assessing pension losses. He held that such evidence was not admissible as a primary basis of assessing future losses, that in making subjective judgments on issues of fact the actuary was usurping the function of the court and thus his evidence was inadmissible. The evidence of an actuary on inflation was not admissible and further, following the House of Lords decision in *Cookson* v *Knowles (1978)*, future inflation could not be taken into account.

However, in certain cases the multiplier/multiplicand method may have to be abandoned. For example, in *Joyce* v *Yeomans (1981)*, the court rejected the method because there were too many imponderables. They did not know how long the plaintiff (aged ten at the accident date) would live or what sort of job he could or would get. All they could do was to award a sum for injury to earning capacity which they considered reasonable. The danger is that with a young person a relatively high multiplier would be employed, and even with the application of a modest multiplicand a substantial compensation figure might result. In a claim under the Fatal Accidents Act damages recoverable are calculated by establishing the extent of the dependant's pecuniary reliance on the deceased and the deceased's future earning expectation. The multiplier will reflect the age and expectation of the working life of the deceased and the expectation of life of the dependants. The expenses the deceased would have spent on himself have to be deducted from the estimated future loss of earnings. While damages may first be calculated as a total figure the court apportion a separate award for each dependant child on the basis of the assumed cost of the child's keep according to the family circumstances. Each case, however, has to be considered on its own merits.

Future medical care

In the event of a serious injury which involves permanent total disablement, the cost of future medical care can be a very large item in the total assessment of damages. Frequently relatives are able and willing to provide nursing care at home, thus avoiding the cost of professional nursing. The criterion, however, is the need for treatment, and the cost of care must be assessed realistically. A wife looking after a husband, although not paid as such is entitled to a reward for her services. In *Lim* v *Camden and Islington Area Health Authority (1979)* out of a total award of nearly £230,000 the provision for the cost of future care was £76,800.

In *Walker* v *John McLean and Sons (1979)* there was an award of £2000 a year for twelve years in respect of future care, housekeeping and nursing. In *C, a minor by his next friend,* v *Wiseman and Another (1981)* a child of twenty-one months was taken to hospital with symptoms of croup. Whilst being examined by doctors he suffered a cardio-respiratory arrest which destroyed his brain beyond repair. The Court of Appeal did not think it right in principle that a baby should have two separate items of damage, one for cost of future care and one for loss of future earnings. The award for the cost of future care was based on a multiplicand of £7850 with a multiplier of twelve. The total award came to about £170,000.

The largest award in recent times was made in the case of *Brown* v *Merton Sutton and Wandsworth Authority (Teaching) (1982)*. The plaintiff aged thirty-eight had become a tetraplegic through negligent medical treatment. She would need a daily housekeeper, a night nurse and a part-time helper during the day for the rest of her life at an estimated cost of £20,000 per annum. She was expected to live until fifty-five or sixty. An agreed multiplier of thirteen produced an award of £260,000 for future care. Other items brought the total award to £414,663.

Loss of expectation of life

The Law Reform (Miscellaneous Provisions) Act 1934 provided that on the death of any person all causes of action, subsisting against or vested in him at the time of his death (except defamation), survive against or for the benefit of his estate. The question arose, could damages be claimed for loss of expectation of life? The case of *Rose* v *Ford (1937)* decided that such an action did survive for the benefit of the estate. The House of Lords held that only moderate sums should be awarded and in *Naylor* v *Yorkshire Electricity Board (1966)* fixed the sum at £500. Since then inflation has increased awards to about £1500*.

Future loss of earnings

When the victim of an accident finds that he can no longer earn his pre-accident earnings the reduction of earning capacity can be ascertained precisely and calculated as an annual sum. It is then possible to form a view of his working life and to apply to the annual sum a figure which represents the number of years' purchase in order to reach a capital figure. In respect of future pecuniary loss the award is assessed by reference to the value of money at the date of trial, for the rate of future inflation is an unknown quantity. The multiplier is fixed in relation to the pre-accident expectation of life, not post-accident expectation of life.

Oliver v *Ashman (1962)* was a decision which operated harshly against badly

---

* See Administration of Justice Act 1982 page 254.

injured victims of accidents with an impaired life expectancy. It provided that if the life expectation was reduced, then in providing compensation for future loss of earnings regard should be paid to the impaired expectation, the difference between that and a normal one being referred to as the 'lost years'.

The main strands in the law as it then stood were:

(a)   The Law Reform (Miscellaneous Provisions) Act 1934 had abolished the old rule *actio personalis moritur cum persona* and provided for the survival of causes of action in tort for the benefit of the victim's estate.

(b)   The decision of *Flint* v *Lovell (1935)* where an item of damages by a living person for loss of expectation of life was allowed for the first time. The House of Lords in *Rose* v *Ford (1937)* decided that a claim for loss of expectation of life survived under the 1934 Act and was not a claim for damages based on the death of a person and so barred at common law.

(c)   The decision of the House of Lords in *Benham* v *Gambling (1941)* that damages for loss of expectation of life could only be given up to a conventional figure, then fixed at £200.

(d)   The Fatal Accidents Act under which proceedings might be brought for the benefit of dependants to recover the loss caused to them by the death of the breadwinner. The amount of that loss was related to the probable future earnings which the deceased would have made during the 'lost years'.

If an action for damages was brought by the victim during his lifetime and either proceeded to judgment or was settled, it appeared that further proceedings could not be brought after his death under the Fatal Accidents Act.

A victim might therefore be placed in a dilemma of having to decide whether to proceed with his claim or allow his dependants to secure a higher award after his death.

However, *Oliver* v *Ashman* was overruled by *Pickett* v *British Rail Engineering Ltd (1979)*. The House of Lords held that a deceased's personal representative could under the Law Reform (Miscellaneous Provisions) Act 1934 maintain a claim to damages for the loss of earnings that the dead person would have experienced for the period of which the accident had robbed him. Mr Pickett had died from mesothelioma of the lung caused by asbestos dust. However, he had brought his claim whilst still alive, and the future loss of earnings award was related to a life expectancy of one year from the date of trial. Mr Pickett appealed against the judgment but died before the appeal was heard. The Court of Appeal did not award any sum for loss of earnings beyond the survival period and the widow appealed to the House of Lords. They held that in the case of an adult wage earner with or without dependants who sued for damages during his lifetime a rule which enabled the 'lost years' to be taken into account was a just principle. The amount to be recovered in respect of earnings in the 'lost years' should be after deduction of an estimated sum to represent the victim's probable living expenses during those years. The principle for recovery lay in the interest which he had in making provision for dependants and others and that he would do out of his surplus. This brought the award into line with what could be recovered under the Fatal Accidents Acts.

An example of an award of damages where the plaintiff did not suffer any deduction for the 'lost years' was *Chambers* v *Karia (1979)*. A married man aged thirty at the date of accident, suffered appalling injuries and was seriously paralysed. He required constant nursing care and heated accommodation at an even temperature. His life expectancy was estimated at twenty years. The judge took fifteen years' of purchase of the loss of future earnings and awarded £68,000 under that head. £40,000 was added for pain suffering and loss of amenity. A further £40,000 was awarded for future nursing and for the wife's services. There were further items including the cost of altering the plaintiff's house. Special damages to date of trial were £14,000 and the total award was £183,650.

Not only has the *Oliver* v *Ashman* principle been overruled regarding living plaintiffs, but as regards the 'lost years' there was no difference between their claims and claims brought by their estates after death.

In *Gammell* v *Wilson* and *Furness* v *B. and S. Massey Ltd (1981)* the estates of two young deceaseds were awarded sums representing the financial loss relating to the deceased's surplus over earnings for the 'lost years' (that is, after deducting from earnings an amount deemed to be spent on living expenses). Kevin Furness aged twenty-one, was killed at work and the estate was awarded £15,547 for the 'lost years'.

However, the judiciary are not happy with the results arrived at, especially where the victims are young men with no established earning pattern. The award of damages in such circumstances becomes speculation rather than justice. The effect on insurers is important for it means that in fatal accident claims where there are no dependants the cost of settlement will be very considerably higher than it has been in the past. For living plaintiffs the awards for future loss of earnings may also be considerably higher*.

Possible future loss of earnings   Occasionally the earnings of an injured worker may be the same after the accident as before. However, as the result of the accident the employee may have a weakened position in the labour market, should the employment be lost at a later date. The possibility of future loss of earnings is something a court may take into account in awarding damages (see *Smith* v *Manchester City Council (1974)*). As explained in *Moeliker* v *Reyrolle (1977)* there must be a real risk of the plaintiff losing his job in the future.

Interest on damages

Section 22 of the Administration of Justice Act 1969 requires the court to exercise its power to award interest on damages for personal injuries or on such part of them as it thinks appropriate, unless the court is satisfied that there are special reasons why no interest should be given in respect of those damages. These provisions confer a discretion on the court to decide what part of an award of damages shall carry interest, the rate of that interest and the period for which it should be given.

---

* Now amended by the Administration of Justice Act 1982. See page 254.

The principle underlying the award of interest is:

When money is owing from one party to another and that other is driven to have recourse to legal proceedings in order to recover the money due to him, the party who is wrongfully withholding that money from the other ought not in justice to benefit by having the money in his possession and enjoying the use of it, when the money ought to be in the possession of the other party who is entitled to its use.

(*London, Chatham and Dover Railway Co* v *South Eastern Railway Co (1893)*). This view was confirmed in *Harbutt's Plasticine Ltd* v *Wayne Tank and Pump Co Ltd (1970)*:

The basis of an award of interest is that the defendant has kept the plaintiff out of his money and the defendant has had the use of it himself, so he ought to compensate the plaintiff accordingly.

Guidelines for the calculation of interest were first laid down in *Jefford* v *Gee (1970)*, but have been subsequently amended by *Cookson* v *Knowles (1978)*, *Pickett* v *British Rail Engineering Ltd (1979)* and *Birkett* v *Hayes (1982)*.

The present position is as follows.

Pecuniary loss to date of trial (special damages)   Interest is awarded from the date of accident to the date of trial. The rate will normally be one half of the rate of interest allowed by the court on the short-term investment account, averaged over the period the interest is awarded.

Future pecuniary loss   As the loss is in the future no interest is awarded.

Non-pecuniary loss (pain, suffering, loss of amenity)   Interest is awarded from the date of service of the writ to the date of trial. *Birkett* v *Hayes (1982)* decided that the rate of interest should be 2%. The figure for pain, suffering and loss of amenities is always assessed at the date of the trial and therefore takes into account any inflationary element up to the date of trial. The court did not consider it reasonable that a high rate of interest should be given on an award which already takes inflation into account. An increase for inflation and payment of interest are separate matters but current rates of interest contain a large inflationary element. The interest even at 2% will not necessarily be awarded for the whole period from the date of service of the writ. If there has been unjustifiable delay after the date of service of the writ the period may be reduced and interest awarded for a lesser time according to the circumstances of the case.

Fatal accidents   Damages have to be assessed in two stages:

(a) *Pecuniary loss up to date of trial:* interest is awarded from the date of accident to the date of trial at one half of the rate of interest allowed by the court on the short-term investment account averaged over the period the interest is awarded.
(b) *Future pecuniary loss:* no interest is awarded as the loss has not yet been incurred.

Agreed settlements   Strictly, interest is only awarded on judgments of the court and an agreed settlement between the parties does not qualify for the payment of interest. However, there is nothing to prevent an agreement between the parties that a settlement between them will take account of interest which might be payable if the case proceeded to court. Where liability is admitted insurers are usually anxious to settle the claim out of court if it is at all possible to do so. Delays and litigation can only increase the final cost of settlement. However, where there is a payment into court by the defendant, interest is now payable on such payments.

As interest on general damages is payable from the date of issue of a writ, there is a tendency for writs to be issued automatically at an early stage and this in itself adds to the insurer's claims costs.

Nervous shock

Under the Law Reform (Personal Injuries) Act 1948 personal injury is defined as including any disease and any impairment of a person's physical or mental condition. To substantiate a claim, therefore, an injury does not have to be merely physical, a mental condition may justify an award for damages.

In *Lynch* v *Knight (1861)* it was said that the law cannot value mere mental pain or anxiety, but if it results in physical harm then compensation can be awarded for that harm just as if it had been caused by a blow. There are many legal cases involving nervous shock and some apparently inconsistent decisions. Shock is merely one form of injury and a cause of action has the same constituents as any other personal injury claim: (a) a duty to the plaintiff to take care; (b) a careless act amounting to a breach of that duty; and (c) resultant injury to the plaintiff.

Bourhill v *Young (1942)* considered the problems of foreseeability and duty. It was held that a driver was not liable for shock unless the injury was within that which he ought to have reasonably contemplated as the area of potential danger which would arise as the result of his negligence.

This ruling was followed in *King* v *Phillips (1953)* where a taxicab driver negligently reversed into a child on a tricycle. The mother, seventy or eighty yards away did not see the accident but heard the child scream. The action for damages for shock suffered by the mother failed; the driver could not have anticipated he would cause the mother the injury complained of and he was therefore not liable to her for negligence.

Nervous shock is an area where the judges set bounds to liability, they decide on the facts of a particular case whether or not the claim is too remote. Until recently, where damages have been recovered, the plaintiff has been at or near the accident at the time or very shortly afterwards. In *Hinz* v *Berry (1970)*, Mr Hinz was parked in a lay-by, a car crashed into his van killing him and injuring the other occupants. Mrs Hinz at the time was nearby picking bluebells. She did not suffer any physical injury but she later suffered extreme mental depression. She was able to recover damages for nervous shock, but not for grief and worry for which damages could not be recovered. Generally a person who was not present at the time of the accident or was told of the accident or saw its results later was outside the ambit of the wrongdoer's responsibility. In *McLoughlin* v *O'Brian (1982)* a mother was told at her home that her family had been in a car accident. On visiting hospital she found that her youngest daughter was dead and that her husband and other children were injured. The Court of Appeal held that the mother's claim for damages for shock was too remote and felt that to allow this type of litigation claim would open the 'floodgates of litigation'.

However, the House of Lords altered this and held that she could recover damages. The illness caused by shock was a reasonably foreseeable consequence of the admitted negligence of the drivers of two lorries involved in the accident. Three elements were considered, the class of persons whose claims should be recognised, the proximity of such persons to the accident and the means by which the shock was caused. It was necessary to distinguish between a family tie, for example, parent and child, husband and wife, and the ordinary bystander. On proximity to the accident it must be close in both time and space. The shock must come through sight or hearing of the event or its immediate aftermath. Lord Wilberforce said it would be illogical to allow the claim where the mother saw the accident occur but to refuse it where she saw her family two hours later at the hospital.

Although the House of Lords did not feel their decision introduced any new principle of law, nevertheless it seems that the boundaries for nervous shock claims have been widened a little and an increase in the number of claimants can be expected.

Disease claims

Industrial disease (and this includes noise-induced hearing loss) claims create many problems for insurers. There is usually a lengthy time-lapse from the exposure to the cause of the disease, to the actual manifestation of the disease. For example, many chemicals have proved to be cancer-producing agents and the hazard of working with asbestos has only been realised more fully in recent years.

An employers' liability policy covers injury or disease caused during an insurance period. With a disease of gradual onset, for example, pneumoconiosis, how is it decided when the disease was caused? It frequently happens that there are many different insurers involved over the period of the alleged exposure. In practice insurers will share any claim in proportion to the period their policies were in force during the period of the exposure to the disease. However, this is not always straightforward. Insurers and insureds may not have records going back for say, twenty years and even when the existence of a policy is known it may not be possible to prove the exact periods it was in force. An insured, despite the maintenance of employers' liability insurance, may thus find himself uninsured for certain periods in the past.

A few insurers may put their policies on to a 'claims made' basis but there is usually a reluctance to accept liability for risks which relate, more correctly, to policies of previous years.

A plaintiff is entitled to make a claim against his employer within a period of three years from the date of knowledge, that is, the date upon which he had knowledge he was suffering from the disease which was due in whole or in part to his employer's negligence or breach of duty.

Employment in an occupation causing the disease does not have to be continuous. For example, exposure to asbestos over a period of months might result in the manifestation of asbestosis many years later. In the intervening years the employee might have been engaged in completely non-hazardous work. The insurers of the employer during the period of exposure are the ones who are responsible; the subsequent insurers would have no liability.

Disease claims have to be investigated thoroughly, and this is another source of difficulty. How is it possible to investigate conditions in a factory of twenty or more years ago? However, it is usually possible to trace employees and to build up a picture of working conditions and practices.

Much will depend on the medical evidence and here there can be conflict of opinion both as regards the cause and severity of the disease. By the time insurers are involved in a claim the disease has usually progressed to the stage where the employee is seriously disabled and may be incapable of ever working again. Such claims are inevitably expensive and from the insurers's point of view have to be paid for out of past premiums which never contemplated either the number or the cost of disease claims today.

However, it does not automatically follow that an employer is legally liable for disease; an employee has to prove negligence against his employer. In *Joseph* v *Ministry of Defence (1980)* a naval dockyard employee suffered from Raynaud's Phenomenon (vibration white finger). For twenty-three years he had been using rivetting and caulking tools. In 1972 he became aware of numbness and whiteness of his finger tips and he collapsed early in 1973. The Court of Appeal found the Ministry was not negligent in failing to warn caulkers and rivetters in its employ about the symptoms of vibration white finger, nor was it negligent in failing to introduce a system of regular medical examinations.

Noise-induced injuries

Although the connection between noise and deafness has been known since the Industrial Revolution, it is only in the more recent past that the relationship has been investigated in detail. Deafness claims produce similar problems to other diseases of gradual onset.

The number of claims for noise-induced hearing loss or industrial deafness has increased tremendously over the past few years. Occupational deafness may occur comparatively quickly, for example, if there is exposure to a high noise level for a short period. More commonly, a person's hearing will be damaged by prolonged exposure at much lower noise levels.

Investigation of noise-induced injuries   Investigation of the sudden injury involves investigations on the lines of any other accident. It is necessary to establish what happened and medical evidence will be available to establish the effect on the claimant's hearing. It has been said that some specialists insist on seeing deafness cases on a Monday after the weekend has allowed the hearing to settle down. The facts of the case will decide to what extent an employer has a legal liability.

Claims arising from long-term exposure, however, present additional problems. It is not sufficient to indicate a mere causal link between the noise exposure and loss of hearing. It is necessary to consider what the employer knew or ought to have known and what practices and remedies he took or ought to have taken over the relevant period in question. It may be difficult to establish what the conditions were in a particular workshop say fifteen or twenty years ago. It is also necessary to distinguish between impaired hearing arising out of the employee's work and impairment arising out of leisure activities, for example, motor cycling or rifle shooting. War service may also have played a part, for example, service in the Tanks Corps or in the Navy. If the noise is continuous an engineer specialising in noise surveys can measure the noise level in a building over a period if there is a doubt about the noise level being the cause of the injury.

The following points are important:

(a) In the light of available technical and general knowledge during the period the employee was exposed to noise, should the employer have taken any or greater precautions?

(b) Has the employer issued protective equipment to the employees, and is the use of such equipment mandatory?

(c) Has production been affected by noise, or is there a high labour turn-over because of noise problems?

(d) Are there notices about precautions against noise posted in the factory?

*Negligence in noise-induced injuries** Generally an employer who disregards the risk of noise-induced hearing loss will be held to be negligent.

In *Berry* v *Stone Manganese and Marine Ltd (1971)* the plaintiff had been employed from 1957 in the defendant's factory where manganese bronze propellors were shaped with pneumatic hammers. The noise levels reached 120 decibels which was accepted by the court as 'bordering on the threshold of pain'. Ear-plugs had been issued but the expert evidence was that the men could not be trusted with the selection of suitable ear-plugs and supervision was necessary. This the defendants failed to give. Ear-muffs were the only answer so far as adequate protection against the noise was concerned and these the defendants ought to have supplied. The plaintiff's action was successful.

Where it is widely known that employees suffer from deafness an employer who takes no notice of the hazard is clearly negligent. However, to what extent is a reasonable employer expected to be aware of the hazards of much lower levels of noise?

There are few statutory requirements, although section 44 of the Woodworking Machines Regulations 1974 provides that if a person is likely to be exposed for eight hours continuously to a sound level of ninety decibels, measures should be taken to reduce the noise to the greatest extent which is reasonably practicable. Furthermore, suitable ear protectors have to be provided and made readily available.

However, an important development was the issue in 1972 by the Department of Employment of a Code of Practice for reducing the exposure of employed persons

---

* The apportionment of damages for liability has become more difficult since *Thompson* v *Smiths Ship Repairers (North Shields) Ltd (1983)*, see *Insurance Law Reports* Volume 8 (George Godwin), because of the additional scientific factor that more damage is done to the hearing in the earlier years in a noisy environment. See footnote on page 268.

to noise (known as the 'Yellow Peril' because of the booklet's colour). This gave the same wording two years earlier.

A breach of statutory regulations may therefore result in a claim. However, the level of sound of ninety decibels seems to be an indication of the level at which a claim based on negligence might succeed. Much will depend upon when the reasonably careful employer ought to have become aware that if his employees were exposed to a high level of noise their hearing might be at hazard and there might be steps which they could or should be taking to eliminate or reduce the hazard. *McGuiness* v *Kirkstall Forge Ltd (1979)* considered this problem. The plaintiff had worked in a forge from 1939 to 1972 in conditions which involved a noise level of 104 decibels. However, it was not proved that the plaintiff's deafness was attributable to exposure to noise in the defendant's employment — there was medical evidence that the plaintiff suffered from a degenerative disorder affecting the middle ear. The judge considered that 1963 was the 'watershed' date after which reasonably careful employers should have been aware of the hazards and that in all the circumstances they were not in breach of duty to the plaintiff*.

In *McIntyre* v *Doulton and Co Ltd (1979)* an employee in a joinery shop claimed for loss of hearing caused by exposure to noise at work. The medical evidence held it was much more probable that the loss of hearing was caused by the plaintiff's use of a shot-gun, and the Court of Appeal dismissed the claim.

The Court of Appeal in *Smith* v *British Rail Engineering Ltd (1980)* considered for the first time the proper measure of damages for loss of hearing. (The defendants had admitted liability.) Except for a short break in 1935 the plaintiff was employed as a plater and rivetter all his working life until retirement in 1978. His hearing began to be affected twenty years previously and it got progressively worse. Tests showed a considerable loss of hearing and the evidence was that there would be further deterioration as he got older. His expectation of life was about twelve years. The trial judge had awarded damages of £3250 and the Court of Appeal substituted £4500. This took into account the plaintiff's age of sixty-seven; but for this the award would have been in the order of £6000 to £7000. The significant matter was that the plaintiff had suffered an interference with the quality of his life for twenty years, had worn a hearing aid for two years and would continue to do so till the end of his life.

In *Bailey* v *ICI Ltd (1979)* a thirty-five year old plaintiff was awarded £8500 for loss of hearing and tinnitus caused by his employer's admitted negligence. There was an average bilateral hearing loss of between fifteen and eighteen decibels. The award comprised £7000 for pain, suffering and loss of amenity plus £1500 for loss of future earning capacity.

The most difficult cases to deal with are those where the sound level is not continuous, although the Code of Practice mentioned above attempts to deal with this. For example, the workman in the plant hirer's workshop who makes his own noise level by straightening excavator buckets with a sledge hammer before welding is such a case. Drivers of some earth-moving equipment are in a similar position.

* According to Thompson's case (see footnote on page 267) a lifetime in noisy work leaves a disproportionately lower amount of damages after the threshold of liability year of 1963. See *Insurance Law Reports* Volume 8.

Recoveries

Where another party either through negligence, breach of statutory duty, or liability assumed under contract is responsible in whole or in part for the bodily injury, then insurers will usually join that other party in the action. Alternatively they will seek recovery after having made their payment to indemnify the insured under the terms of their policy.

A typical example is where a contractor is liable, either at common law or under contract, for the negligent acts of a subcontractor. Having paid the claim the insured's rights against the subcontractor will be transferred to the insurers.

Subrogation, the right which one person has of standing in the place of another and availing himself of all the rights and remedies of that other whether already enforced or not, is a common law principle. An express subrogation condition is inserted in liability policies which empowers the insurers to exercise the subrogation rights even before they have settled their insured's claim.

Recovery under contract

The implications of liabilities assumed under contract have been discussed in chapter 1. The Unfair Contract Terms Act 1977 states that a person cannot by reference to any contract term or to a notice given to persons generally or to particular persons exclude or restrict his liability for death or personal injury resulting from negligence.

However, there appears to be an exception in the case of indemnity clauses in contracts with non-consumers (see later and also chapter 1). 'Contract term' is not defined in the Act but according to the Law Commission it 'bears its natural meaning of any term in any contract (and is not limited to terms in a contract between the instant parties)'. Thus liability to an employee or to a third party for the consequences of negligence resulting in death or injury cannot be avoided and the injured person will be able to recover compensation. Even where a person agrees to or is aware of an exclusion or restriction of liability for negligence this is not of itself to be taken as indicating voluntary acceptance. Voluntary acceptance of risk (or *volenti non fit injuria*) is a general defence in tort although its application has been resisted strongly by the courts in recent years. In any event it has little application in the master/servant relationship.

Where a contract term does refer to 'negligence' and the contract is where one party deals as a consumer, can that party be made to indemnify another person in respect of liability that may be incurred by the other for negligence or breach of contract? Apparently liability can be transferred provided the contract term satisfies the requirement of reasonableness. An exemption clause in respect of death or personal injury is void but a clause requiring one person to indemnify another against such liability is valid if reasonable. It lies upon the party seeking to enforce the indemnity clause against the consumer to show that it is reasonable.

A party to a contract 'deals as a consumer' in relation to the other party if:

(a) he neither makes the contract in the course of a business nor holds himself out as doing so; and

(b)   the other party does make the contract in the course of a business; and

(c)   in the case of a contract governed by the law of sale of goods or hire purchase, the goods passing under or in pursuance of the contract are of a type ordinarily supplied for private use or consumption.

On the other hand the Act does not appear to apply to indemnity clauses between non-consumers. Commercial contracts between two equal parties can contain effective clauses either excluding or restricting liability for negligence in tort or in breach of a contractual duty of care. The test of reasonableness may have to be applied but businessmen entering into contracts are presumed to be capable of apportioning the risks and deciding their intentions.

Many contract terms are a combination of both indemnity and exemption clauses and the issues involved are frequently not straightforward. Nevertheless, it is always worth while to examine contract clauses in detail to ascertain whether there can be a right of recovery under contract against another party.

Recovery from employees

Many accidents are caused by the negligence of fellow employees. The employer is, of course, responsible for the torts committed by his servants but there is nothing in law to prevent an employer from joining a negligent worker in proceedings with the object of obtaining an apportionment of damages.

Employers' liability insurers have entered into a formal agreement throughout the insurance market not to institute proceedings against a negligent employee unless the weight of evidence clearly indicated collusion or wilful misconduct of the employee. In *Morris* v *Ford Motor Co: Cameron Industrial Services (Third Party): Robert (Fourth Party) (1973)* an attempt was made to circumvent this agreement. The court said:

Where the risk of a servant's negligence is covered by insurance his employer should not seek to make that servant liable for it, at any rate the court should not compel him to allow an indemnifier to use his name to do so by subrogation.

It is not therefore a practical proposition to attempt recovery or contribution from a negligent employee.

Employer's Liability (Defective Equipment) Act 1969

This Act imposes a strict liability on the employer in respect of defective equipment. Prior to the passing of the Act an employer was not necessarily liable for defective tools. For example, in *Davie* v *New Merton Board Mills Ltd (1959)* the plaintiff lost an eye through the breaking of a tool supplied by an old-established firm of tool-makers. The defect was not discoverable on inspection and it was held that the employers were not liable, they had fulfilled their duty to take reasonable care and they were not responsible for the negligence of the manufacturers.

This situation could create hardship on an employee if the manufacturer of a defective tool could not be traced. The Act provides that where the defect in equipment is attributable wholly or partly to the fault of a third party (whether identified or not) the injury shall be deemed to be also attributable to negligence on the part of the employer (whether or not he is liable apart from the provisions of the Act). The employee must prove that the defect in the equipment is the fault of a third party — fault means negligence, breach of statutory duty or other act or omission which gives rise to liability in tort. The practical effect is that the employer has to take over liability for the negligence of the manufacturer and compensate his employee for the injury suffered. However, having paid a claim the employer has the right of recovery or contribution from the negligent manufacturer. If traced the manufacturer would probably be joined in the action before the claim was settled. This, however, is not the employee's problem, his action lies against the employer. The right of recovery from the negligent manufacturer is an important aspect for the employer's insurers.

Recovery from employers

Many employers' liability policies contain a condition on the following lines:

The indemnity provided. is deemed to be in accordance with such provisions as any law relating to compulsory insurance of liability to employees in Great Britain, Northern Ireland, the Isle of Man, or the Channel Islands may require but the insured shall repay to the insurers all sums paid by the insurers which the insurers would not have been liable to pay but for the provisions of such law.

It is not possible to repudiate liability under an employers' liability policy because of a breach of certain policy conditions. However, a breach may seriously prejudice the insurers and in such an event they may wish to invoke this condition. Insurers will handle the claim to the best of their ability and indemnify their insured by paying the employee the compensation to which he is legally entitled. They are then entitled to seek recovery from their insured for the amounts they have paid which they would not have paid but for the requirements of the compulsory insurance laws. Recovery may be sought for all the payments made or for just a part of them. For example, the insurers may be quite satisfied about the position on liability but unnecessary additional legal costs may have been incurred because of obstruction, delay or lack of co-operation on the part of the insured. They could then seek recovery for the amount of these additional legal expenses.

In practice these circumstances arise only rarely. In any event it is usually very difficult to secure recovery. An insured will resent the fact that his policy liability is being denied to him and will resist strongly attempts made to seek reimbursement. Where the amount is small insurers will not usually consider the exercise to be worth while and if their insured is so unsatisfactory they will relieve themselves of the risk at renewal date. Where the amount is large, even greater problems may arise and if the insured is a 'man of straw' inability to pay is a practical and usually insurmountable problem.

Appendix 1

Specimen Contractors' Combined Insurance Policy and Conditions

Whereas, the Insured, carrying on the Business stated in the Schedule and no other for the purposes of this insurance, by a proposal and declaration which shall be the basis of this Contract and is deemed to be incorporated herein has applied to the ALLSTATE INSURANCE COMPANY LIMITED (hereinafter called the "Company") for the insurance hereinafter contained and has paid or agreed to pay the premium as consideration for such insurance

Now this Policy Witnesseth that during the Period of Insurance or during any subsequent period for which the Company may accept payment for the continuance of this Policy and subject to the terms, exceptions and conditions contained herein or endorsed hereon, the Company will indemnify the Insured as hereinafter specified.

Section 1 — Employers' Liability

If any person who is under a contract of service or apprenticeship with the Insured (hereinafter called "an employee") shall sustain bodily injury or disease arising out of and in the course of employment by the Insured in connection with the Business described in the Schedule the Company will indemnify the Insured against all sums for which the Insured shall be liable at law for damages in respect of such injury or disease.

Avoidance of Certain Terms and Right of Recovery

The indemnity granted by Section 1 of this Policy is deemed to be in accordance with the provisions of any law relating to compulsory insurance of liability to employees in Great Britain, the Isle of Man or the Channel Islands.
But the Insured shall repay to the Company all sums paid by the Company which the Company would not have been liable to pay but for the provisions of such law.

Section 2 — Public Liability

The Company will indemnify the Insured against all sums for which the Insured shall be legally liable to pay in respect of

(a) Accidental bodily injury to or illness of any person

(b) Accidental loss of or damage to property

(c) Accidental obstruction or trespass

happening or caused in connection with the Business during the Period of Insurance but not exceeding the Single Accident Indemnity Limit specified in the Schedule in respect of any one accident or number of accidents occurring in connection with or arising out of any one event.

Exceptions to Section 2

Except so far as the Company shall by endorsement hereon have agreed, the indemnity expressed in this Section shall not apply to or include

1. liability in respect of bodily injury to or illness of any person who is under a contract of service or apprenticeship with the Insured where such injury or illness arises out of and in the course of the employment of such person by the Insured.

2. liability arising out of the ownership, possession or use by or on behalf of the Insured of

(a) any locomotive, aircraft, watercraft exceeding 20 feet in length or hovercraft.

(b) any mechanically propelled vehicle and any trailer or appliance whilst attached thereto.

   (i) used in any circumstance necessitating compulsory third party insurance under statute (other than whilst operating as a tool of trade).

   or

   (ii) more specifically insured under any other policy of insurance.
The indemnity granted hereby does not apply to any risk which has to be insured to comply with the requirements of the Road Traffic Acts.

3. liability for loss of or damage to the contract works, materials, tools, plant, appliances and temporary buildings for which the Insured is responsible.

4. liability in respect of loss of or damage to property

(a) belonging to the Insured

(b) being that part of any property worked upon where the loss or damage arises out of such work.
Subject always to Exception 3 above Exception 4(b) does not apply if loss or damage is caused by fire, explosion or by removal or weakening of support to any land, property or building

5. the making good, replacement or reinstatement of work, materials or goods supplied, installed or erected.

6. liability arising from loss or damage due to theft unless caused by negligence of the Insured or any employee of the Insured.

Motor Contingent Liability

The Company will also indemnify the Insured within the terms of Section 2 of this Policy for legal liability for injury, illness, loss or damage (as defined herein) caused by or through or in connection with any motor vehicle (not belonging to or provided by the Insured) being used in the course of the Business.

Provided always, the Company shall not be liable for

(a) loss of or damage to any such vehicle

(b) any claim arising whilst the vehicle is

    1) engaged in racing, pace-making, reliability trials or speed testing.

    2) being driven by the Insured

    3) being driven with the general consent of the Insured or of his representative by any person who to the knowledge of the Insured or of such representative does not hold a licence to drive such vehicle unless such person has held and is not disqualified for holding or obtaining such a licence.

    4) used elsewhere than in Great Britain, the Isle of Man or the Channel Islands.

So far as concerns loss or damage to property the liability of the Company is limited to £100,000 in respect of any one claim or number of claims arising out of any one event.

Defective Premises Act 1972

The indemnity granted by Section 2 of this Policy shall apply also with respect to any legal liabilities incurred by the Insured by virtue of Section 3 of the Defective Premises Act 1972 in connection with premises which have been disposed of by the Insured.

The Company shall not be liable hereunder:—

(a) for the cost of remedying any defect or alleged defect in the premises disposed of.

(b) if the Insured is entitled to indemnity from any other source.

Property in the Ground Precautions

The indemnity granted by Section 2 of this Policy in respect of damage to property in the ground is operative on the express understanding and condition that:-

1. the Insured shall advise all owners of underground services (Post Office, Electricity, Gas, Water and Sewerage Authorities) of the intention to commence operations giving full details of the work to be done.

2. the Insured shall inspect available plans of cables, pipes and all other underground services before commencing operations.

3. the Insured shall comply fully with Section 26 of the Public Utilities Street Works Act 1950.

Fire Precautions

It is hereby declared and agreed that where electric, oxy-acetylene, or similar welding or cutting plant or blow lamps or torches are used away from the Insured's premises the Insured shall

(a) wherever possible appoint a responsible employee to act as a fire watcher

(b) arrange for portable fire extinguishing appliances to be available at any place where work is to be carried out

(c) not leave lighted equipment or naked lights unattended.

(d) after completion of each period of work examine the immediate vicinity and the area on the other side of any wall or partition to ensure there is no risk of fire.

Clauses applicable to Sections 1 or 2 of the Policy

Costs

The Company will also

1. Pay all costs and expenses incurred with its written consent.

2. Pay all the Solicitor's fees incurred with its written consent for representation of the Insured at any Coroner's inquest or fatal enquiry or proceedings in any Court arising out of any alleged breach of a statutory duty resulting in bodily injury or disease which may be the subject of indemnity under this Policy.

Contractual Liability

Under Sections 1 and 2 of this Policy (subject to the terms and conditions of these Sections and the general exceptions of the Policy) the Company will indemnify the Insured in respect of liability for injury illness loss or damage (as defined herein) under indemnity clauses forming part of any contract to which this Policy applies.

Provided that the Company shall not be liable hereunder

1. unless the Company has the sole conduct and control of all claims covered by this extension.

2. for loss of or damage to property hired leased or rented where liability would not have attached in the absence of a hire lease or rent agreement.

Work Abroad

Notwithstanding anything contained in General Exception 1 of the Policy, the indemnity granted under Sections 1 and 2 of the Policy (subject to its terms conditions and limitations) shall apply also with respect to liability arising out of the visits abroad of directors, partners, principals or employees (other than those engaged in manual labour).

Labour Masters

It is agreed that for the purposes of this Policy the persons specified below engaged by the Insured in connection with the Business shall be deemed to be employed by the Insured under a contract of service or apprenticeship and for premium adjustment purposes the Insured shall declare all payments made to them.

(a) labour masters and persons supplied by them

(b) persons employed by labour only subcontractors

(c) self employed persons

(d) any person hired or borrowed by the Insured under an agreement by which the person is deemed to be employed by the Insured

Legal Defence

Irrespective of whether any person has sustained bodily injury or disease the Company will, at the request of the Insured also pay the costs and expenses incurred in defending any director, manager, secretary or employee of the Insured in the event of such a person being prosecuted for an offence under the Health and Safety at Work etc. Act, 1974.

The Company will also pay the costs incurred with its written consent in appealing against any judgment given.

Provided that:

(1) the offence was committed during the Period of Insurance.

(2) the liability of the Company in any one Period of Insurance shall not exceed £1,000 in respect of each director, manager, secretary or employee prosecuted.

(3) the indemnity granted hereunder does not

(a) provide for the payment of fines or penalties.

(b) apply to prosecutions which arise out of any activity or risk excluded from the Policy.

Subject otherwise in all respects to the Policy terms, conditions and limitations.

Business

Reference to Business in this Policy shall include

1. the provision and management of canteens, clubs, sports, athletic, social and welfare organisations for the benefit of the Insured's employees.

2. first aid, fire and ambulance services.

3. private work carried out by an employee of the Insured (with the consent of the Insured) for any director, partner or employee of the Insured.

Indemnity to Other Persons

Subject to the terms exceptions and conditions of this Policy and at the request of the Insured the Company will indemnify

1. any director, partner or employee of the Insured in respect of any claims for which the Insured would be entitled to indemnity under the Policy if claims were made against the Insured.

2. any director, partner or employee of the Insured, for whom with the consent of the Insured an employee is undertaking private work.

3. any officer or member of the Insured's canteens, clubs, sports, athletic, social or welfare organisations and first aid, fire and ambulance services

Provided that such persons shall as though they were the Insured observe, fulfil and be subject to the terms exceptions and conditions of this Policy in so far as they can apply.

Section 3 — Loss of or Damage to Contract Works

The Company will indemnify the Insured against loss and/or damage from whatsoever cause occurring during the Period of Insurance to:

(a) the Contract Works consisting of the permanent and/or temporary works erected or to be erected in performance of the contract(s) and all materials for use in connection therewith

(b) tools, plant, equipment and temporary buildings the property of the Insured or for which he is responsible

(c) personal effects and tools of the Insured's employees in so far as such items are not otherwise insured.

Whilst on the contract site(s) or in transit thereto or therefrom by road, rail or inland waterway.

The liability of the Company shall not exceed the Indemnity Limits specified in the Schedule but subject to these Limits the Company will also indemnify the Insured against.

1. Liability for loss and/or damage to the Contract Works specified in (a) above where such loss or damage

(i) happens within the Period of Maintenance and arises from a cause occurring prior to the commencement of the Period of Maintenance

(ii) is caused by the Insured during the Period of Maintenance in the course of any operations carried out for the purpose of rectifying any defects

The Period of Maintenance shall be the period stated in the Contract Conditions but not exceeding twelve months.

2. Architects', Surveyor's Legal and other fees for estimates, plans, specifications, quantities, tenders and supervision necessarily incurred in the reinstatement following loss or damage to the property insured by any peril hereby insured against at a percentage in accordance with the authorised scale of the appropriate body or institute.

This extension shall not apply to fees for preparing any claim hereunder.

3. Costs and expenses necessarily incurred by the Insured with the consent of the Company in

(i) removing debris

(ii) dismantling and/or demolishing

(iii) shoring up or propping

of the property insured hereunder following loss or damage by any peril hereby insured.

4. Liability which falls on the Insured under Clause 14(2) of the Standard Form of Building Contract or Clause 54(3) of the Standard I.C.E. Conditions in respect of loss or damage to materials and goods for incorporation in the works whilst temporarily stored away from the Contract Site(s)

Provided that (i) in the event of an increase in the value of any contract the Indemnity Limit specified in the Schedule of the Policy under item (a) Contract Works, shall be increased automatically by an amount not exceeding 20% (twenty per cent) (subject always to adjustment of premium under the terms of Condition 7).

(ii) Any sum or sums that may from time to time be paid by the Company by reason of this indemnity shall be accounted in diminuition of the Indemnity Limit so that in the event of any subsequent claim or claims occurring during the Period of Insurance and/or the Period of Maintenance the liability of the Company under this Policy shall not exceed in total the amount of the Indemnity Limit unless such diminuition of the Indemnity Limit be reinstated by payment of an agreed additional premium.

(iii) Buildings in course of erection are deemed to be constructed of brick, stone or concrete and roofed with slates, tiles, concrete, asphalt, metal or sheets or slabs composed entirely of incombustible mineral ingredients.

(iv) The sum insured in respect of tools plant equipment and temporary buildings is declared to be Subject to Average and if the tools plant equipment and temporary buildings covered hereunder, shall at the time of any loss and/or damage insured by the Policy be of greater value than such sum insured then the Insured shall be considered as being his own insurer for the difference and shall bear a rateable share of the loss accordingly.

Exceptions to Section 3

The Company shall not be liable to indemnify the Insured in respect of:

1. (a) the cost of repairing, replacing or rectifying property which is defective in material or workmanship
(b) loss or damage due to fault, defect, error or omission in design plan or specification

2. loss or damage due to wear and tear or gradual deterioration.

3. loss of or damage to plant due to its own explosion, mechanical or electrical breakdown, failure, breakage or derangement, but this exception does not relate to damage to other parts of the plant arising as a consequence of such breakdown or derangement.

4. damage to tyres by application of brakes or by punctures cuts or bursts.

5. loss of or damage to the Contract Works or any part thereof after such works or such part have been or has been completed and delivered up to the Principal or taken into use or occupation by the Principal except to the extent that the Insured may remain liable:
(i) under the maintenance conditions of the contract
(ii) during a period not exceeding fourteen days after the issue of a Certificate of Completion and which is the responsibility of the Insured to insure.

6. loss of or damage to locomotives, aircraft, watercraft exceeding 20 feet in length or hovercraft.

7. loss of or damage to tools, plant and equipment belonging to Subcontractors or their employees

8. consequential loss of any nature whatsoever

9. loss destruction or damage occasioned by pressure waves caused by aircraft or other aerial devices travelling at sonic or supersonic speeds

10. Cash, bank notes, cheques, stamps or securities for money

11. property more specifically insured

12. loss of property unless this is identifiable by the Insured with an occurrence which has been the subject of notification to the Company under the terms of Condition 1.

13. loss or damage which is the responsibility of the Principal under the Conditions of Contract.

Clauses applicable to all sections of the Policy.

Excess Clause

If any payment made by the Company under any Section of this Policy shall include an amount for which the Insured is responsible hereunder such amount shall be repaid to the Company forthwith. The expression "claim" shall mean a claim or series of claims arising out of one event.

Indemnity to Personal Representatives

In the event of the death of the Insured the Company will in respect of liability incurred by the Insured indemnify the Insured's personal representatives under the terms of this Policy provided that such personal representatives shall as though they were the Insured, observe, fulfil and be subject to the terms, exceptions and conditions of this Policy in so far as they can apply.

Indemnity to Principals

Where any contract or agreement entered into with any Principal so requires the benefits of this Policy shall apply jointly to the Insured and the Principal (named in the Schedule hereto) and the Company will indemnify the Principal within the terms of this Policy for any claim resulting from injury, illness, loss or damage (as herein defined) where such injury, illness, loss or damage occurs during the currency of the Policy and arises out of, in the course of or by reason of the carrying out by the Insured and/or his Subcontractors of work for which an indemnity is provided by this Policy.

Provided that

(a) the Insured shall have arranged with the Principal for the conduct and control of all claims for which the Company may be liable by virtue of this indemnity to be vested in the Company.

(b) the Principal shall observe fulfil and be subject to the terms and conditions of this Policy in so far as they can apply.

(c) claims made by the Principal shall be treated as though the Principal were not insured by this Policy.

General Exceptions

The Company will not be liable under this Policy

1. for injury, illness, loss or damage caused elsewhere than in Great Britain, the Isle of Man, or the Channel Islands.

2. (i) loss or destruction of or damage to any property whatsoever or any loss or expense whatsoever resulting or arising therefrom or any consequential loss or
(ii) any legal liability of whatsoever nature
directly or indirectly caused by or contributed to by or arising from

(a) ionising radiations or contamination by radioactivity from any nuclear fuel or from any nuclear waste from the combustion of nuclear fuel

(b) the radioactive, toxic, explosive or other hazardous properties of any explosive nuclear assembly or nuclear component thereof.

In respect of bodily injury to or disease of any employee this Exception shall apply only when the Insured under a contract or agreement has undertaken to indemnify any Principal or has assumed liability under contract for such bodily injury or disease.

3. Under Section 2 and 3 for any consequence of war, invasion, act of foreign enemy, hostilities (whether war be declared or not) civil war rebellion, revolution, insurrection or military or usurped power.

4. for any sums which may become payable under Contract as fines or penalties for delay or non-completion of the works.

In Witness whereof this Policy has been signed on behalf of the Company

Allstate Insurance Co. Ltd.
Head Office (and Registered Office),
Marsland House,
Marsland Road,
Sale,
Cheshire, M33 3AQ.
Registered Number 80623, England.

Managing Director

Conditions of the Policy

This Policy and the Schedule shall be read together and any word or expression to which a specific meaning has been attached in any part of this Policy or the Schedule shall bear such meaning wherever it may appear.

1. The Insured or his legal representatives shall give notice in writing to the Company as soon as possible after the occurrence of any accident, loss or damage with full particulars thereof. Every claim, notice, letter, writ or process or other document served on the Insured shall be notified or forwarded to the Company immediately on receipt. Notice in writing shall also be given immediately to the Company by the Insured of impending prosecution inquest or fatal inquiry in connection with any such occurrence as aforesaid. In the case of loss or damage by theft or malicious act the Insured shall also give immediate notice to the Police.

2. No admission. offer, promise, payment or indemnity shall be made or given by or on behalf of the Insured without the written consent of the Company which shall be entitled if it so desires to take over and conduct in the name of the Insured the defence or settlement of any claim or to prosecute in the name of the Insured for its own benefit any claim for indemnity or damages or otherwise and shall have full discretion in the conduct of any proceedings and in the settlement of any claim and the Insured shall give all such information and assistance as the Company may require.

3. The Insured shall take reasonable precautions to prevent accidents, loss or illness and to safeguard the property insured.

4. In the event of misrepresentation, misstatement or omission in any material particular or if the circumstances in which the Insurance was entered into shall be altered or if the risk shall be materially increased, the Company shall not be liable under this Policy unless notice of such alteration or increase of risk has been given to the Company and the written consent of the Company thereto obtained.

5. Any claimant under this Policy shall at the request and at the expense of the Company do and concur in doing and permit to be done all such acts and things as may be necessary or reasonably required by the Company for the purpose of enforcing any rights and remedies or of obtaining relief or indemnity from other parties to which the Company shall be or would become entitled or subrogated upon its payment for or making good any loss or damage under this Policy whether such acts and things shall be or become necessary or required before or after their indemnification by the Company.

6. Under Section 2 of this Policy the Company may at any time by notice in writing to the Insured limit their liability in respect of any claim or claims arising out of any one accident or number of accidents occurring in connection with or arising out of any one event to the amount of the Single Accident Indemnity Limit together with the costs incurred with the consent of the Company up to the date of the said notice and thereupon the Company will be liable to pay only the amount of the Single Accident Indemnity Limit (after deduction of any sum or sums already paid as compensation) and the said costs upon completion of the settlement or other final disposal of the said claim or claims.

7. The premium for this Policy is based on estimates provided by the Insured and the Insured shall keep an accurate record containing all particulars relative thereto. The Insured shall at all times allow the Company to inspect such record and shall supply such particulars and information as the Company may require within one

month from the expiry of each Period of Insurance and the premium shall thereupon be adjusted by the Company (subject to the minimum premium chargeable for the risk being retained by the Company).

8. The Company may at any time cancel the Policy by sending seven days notice of termination by Registered Letter to the last known address of the Insured and in such event the Premium shall be adjusted in accordance with Condition 7. Such determination shall be without prejudice to any rights or claims of the Insured or the Company prior to the expiration of such notice.

9. The Company is not to be called upon in contribution under this Insurance and is only to pay any loss hereon in and so far as this is not otherwise recoverable under any other insurance.

10. If any difference shall arise as to the amount to be paid under this Policy (liability being otherwise admitted) such difference shall be referred to an Arbitrator to be appointed by the parties in accordance with the Statutory provisions in that behalf for the time being in force. Where any difference is by this Condition to be referred to arbitration the making of an Award shall be a condition precedent to any right of action against the Company.

11. The due observance and fulfilment of the terms, provisions, conditions and endorsements of this Policy in so far as they relate to anything to be done or complied with by the Insured and the truth of the statements and answers in the proposal and declaration shall be conditions precedent to any liability of the Company to make any payment under this Policy.

Specimen Contract Guarantee Bond

Contract Guarantee Bond

Bond Number CG

Amount £

𝕶now all men by these Presents,

That We

(hereinafter called 'the Contractor/s')
and the Allstate Insurance Company Limited,
whose Registered Office is situated at Marsland House,
Marsland Road, Sale, Cheshire M33 3AQ,
(hereinafter called the 'Sureties') are jointly and
severally held and firmly bound to

(hereinafter called 'the Corporation') in the penal sum of

sterling to be paid to the Corporation for which payment
to be well and truly made we bind ourselves and each of us
and any of us our and each and any of our successors
and assigns heirs executors and administrators jointly
severally and respectively firmly by these presents
and sealed with our seals.

Dated this   day of
One Thousand Nine Hundred and

Whereas by a certain Contract or Agreement bearing
the date of the   day of
One Thousand Nine Hundred and
and made between the Contractor/s of the one part
and the Corporation of the other part the Contractor/s
contracted with the Corporation for the execution of
certain works in connection with

for such sum as may become payable under the
aforesaid Contract or Agreement.

Now therefore the condition of the above written Bond is
such that if the Contractor/s their assigns executors or
administrators shall duly perform and observe all
the stipulations agreements and provisions contained
in the said Contract and on their part to be performed
and observed then the above written Bond shall be void
and of no effect or otherwise shall remain in full
force and virtue. Any alterations which may be made
by agreement between the Contractor/s their
executors or administrators and the Corporation or
their assigns in the terms of the said Contract
conditions stipulations and provisions and the work
or plant to be done or provided thereunder or otherwise in relation thereto, or the giving by the Corporation
or their assigns of any extension of time for performing
the said Contract, or any of the stipulations therein
contained and on the part of the Contractor/s to be
performed, or any other forgiveness or forbearance on
the part of the Corporation or their assigns to the
Contractor/s their executors or administrators shall
not in any way release the Sureties from their liability
under the above-written Bond.

And provided also that the Sureties shall be released and
discharged from their obligations under the above
written Bond upon the happening of whichever shall
last occur of the following events (hereinafter called
'the relevant event') namely the issue of a certificate
of practical completion of the works by the Architect
or of substantial completion of the works by the Engineer
as the case may be or upon the subsequent expiration
of any defects liability period or period of maintenance
provided by the said Contract and if any suits at law
or proceedings in equity are brought against the
Sureties to recover any claim hereunder the same must
be instituted within six months after the date of the
relevant event and thereafter shall be absolutely barred.

Signed Sealed and Delivered
by the said

in the presence of

The Common Seal
of the Allstate Insurance Company Limited
was hereunto affixed in the presence of

Director

Secretary

Appendix 3

Specimen Counter Indemnity Form

Counter Indemnity Form

𝔐emorandum of 𝔄greement made the _____

day of _____ One thousand nine hundred

and _____ between the Allstate Insurance Company Limited whose registered office is at Marsland House, Marsland Road, Sale, Cheshire M33 3AQ (hereinafter called "the Surety") of the one part

and _____

_____

(hereinafter called "The Insured")

and _____

_____

(hereinafter called "the Indemnitor/s") of the other part

𝔄nd 𝔚hereas _____

(hereinafter called "the Obligee") having required the insured to enter into a bond with surety in

respect of _____

_____

(hereinafter called "the said works") in the amount of

£ _____ the Insured have applied to the Surety to be bound with them as surety as aforesaid which the Surety has agreed to do in consideration of the payment to the Surety of the required premium and of the Insured and the Indemnitor/s entering into such an agreement as is hereinafter contained.

𝔑ow in consideration of the premises it is hereby agreed between the parties hereto as follows:

1 The said Insured and/or the Indemnitor/s shall pay to the Surety all sums of money which the Surety from time to time may have paid in respect of the obligations of the Surety under the said Bond and all costs charges and expenses which the Surety may have expended or been put to in respect of any claim or purported claim under the said Bond and all other charges and expenses which the Surety may have incurred by reason of their having become so bound as surety as aforesaid.

2 In the event of failure on the part of the Insured to complete the said works or of any default on their part by reason of which the said works are terminated the Insured and the Indemnitor/s hereby grant authority to the Obligee and the Surety to use and expend any moneys due to the Insured from the Obligee by way of retention or otherwise at the time of such failure or default towards the cost of completing the said works and in such event the Insured and the Indemnitor/s waive their claim to receive all or part of such moneys.

3 In the event of any claim being made against the Surety or the Insured under the said Bond or the said works it shall be lawful for the Surety to effect any compromise or settlement of the same with or without reference to the Insured and the Indemnitor/s upon such terms as the Surety may consider expedient and any moneys which the Surety may pay in settlement or compromise of any such claim shall be deemed expenses within the meaning of Clause 1 hereof.

4 For the due performance hereof the Insured and the Indemnitor/s (and each of them) hereby (jointly and severally) bind $\frac{himself}{themselves}$ $\frac{his}{their}$ executors successors administrators and/or assigns.

Form of Subcontract

Designed for use in conjunction with the ICE General Conditions of Contract: Fifth Schedule

Example 1: Insurances

Part I: Subcontractor's insurances

Employer's liability – unlimited in amount.
Third party (public) liability – minimum on any one occurrence £1,000,000 (number of occurrences unlimited).
Plant and equipment – to the full value.
Subcontract works – to the full value.

The Subcontractor shall effect the above insurance in such a manner in the name of himself and such others as are required under the Main Contract that he assumes in respect of the Subcontract all the obligations and liabilities required of the Contractor under the Main Contract.

Part II: Contractor's policy of insurance
No benefit under the Contractor's policies of insurance is available to the Subcontractor.

Example 2: Insurances

Part I: Subcontractor's insurances

Employer's liability
Plant and equipment

Part II: Contractor's policy of insurance
The Subcontractor shall have the benefit of the Contractor's 'Contractor's All Risks' and 'Public Liability' insurance (subject to policy excesses).
    Note: The public liability insurance will be included where it is part of the contractor's all risks policy.

Summary of Responsibility for the Works and Liability Insurance Table

Risks to the works

Loss or damage to the works up to certificate of completion date or arising from work performed during the maintenance period

	Employer		Contractor	
	Contract	*Insurance*	*Contract*	*Insurance*
1 War and kindred risks plus riot	(a) ⎫ Sustains loss	1 No cover except riot in certain areas	(a) ⎫ No responsibility	(c) CAR policy
2 Nuclear and sonic waves risks	(b) ⎬	2 No cover	(b) ⎬	
3 Use or occupation by employer	⎭	3 Employer to arrange	⎭	
4 Fault defect error or omission in design (other than contractor's design; see 7 below)		4 Employer can make sure that engineer arranges professional indemnity (no non-negligent) cover		
5 Materials or workmanship not in accordance with the contract	No responsibility		(e) Liable ⎱ but no liability to insure	
6 All other causes	No responsibility		Liable ⎰	CAR policy
7 The contractor's design	No responsibility		Liable	Can insure professional indemnity of design department but not non-negligent design

	Employer		Contractor	
	Contract	Insurance	Contract	Insurance
Public liability losses				
1 Damage to crops on site; liability for use or occupation of land or interference with easements, right to construct the works	Liable	Public Liability to a limited extent	Not liable	
2 Unavoidable result of construction of the works	Liable	No cover	Not liable	
3 Negligence or breach of statutory duty of employer, etc.	(f) Liable	(f) Public Liability	Not liable	
4 Negligence of contractor	Not liable		(f) Liable	(f) Contractor's public liability
5 Negligence of subcontractor not employed by contractor	(d) Liable	Public Liability	Not liable	
Injury to employees of contractor/ subcontractor				
1 Negligence of contractor	Not liable		Liable	Contractor's employers' or public liability
2 Negligence of employer, etc	(f) Liable	(f) Public liability	Not liable	
3 Negligence of subcontractor	Not liable		(d) Liable	Contractor's employers' or public liability

For the purposes of this table the phrases 'act or neglect' and 'act or default' are taken to mean negligence for the sake of simplicity but see chapter 1 and the chapters on clauses 22 and 24 for a further discussion of these phrases

Note: where a policy is named as applying this is always subject to the terms exceptions and conditions of that policy.

(a) Usually Goverment pays compensation
(b) Nuclear operator and Goverment have statutory liability for nuclear risks and possible responsibility for sonic waves
(c) Excluding the actual faulty part
(d) May be able to recover from subcontractor
(e) Possible recovery from supplier of faulty materials
(f) Contract does not exclude nuclear risks but policy does

The Presentation of CAR and Other Construction Insurance Damage Claims

Some contractors tend to perform poorly in their method of claims presentation under the CAR policy. CAR or other property damage claims should be presented under clear and precise headings of claim supported by authenticated daywork sheets or other approved site documents and plant and material allocation forms which will enable claim settlements to proceed smoothly. Poor presentation can involve delays due to misunderstandings with insurers or loss adjusters.

The remedial works must be the subject of as detailed a specification as possible and a bill of quantities. At the earliest possible stage daywork areas must be indicated and a basis of costing agreed. Valuation as 'daywork' arises when additional or substituted work cannot be valued by measurement. See clause 13.5.4 of JCT 1980 edition or clause 52(3) of the ICE Conditions.

The cost of remedial works is dependent, *inter alia*, on the time involved. Thus it is necessary to have a programme of works arranged from the start. This may involve site demolition and clearance before the remedial works proper can commence. Specialist subcontractors may undertake this type of work as it may be more economic from a direct cost viewpoint, leaving the main contractor free to work on other areas of the contract.

It is relevant to remind the contractor that clause 21 requires him to insure the permanent works, the temporary works (including unfixed materials, etc.) and the constructional plant to their full value. Furthermore under clause 8 the contractor is required to construct complete and maintain the works etc. Therefore the adequacy of the insurance is largely his problem.

The costs submitted, so far as they relate to the repair or reinstatement of the permanent or temporary works, will consist basically of labour costs, material and plant costs and overheads. Incidentally it is usually the loss adjuster who recommends the payments on account. The payment being in the joint names of the employer and contractor.

Labour costs

The labour costs are supported by daywork sheets which record the type of labour employed, the number of hours entered for each class of labour, usually the tasks performed, a rate and the extent to which that rate is varied by on-costs such as bonuses, subsistence or lodging, travelling etc. The daywork sheets should be authorised by a responsible official on site and/or should as far as possible have been approved by the loss adjuster during interim site visits.

On general building work there can be two methods of calculating the main contractor's labour costs. The first is the unit rate included in the original main contract bill of quantities, used for tender purposes. This method can be used where the damaged areas are easily recognised as part of the original bill. Where the items of damage were not included in the original bill there is no alternative but to adopt a daywork basis or agree a unit figure. In the case of hourly-paid labour questions sometimes arise concerning waiting time for either the arrival of materials, or plant, or just for an improvement in the weather. Provided the labour has been properly and reasonably brought to the site and cannot be employed elsewhere either on-site or at other nearby sites such costs are usually recoverable. Overtime working would not normally be paid by insurers on the grounds that the insured benefits in that he can avoid other liabilities, but in some circumstances such overtime has the effect of reducing on-costs such as travelling and lodging by reducing the repair period overall and this may be recoverable under the policy. Also there may be other acceptable reasons why overtime payments should be allowed.

The Conditions of Contract used in civil engineering work (revised 1979) 5th edition) make reference to daywork. Clause 52(3) indicates reference to the 'Schedules of dayworks carried out incidental to Contract Work' issued by The Federation of Civil Engineering Contractors. There are very detailed rules on how labour rates are to be built up from the productive working hours and the base rates. The amount of wages are defined as:

Wages, actual bonus paid, daily travelling allowances (fare and/or time) tool allowance and all prescribed payments including those in respect of time lost due to inclement weather paid to workmen at plain time rates and/or at overtime rates.

All payments shall be in accordance with the Working Rule Agreement of the Civil Engineering Construction Conciliation Board for Great Britain Rule Nos I to XXIII inclusive or that of other appropriate wage-fixing authorities current at the date of executing the work and where there are no prescribed payments or recognised wage-fixing authorities, the actual payments made to the workmen concerned.

Material costs

The cost of materials should be straightforward provided that the nature and amount of the materials is in accordance with the particular loss or damage. The claim must be supported by invoices. The main contractor is permitted to charge the invoiced price of all materials after deducting VAT including delivery to site plus 12½%. The question of VAT in the construction industry is not as complex as it was before the March 1984 budget. As a general guide, new construction work is zero rated, but alterations (from 1 June 1984) repairs and maintenance are at the standard rate. VAT, when charged, is not normally recoverable from the insurer as it can be set off by the insured. Further details on VAT will be found in chapter 20 on Contract Works Claims. Reference should be made to the end of this appendix for some details about the National Economic Development Office (NEDO) Price Adjustment Formula, which is used throughout the construction industry for up-grading prices to current value. The NEDO statistics are published in most of the construction industry journals and these are used to act as multiplication factors to change the rates quoted for particular items, in any specific category of building work from the date of origin to the date on which the work was carried out.

Overheads

The most common overheads are site overheads which are included as a provisional sum in the bill of quantities, i.e. sums which cannot be entirely foreseen, defined or detailed at the time of tendering. They include safety aspects, sick pay, transport for workers, safeguarding the works materials and plant etc.

Another type of overhead arises in the form of administration in placing and processing orders for replacement materials and is added to the invoice value. Although there are no guidelines concerning overheads in the building industry there are in the civil engineering industry.

There is a percentage addition, currently 133%, to the amount of wages paid to workmen which provides for statutory charges and other charges including:

- National Insurances and Surcharge
- Normal Contract Works, Third Party, and Employers' Liability Insurances
- Annual and Public Holidays with Pay
- Non-contributory Sick Pay Scheme
- Industrial Training Levy
- Redundancy Payments Contribution
- Contracts of Employment Act
- Site supervision and staff including foremen and walking gangers, but the time of the gangers or charge hands working with their gangs is to be paid for as for workmen.
- Small tools — such as picks, shovels, barrows, trowels, hand saws, buckets, trestles, hammers, chisels and all items of a like nature.
- Protective clothing
- Head Office charges and profit

All *labour only* subcontractors' accounts to be charged at full amount of invoice (without deduction of any cash discounts not exceeding 2½%) plus 64%.

Subsistence or lodging allowances and periodic travel allowances (fare and/or time) paid to or incurred on behalf of workmen are chargeable at cost plus 12½%.

The charges referred to above are as at date of publication and will be subject to periodic review.

Claims submitted to insurers will usually include profit, and provided the insured contractor can show that had his resources not been employed on the repair and replacment work they would have been employed elsewhere, this profit is justified. Associated with this aspect is the cost of supervising and material staff involved in the investigation of the extent of the damage and the appropriate method of repair, and these costs are normally considered justified. Incidentally, the cost of presenting the claim is not justified.

The National Economic Development Office Price Adjustment Formula

Estimates and quotations in the construction industry are usually related to a specific date and are held only for a short period. One of the most convenient methods of amending these prices to suit changes which occur through the passage of time is the use of the monthly bulletin *Construction Indices* for use with the National Economic Development Office Price Adjustment Formula (usually referred to as the NEDO formula for short). What happens is that the HMSO through the Working Group on Building Indices Property Services Agency, Croydon, publishes monthly a bulletin in most of the construction industry journals. The index is published in 49 work categories covering demolition, concrete, asphalt, steelwork etc. and seven appendices covering building, labour and plant. The indices are published provisionally in the month following the date of the index, and they remain unchanged until replaced by a firm index number after about 3 months. The current series takes June 1976 as its base. Indices can be extended regressively to 1970. There is a complete book on how the price adjustment formula works, but in a word the prices are increased on the basis of the difference between the indices over the period of time in accordance with the movement of the index numbers.

As far as the indices for the price adjustment formula are concerned, there are the First Series which uses a base of 1970 as 100 which is primarily for civil engineering work and that is often referred to as the Baxter index and then there is the Building Works Series Two which uses a base line of June 1976 of 100 and this is often referred to as the Osborne Index. In both cases the description of the trade is related to the nature of the work.

Table of cases

Index